ADVANCED BIOSTATISTICS

AF655961

ADVANCED BIOSTATISTICS

K. C. Bhuyan
Professor of Statistics
American International University
Dhaka, Bangladesh

New Central Book Agency (P) Ltd
LONDON
HYDERABAD ERNAKULAM BHUBANESWAR
NEW DELHI KOLKATA PUNE GUWAHATI

NCBA

REGD OFFICE

8/1 Chintamoni Das Lane, Kolkata 700 009, India
email: ncbapvtltd@eth.net

OVERSEAS

NCBA (UK) Ltd, 149 Park Avenue North, Northampton, NN3 2HY, UK
email: ncbauk@yahoo.com

EXPORT

NCBA Exports Pvt. Ltd, 212 Shahpur Jat, New Delhi 110 049
email: ncbaexports@ncbapvtltd.com

BRANCHES

212 Shahpur Jat
New Delhi 110 049
email: ncbadel@ncbapvtltd.com

Shop Nos. 3 & 4, Vinayak Towers
681/B Budhwar Peth, Appa Balwant Chowk
Pune 411 002
email: ncbapune@ncbapvtltd.com

Shop No. 15, Bharati Towers, Ground Floor
Forest Park
Bhubaneswar 751 001
email: ncbabub@ncbapvtltd.com

House No. 3-1-315, 1st Floor
Nimboliadda, Kachiguda, **Hyderabad** 500 027
email: ncbahydb@ncbapvtltd.com

GSS Shopping Complex, 1st Floor
Opposite College Post Office, Convent Road
Ernakulam 682 035
0484-4021054/53

Radhamayee Complex (Second Floor)
Opp. SBI, Panbazar, Jaswanta Road
Guwahati 781 001
email: ncbaguwahati@ncbapvtltd.com

ADVANCED BIOSTATISTICS • K. C. Bhuyan

© Copyright reserved by the Author

Publication, Distribution, and Promotion Rights reserved by the Publisher

All rights reserved. No part of the text in general, and the figures, diagrams, page layout, and cover design in particular, may be reproduced or transmitted in any form or by any means—electronic, mechanical, photocopying, recording, or by any information storage and retrieval system—without the prior written permission of the Publisher

First Published: March 2015

PUBLISHER AND TYPESETTER
New Central Book Agency (P) Ltd
8/1 Chintamoni Das Lane, Kolkata 700 009

PRINTER
New Central Book Agency (P) Ltd
Web-Offset Division, Dhulagarh, Sankrail, Howrah

COVER DESIGNER
Soumen Paul

COVER PRINTER
Techno Process, Kolkata

TECHNICAL EDITOR
Dipan Roy

PROJECT TEAM
Subir Pal, Prodip Baidya, Pradip Biswas, and Prabhat Jas

ISBN: 978 81 7381 954 4

Price: ₹ 245.00

Dedicated to

My Granddaughter

Neelormi

Contents

Preface

Biostatistics is a multivariate subject. In it the different aspects of life and living organisms are studied through many controlled and uncontrolled experiments. The experiments are conducted in the field of Biological Sciences, Microbiology, Psychology, Genetics, Medical Sciences, etc. The experimental results are analysed and interpreted using statistical tools. The analytical procedure for different aspects of life sciences is the basis of biostatistics and advanced biostatistics.

This advanced biostatistics book is for the students of undergraduate and graduate classes in the field of biological sciences, medical sciences, and specially for the students of Masters in Public Health of different universities. The basic idea of statistics and its different measures are explained herein with examples so that the readers can grasp the subject matter easily. The procedure to infer about the population unknown constant (parameter) on the basis of sample observations has also been discussed in the book. The book contains 13 chapters in all — covering all topics related to biosciences including some aspects of demography. Seven APPENDICES given will, of course, be of immense help to the students and readers alike.

I convey my gratitude and sincere thanks to Professor Dr. T. Islam, Dean, Faculty of Arts and Social Sciences, American International University, Bangladesh, who encouraged me to write such a book.

Finally, I am thankful to Mr. Amitabha Sen, Managing Director, New Central Book Agency (P) Ltd, and the project team-members of the publisher — for their initiative to complete the book successfully.

Any suggestion from any corner to improve this book will be highly appreciated.

26 January 2015 **K. C. Bhuyan**

Chapter 1

Introduction

1.1 Meaning and Definition of Statistics

Statistics is as old as human civilization. In an ancient time it was called "*Science of King*" as the king or the ruler of a state used it to collect information on social and economic aspects of the state as well as its inhabitants. The statistical work was started in U.K. in 1532 using the data of weekly death record. In 1539 the information on marriage and death were collected in France. In 1693 the birth and death rates and yearly cost of living index were calculated. These information on birth, death, marriage and cost of living index were used to explain the state affairs and these were used by the planners in framing Government policies. Thus, it seems that the word "*Statistics*" has been derived from the latin word "*Status*" or the French word "*Statistique*", each of which means a political state. It means that statistics is used as a mirror through which one can see the picture of the political and economical status of the country. However, in recent times, the scope of collection and compilation of facts and figures gradually extended to many fields related to human life and even to the field of animal life for the well-being of the country through intelligent decisions. As our world grows in complexity, it becomes difficult to make intelligent decisions without basic information prevailed in the society. So, we need a scientific technique of collection, compilation, analysis and interpretation of information, where information are collected according to some pre-determined objective. The interpretation is based on a method or technique by which collected information are compiled, analysed, numerically expressed so that comprehensive view of the collected information are depicted. This comprehensive depicted information will work as a mirror through which one can see the socio-political and socio-economical status of the country. Thus, statistics is used in different fields of inquiry and hence, it is not only the "*Science of King*" rather it is the "*King of Science*".

The word "*Statistics*" has three different meanings

First: Statistics is the quantitative information of any scientific inquiry. For example, the statistics of birth and death, the statistics of health management, the statistics of hospital management, the statistics of scientific development in the production sector of medicine, the statistics of scientific development in treatment, the statistics of digital changes in different spheres of life, the statistics of development in information technology, the statistics of input and output of industry, the statistics of loss and profit of different industries, the statistics of manpower, etc.

Second: It is the scientific technique of collection, compilation, presentation, analysis and interpretation of both qualitative and quantitative information. For example, the information on adoption of family planning and number of ever born children can be collected from currently married women, information of hospital management can be collected from administrative set-up of different hospitals, etc. Collected information can be analysed and interpreted to know about the growth of population and pros-and-cons of hospital management.

Third: It is also used in plural sense to explain the unknown characteristic of the units under study by a function of collected information, when information are collected from a representative part of units under investigation. Such function is called statistic and the word 'statistics' is the plural of the word 'statistic'.

Definition of Statistics

From the meaning of "*Statistics*" it may be mentioned that simple or isolated item cannot be termed as statistics unless it is a part of aggregate of facts relating to any particular field of scientific inquiry. The aggregate of facts is a part of applied mathematics. According to R.A. Fisher, "statistics" is the branch of applied mathematics, where the mathematics is used to aggregate the collected information (data). Besides this, the different writers defined the word "Statistics" in different ways.

Definition of Webster

"Statistics are the classified facts representing the conditions of the people in a state, specially those facts which can be stated in numbers or in tables of numbers or in tabular or classified arrangement."

Bowely's Definition

"Statistics are numerical statement of facts in any department of inquiry placed in relation to each other."

Definition of Horace Secrist

"Statistics may be defined as the aggregate of facts affected to a marked extent by multiplicity of causes, numerically expressed, enumerated or estimated according to a reasonable standard of accuracy, collected in a systematic manner for a pre-determined objective and placed in relation to each other."

Definition of Selligman

"Statistics is the science which deals with methods of collecting, classifying, presenting, comparing, interpretating numerical data collected to throw some light on any sphere of inquiry."

Biostatistics

Since statistics deals with data collected from any investigating unit, living or non-living, it may deal with data of unit (object) in the field of life science, viz., in the field of human beings, animal life, botanical elements and/or any living organism. The data provide number of elements dealt with, the relationship of various characteristics of elements or units. The scientific method of studying the different characteristics of units in life science is known as biometry. It is also called biostatistics.

Sir Francis Galton (1822–1911) studied the changes in different characteristics from generation to generation and formulated the rule of change in the characteristics. He studied the changes in the height of offsprings with respect to the changes of height of parents. This study enabled him to develop the regression line.

Kearl Pearson (1857–1936) developed researches in the field of life sciences. Later on, research activities were performed to study the characteristics in the field of botany, animal science, psychology, medical science, social science, educational science, political science, economics, commerce, agricultural science, etc. Currently statistics are applied in the field of computer science and information technology, engineering and environmental science. Whenever it is applied or whatever is the application, statistics needs information (data) of different investigating units/individuals. The information are collected according to some pre-determined objective. Some of the objectives may be as follows:

(i) To study the knowledge and attitude towards breast feeding of babies.

(ii) To study the impact of socioeconomic factors on differential fertility according to sex preference

(iii) To study the knowledge regarding child health care, child mortality and fertility

(iv) To study the attitude of doctors and nurses in improving hospital management, etc.

To conclude about any of the objective mentioned above or any other objectives in mind we need information (data) from the units/items/individuals which are under investigation. The information are to be collected from all the units or from a representative part of the units. Let us now discuss the terms related to data and let us discuss from whom, by whom and how data are collected.

1.2 Terms Related to Data

Population

By population we mean total of units/items/individuals which are under investigation according to some pre-determined objective and are available in a specified area at a specified time period. For example, if the objective is to study the attitude of doctors in improving the hospital management in future, all the doctors working currently in all hospitals in all area/country/district/city will constitute a population of doctors. Again, if the objective is to study the hospital management of a district, all the available hospitals in the district in a specific period will constitute a population of hospitals. If the objective is to study the socioeconomic characteristics of the patients visiting an outdoor of a hospital, all the patients visited the hospital in a month or in a specific period of time will constitute the patients population. Other examples of population are all pharmaceutical industries of a country in a year, all nurses working in different hospitals or clinics in an area in a year, ever married women of a state/district/region in a year, etc.

Each of the population mentioned above has a finite population size, say N. This N is known as population size. Thus, the population discussed above is called finite population. Again, there is another type of population, viz., number of insects available in an agricultural plot, bacteria population in a rotten object, fishes in a sea, etc. These latter population have no finite number of units and hence these are known as infinite population.

Sample

A representative part of population units which are under investigation constitutes a sample. For example, if outdoor patients of a hospital in a month constitute a population, then the patients visited in some selected days may constitute a sample. If nurses working in a hospital constitute a population, then nurses working in a particular hour may constitute a sample. The patients of selected days and the nurses working in particular hours will be the representative part of the population of patients and the population of nurses, respectively, if they are selected by a ramdom process. The sample size is denoted by $n (n \leq N)$.

Sampling unit

The population units which are under investigation for a pre-determined objective are known as sampling units.

Data

The recorded characteristics or information collected either from population units or from sample units are known as data. For example, number of patients admitted in a hospital in a day, number of patients released from the hospital in a day, number of patients died in a day, time spent by a specialist doctor for the patients, etc.

The characteristics to be measured depend on the objective of the study. For example, if the objective is to study the knowledge of child health care, child mortality and fertility, then some important characteristics to be recorded are: age of mother (x_1), education of mother (x_2), duration of marriage (x_3), number of ever born children (x_4), family income (x_5), breastfeeding period (x_6), number of dead children (x_7), care taken during pregnancy (x_8), proper vaccination is done or not (x_9), etc. These characteristics vary from unit to unit. So, these are called variable.

Variable

The characteristic which vary from unit to unit is called variable. Variable is of two types, viz., (a) qualitative variable, and (b) quantitative variable.

Qualitative variable

The variable which cannot normally be measured by numerical figure is known as qualitative variable. For example, occupation of persons, type of disease of patients, qualification of doctors, rank of nurses, etc. More examples of qualitative variables are x_8 and x_9 mentioned above. The qualitative variable is also known as attribute.

Quantitative variable

The variable which is measured by numerical figure is known as quantitative variable. For example, age of patients, height of patients, weight of patients, blood pressure level of patients, blood sugar level of patients, duration of disease of patients, number of patients admitted in different days, etc. More examples of quantitative variables are $x_1, x_2, x_3, x_4, x_5, x_6, x_7$ mentioned above. The quantitative variable is again of two types, viz., (a) discrete variable, and (b) continuous variable.

Discrete variable

The variable which is measured only by integral value is known as discrete variable. For example, number of ever born children of mothers, number of dead children of mothers, number of patients admitted in a day in a hospital, number of doctors/nurses working in hospitals, etc. This variable ranges from 0 to ∞.

Continuous variable

The variable which is measured by integral as well as by fractional values is called continuous variable. For example, age of patients, height of patients, body temperature of patients, cholesterol level of patients, level of s. urea of patients, level of uric acid of patients, etc. This variable ranges from $-\infty$ to ∞.

It has already been mentioned that data are collected from sampling units. All the units in the population may be investigated for data collection or a representative part of population units may be selected by a random process and data can be collected from the randomly selected units. Accordingly, the recorded variable is termed as (a) non-random variable, and (b) random variable, respectively.

Random variable

If the values of variable are observed from different units selected by random process, then the variable is called random variable. For example, let there be $N = 500$ patients in a hospital and $n = 100$ patients are selected by any ramdom process. Consider that the age of these 100 patients are recorded in an investigation. In this case the variable age is random variable.

Non-random variable

If a variable is observed from all population units, then the variable is called non-random variable. For example, number of dead patients in different days of a year is a non-random variable. number of admitted patients in different days of a year, number of babies born in different days in a clinic in a year, etc.

It has already been mentioned that data can be collected from the all population units or from some selected units. Accordingly, the process of data collection is, mainly, of two types, viz., (i) census method, and (ii) sample survey.

Census method

If data are collected from all population units under investigation, then the technique of data collection is known as census method. The census method is suitable if resources in terms of money, time, trained investigators, efficient supervising authority, etc. are available. However, inaccuracy in data may creep in due to shortage of trained man power.

Sample survey

If data are collected from some selected units under investigation, then the technique of data collection is known as sample survey. It is done to avoid the resource constraint. The quality of data can also be ensured if these are collected through sample survey as it is possible to conduct the survey employing proper trained personnels.

1.3 Collection of Data

The data are needed to guess about the population characteristic. So, accuracy of data is very important and these are collected through an organized way. For data collection the following steps are followed:

(i) The objective and scope of the inquiry must be well defined.

(ii) The population from which the data are to be collected should be well defined so that the sampling units to be investigated can be identified properly before the start of survey work.

(iii) *Method of data collection*: Data can be collected either by census method or by sample survey method. **The method depends on the resources available for the survey work and** on the type of inquiry.

(iv) *Nature and type of inquiry*: **The nature of inquiry depends on availability of resources** for the survey. It also depends on objective and scope of inquiry. The different types of inquiries are:

(a) Government inquiry or semi-government inquiry or private inquiry

(b) Confidential or non-confidential inquiry,

(c) Regular or irregular inquiry,
(d) Prospective or retrospective survey,
(e) Direct or indirect inquiry,
(f) Primary or secondary inquiry.

(v) *Decision regarding the units of the variable under investigation*: Height of a person can be measured in inch or in centimetre, weight of a person can be measured in pound or in kilogram, blood sugar level can be measured in mmol/l or in mg/dl. Since different units can be used for same variable, a decision is needed for the units to be used for a variable before start of the survey.

(vi) *Mode of measurement*: The variable age can be measured in integral as well as in fractional values. Income, working hours, height, weight, blood sugar level, etc. can be measured in integral and in fractional values. So, a decision is needed regarding the measurement of the values. Uniform technique must be applied in recording the values of a variable from all sampling units.

(vii) *Mode of data collection*: The data are collected by different methods. These are (a) direct interview by the investigator, (b) collection of data by local authority, and (c) mail questionnaire.

(viii) *Decision regarding the level of accuracy of the survey result*: After finalizing the objective and plan of the survey, it needs a decision in executing the plan. For the proper execution of the plan, the following steps are followed:
(a) To set up the administrative organization,
(b) Preparation of the frame of the population units and selection of sample,
(c) Preparation of questionnaire and pre-test of questionnaire through a pilot survey, if possible,
(d) Selection of investigator of enumerator,
(e) Training of investigator and supervisor,
(f) Control of error and minimization of non-response,
(g) Data edition,
(h) Presentation of data after classification and tabulation,
(i) Statistical analysis,
(j) Preparation of survey report.

Investigator

The person who is employed to collect information from sampling units is known as investigator. The investigator works under the supervision of a person, called supervisor. Supervisor and investigator must be well trained so that distorted information cannot be recorded.

Questionnaire

A set of questions related to the information needed for the survey, where information are to be collected from the sampling units constitutes a questionnaire. The following aspects are to be considered in preparing a questionnaire:

(a) The size of questionnaire in respect of number of questions should be minimum as far as possible,
(b) The questions in a questionnaire should be easy, self-explanatory, understandable, short in nature, fine in language and non-complex,
(c) The questions should be related to the objective of the survey,

(d) The word in any question should not be of multiple meaning,
(e) Question affecting the personal belief of a unit must be avoided,
(f) Question is formulated for definite answer or choice from multiple answers,
(g) Pre-test of the questionnaire should be made,
(h) Questionnaire should be well designed and care must be taken so that the collected data can be mechanically analysed.

The method of data collection has been discussed. The collected data are of two types, viz., (i) primary data, and (ii) secondary data.

Primary data

Data which are collected from sampling units by personal interview or by mailed questionnaire are called primary data. The primary data are not usually stored in a systematic fashion and hence, these are known as raw data. For example, let us consider the amount of fasting blood sugar (F.B.S., mmol/l) of some persons recorded in first 15 minutes of a working day. The amount of F.B.S. are as follows:

Amount of F.B.S. (x, mmol/l): $6.2, 5.8, 6.6, 7.8, 8.4, 7.5.$

These information of F.B.S. are primary data. However, primary data can also be presented in a systematic fashion. Let us consider another set of data on s. uric acid (x mg/dl) of some males and females:

S. uric acid of males : $3.5, 4.6, 3.8, 4.2, 4.5, 6.6, 6.8, 6.4, 5.3, 5.9, 6.4, 6.6$

S. uric acid of females : $2.5, 2.8, 3.9, 5.0, 5.2, 5.6, 4.8, 3.8, 3.6, 5.2, 4.7, 2.9$

These information can be presented in a tabular form as shown below. The data are shown in systematic fashion and hence, are called classified data. However, these are primary data.

Class interval of s. uric acid	Number of males	Number of females
< 3	–	2
3–5	7	6
5–7	5	4
Total	12	12

Secondary data

If data are collected from official record or from published work, these are known as secondary data. For example, data used from census report or from "World Fertility Survey" report are known as secndary data.

1.4 Statistical Method

The statistical methods are techniques or principles with the help of which numerical data are systematically treated so as to present a comprehensive view of the variation of observations recorded from different sampling units. The methods are used to simplify the data in such a way that their fundamental features throw light on the whole of the data set. It also helps to study the changes in the values of variables and to identify the causes of changes so that one can formulate a relationship between two or more interrelated variables. For example, the height of a human being is casually linked with his race, his ancestry, his habitation, his diet during youth, his age, etc. Here statistical methods can express the changes in height and formulate a relationship of height with causes of changes.

As a further instance, we can mention the study of progress of family planning activities. The progress depends on adoption of family planning methods by couples. Again, adoption depends on education of couples, occupation of couples, family income of couples, level of fertility of couples, level of child mortality of couples, etc. These information can be collected from different sampling units and these will vary from sampling unit to sampling unit. But various information of sampling units are not of our interest. The information of proportion of couples adopting family planning methods, the average number of children per couple, the level of education of husband and wife, etc. will give a comprehensive picture about progress of family planning activities. These comprehensive picture can be observed through statistical methods. Thus, statistical methods reduce complex data to a simple representative form easily adaptable to human mind. It also helps to compare the performance of similar activities of two or more regions, if proper data are collected from the regions. The major aspects of statistical methods are:

(a) To enable realization of magnitudes,
(b) To enlarge individual experience,
(c) To compare the simplified data and to measure their relationship,
(d) To guide the information of policy,
(e) To verify the laws of other sciences.

1.5 Importance of Biostatistics

The statistical methods depict the clearer picture of any problem affecting the welfare of mankind. The importance of biostatistics is thus pointed out as follows:

(i) Statistics of birth, death, marriage, migration, etc. are important for the plan of welfare of the people

(ii) Statistics of family planning adoption, service provided for family planning activities, current fertility level, child mortality level, level of education, etc. are important for sound population policy,

(iii) Statistics of child health care, health care of pregnant mothers, neonatal death, etc. are needed for planning improved family planning activities and child health care system,

(iv) Statistics of current number of doctors, nurses, health care units, hospitals, clinics, system of service provided in hospitals/clinics, etc. are needed for planning efficient hospital management,

(v) Statistics of production, import, distribution of medicine are also needed to improve health care system.

1.6 Scope of Biostatistics

With the development of advanced statistical methodology, analytical technique and with improved computer facilities, statistics is viewed as a sound technique for the analysis of numerical data and for drawing a valid conclusion about the population characteristics, where the charactristics are affected by multiplicity of causes. Numerical data are available from human population, existing health management, existing medical services and from production and distribution of medicine. The specific points of scope of biostatistics are as follows:

(i) Information on human aspects are collected and analysed for the development of plan for welfare state and for sound population policy. These information are available from vital registration system. Census data may also provide information on human aspects.

(ii) Mortality statistics, incidence of diseases, number of deaths by age, maternal mortality, disease specific deaths, age-specific deaths, neonatal and post-neonatal mortality are of paramount interest to health authority.

(iii) Data related to number of beds, number of doctors, number of nurses and number of other staffs help in planning proper hospital management.

(iv) Promotion of family planning activities depends on current status of family planning adoption and system of work to motivate couples for family planning adoption.

(v) Development of pharmaceutical industry needs information on production, consumption and distribution of different medical products in the country.

(vi) National policy on import, export of medicine and medical instruments needs data on production and distribution of output of the existing industries and current demand of the products.

(vii) For improved medical services data related to existing medical services and sources of diseases and present system of control of diseases are needed.

1.7 Limitation of Biostatistics

Despite of its wide application in medical science, biostatistics has some limitations which restrict its scope and utility. These limitations are:

(i) Qualitative character in medical and biological science is very prominent. But, except some classification by qualitative character the variable cannot be used for mathematical treatment. However, the qualitative variable is used for mathematical treatment if its values are scored by assigning appropriate number.

(ii) Biostatistics deals with comment for population units. So, it needs mass data instead of data from single unit of the population.

(iii) Analysing biostatistical data statistical inference is drawn with certain level of accuracy based on probability theory. Nothing is concluded with 100% confidence.

(iv) Biostatistics cannot prove anything. It plays an auxiliary role to summarise a fact.

(v) Though it can be reduced by efficient planning of the survey, yet biostatistical measurement shows an error due to the difference of population in true value and estimated value from sample survey result.

1.8 Research Methodology

We have discussed the methods of data collection with their scope and limitations. But data are needed to do the research according to the pre-determined objective. Let us discuss, in short, the methodology of research.

Research method

Technique to verify the existing knowledge on any field of inquiry, specially in the field of bioscience and to verify any proposition regarding the parameter assumed by the researchers is known as research method. The objective of the research is to answer the question which will arise whether existing knowledge is valid for all population under study or for all seasons or for any time period. Therefore, research method is a continuous process of investigation by a person or by a group of persons at any time or over a time period.

Research methods in bioscience are concerned with exploring, describing and explaining medical phenomenon involved in health services or in health sciences. The continuous study in the field of health/medical sciences usually helps to correlate the factors creating the health hazard in the country and helps the health wrokers/planners to take measures for improving the health status of human being in the country.

Type of research

It has already been mentioned that the research may be conducted as a continuous process or it can be done at a particular time period. Accordingly, the research may be of two types, viz., (a) *retrospective*, and (b) *prospective*. Retrospective research is done as a continuous process of study starting at a particular time. For example, fertility and child mortality have been declining since long in our country. One can study the trend in fertility level or trend in child mortality level. The trend can be observed, if study is done continuously starting from a particular point of time and continues for certain period in future. This type of study is retrospective study, where individuals/units under investigation are followed up. For cancer patients the study is continued up to the death of patients after the detection of disease. It is also an example of retrospective study.

If it is known that child mortality is declined due to improved health services to mothers of childbearing ages, specially, during their pregnancy, how further reduction of child mortality can be achieved is of interest of investigation, then some selected couples can be investigated to identify the factors which reduce the child mortality. This type of study is prospective one.

The research, whether it is prospective or retrospective, is again of two types, viz., (i) *theoretical research*, and (ii) *empirical research*. Theoretical research is usually done in the field of physical sciences. In medical sciences also theoretical research can be done by the bioscience researchers. For example, cancer and AIDS are two diseases which are taking life sooner or later. Medical researchers have been working since long to discover drugs to control these diseases. As a mode of research, they are making proposition to discover medicine. The propositional status or continuous effort of making propositions is a theoretical aspect of their research. This type of research is also a retrospective research.

The theoretical aspect of the research needs to be implemented by empirical investigation. The effectiveness of a discovered medicine can be studied by empirical investigation. The investigation can be done either by prospective study or by retrospective study. The theory, thus, broadens the scope of empirical findings. The empirical research and theory should proceed hand in hand for development of knowledge. A researcher may choose to begin from either empirical research or choose to begin from the theory. Wherever he chooses to begin his research, he needs to establish, at some stage, the link between his empirical research and the theory underlies it. Only the empirical research without relevance to theory has limited application. On the other hand, if the research deals with theory development without empirical investigation to verify the suitability of application of the theory, the theory will not be of much practical value.

The empirical research can be classified into 3 classes, viz., (a) exploratory, (b) descriptive, and (c) explanatory. The exploratory research is mainly done when little knowledge about prevailing situation works as a drawback of the research. This exploratory research enables the researchers to undertake and to formulate further research plan so that suggestion can be given to the health planners for the welfare of the human beings of the country. Little knowledge or even no knowledge about spreading HIV viruses is a serious problem in controlling the virus. In such a situation some exploratory studies can be performed for further development in the field of research to control AIDS.

Descriptive research is done when much is known about a problem. It is well known that hypertension of people aged 40-50 years leads to increase the mortality level. In such a situation a descriptive study can be taken to evaluate the present situation of patients of hypertension so that further suggestion can be made for the reduction of mortality level of people aged 40-50 years suffering from hypertension. Also, research can be done to identify the causes of more death due to hypertension of people aged 40-50 years. This type of research to identify the causes behind an event is exploratory research.

Whatever be the type of research, it is conducted under some organized steps. The important steps to conduct a research, specially the empirical research, are as follows:

(i) The planning and formulation of research problem,
(ii) The study of current literature related to the problem identified,
(iii) Identification of population and selection of sample,
(iv) The selection of variables and preparation of questionnaire according to proposed research problem,
(v) Implementation of investigation,
(vi) Methods of data collection,
(vii) Decision regarding measuring devices of variables,
(viii) Analytical procedure of data,
(ix) Report writing including interpretation and conclusion.

Sometimes the current literature enables the researcher to formulate a research plan including the objective of the study. At the planning stage the population is identified and mode of data collection is decided. After that the main task is to decide about how the research data can be collected.

The data are the information of variables, where the variables are of two types, viz., (a) *qualitative variable*, and (b) *quantitative variable.* Whatever be the variables, the measuring devices of them are different. These are discussed in the next chapter.

EXERCISE – 1

1. (a) What do you mean by biostatistics? What are the sources of biostatistical data?
   (b) Mention some of the objectives of studying biostatistics.
   (c) Write down some examples of population in the field of biostatistics.
2. (a) Explain the method of collection of biostatistical data.
   (b) What are the sources of biostatistical data?
   (c) Discuss the importance of biostatistics.
3. (a) Discuss the limitations of biostatistics.
   (b) What are the different types of research in the field of biostatistics?
   (c) What do you mean by empirical research?
4. (a) Mention one of the objectives of biostatistical study in a specified population.
   (b) Explain the different steps to complete your research according to your objective.
5. (a) What is research methodology in bioscience?
   (b) What are the different types of empirical research? Explain each type with example.
6. Explain different steps in conducting an empirical research.

7. Write short notes on:
   (a) Population in bioscience,
   (b) Sample
   (c) Variable,
   (d) Retrospective and prospective study,
   (e) Data,
   (f) Primary and secondary data,
   (g) Census and sample survey,
   (h) Scope of biostatistics.

Chapter 2

Presentation of Data

2.1 Introduction

The primary data which are collected through survey are called raw data. The raw data are not always suitable for proper statistical analysis. For analytical purpose and for the purpose of comprehensive idea about the population under investigation the data are usually ordered, classified, tabulated and presented by graphs and diagrams. The process of ordering, classifications and tabulation is called condensation of data. The classified data are presented by graphs and diagrams. The different aspects of condensation and graphical representation of data are discussed in the present chapter. Before condensation, the levels of measurement of variables, specially if the variables are qualitative in nature, are needed as statistical procedures require the quantified data.

2.2 Levels of Measurement

The different levels of measurement are:

(i) Nominal measures,
(ii) Ordinal measures,
(iii) Interval measures,
(iv) Scoring.

Nominal measures

If the variable is qualitative in nature, the data observed from different sampling units are identified by some number or by some notation. For example, the variable measured is fever, it may be malaria, typhoid, etc. This type of fever needs to express in quantity for using this variable in statistical analysis. Let us take malaria = 0 and typhoid = 1, and so on. Here the disease fever is expressed in number. This measurement of variable is known as nominal or nominal measures.

Ordinal measures

The value of qualitative variable, even the value of quantitative variable, is sometimes recorded according to rank. For example, the service of the hospitals is evaluated by a committee, where each member assigned a number on the basis of his evaluation of service. Later on, the service level is evaluated on the basis of the evaluation by all members and, for example, the service level is evaluated by 1, if the service is best, by 2, if the service is medium, and so on. This type of measurement of the variable is known as ordinal scaling. Again, the nurses working in a hospital may be ranked by 1, 2, 3, ... according to their service length. Here quantitative variable service length is measured by ranks 1, 2, 3, ... and this measurement is known as ordinal measurement.

Interval measures

If the value of the variable is measured by ordinal scale and if the distance between any two values of the variable is same, then the measurement of the values of the variable is known as interval measurement. For example, the blood pressure level of persons is measured by 60–70, 70–80, 80–90, and so on. Such measurement is known as interval measurement.

Scoring

Sometime the unit under study is characterized on the basis of values of several variables, qualitative and/or quantitative. For example, the admitted patients in a hosital are characterized as a patient in critical condition, as in danger level and as in out of danger level. These levels depend on many characters of the patients, viz., age, weight, blood pressure level, blood sugar level, level of s. urea, level of s. potassium, length of suffering from disease, treatment received earlier, medicine used, etc. Here observing the values of all the variables the patient can be scored by 1, if he is in critical condition, scored by 2, if he is in danger level and scored by 3, if the patient is out of danger. This type of measurement is known as scoring.

2.3 Data Condensation

As a step of condensation the primary data can be ordered and classified, and classified data can be presented by table. The tabulated data can be presented by graphs and diagrams. Let us discuss the ordering of data. Then ordered or unordered or classified data can be presented by (i) table, and/or by (ii) graphs and diagrams. Let us start with the ordering of data.

Array

The arrangement of values of quantitative variable either in ascending or in descending order is called array. For example, let us consider a set of primary data, where data are for the variable s. urea of some persons:

$$\text{S. urea } (x \text{ mg/dl}) : 18, 22, 12, 15, 20, 27, 30, 25, 40, 38, 37, 17.$$

These data can be arranged in ascending or descending order as follows:

$$\text{Ascending order } (x) : 12, 15, 17, 18, 20, 22, 25, 27, 30, 37, 38, 40.$$
$$\text{Descending order } (x) : 40, 38, 37, 30, 27, 25, 22, 20, 18, 17, 15, 12.$$

The above arrangements of data are known as array.

Classification

The raw data are not suitable to be used in statistical analysis as these are so scattered from each other. For comprehensive presentation, and for statistical analysis the raw data can be presented in a tabular form, where table contains rows and columns used for showing the number of observations according to two characteristics under study. The characteristics may be qualitative or quantitative. The body of the table contains number of observations of each level of characters. The method of division of observations into several classes is necessary so that the observations of each class indicate the frequency of observations of homogeneous character. According to Secrist, "Classification is a technique by which data are divided into several related classes or arrangement of data is made consecutively according to some characteristics so that the observations of a class are homogeneous in nature. One of the levels of characters is shown in rows and the levels of other characters are shown in columns. The characters may be qualitative or quantitative. Thus, we have two types of classification, viz., (i) qualitative classifications, (ii) quantitative classification. The classification of data can be shown by example.

Let us consider a set of primary data shown in Table 1.

Table 1: The information collected from some persons suffering from hypertension

Sl. No.	Sex of persons	Age in years	Level of education	Occupation	Income/month in '000 Rs.	Level of blood pressure	Medicine used
1	1	45	16	1	35	92	2
2	1	58	10	2	50	95	2
3	1	35	16	1	20	88	1
4	0	42	10	0	25	85	1
5	1	54	8	2	48	105	2
6	1	45	5	2	50	105	2
7	0	38	4	0	25	90	1
8	0	44	8	0	20	95	1
9	0	52	10	0	35	90	1
10	1	45	12	2	50	100	2
11	1	65	16	1	40	98	2
12	1	60	16	1	32	95	2
13	1	38	14	1	35	80	0
14	1	42	14	2	40	88	1
15	0	46	14	1	28	85	1
16	0	40	10	1	15	87	1
17	0	35	10	0	40	90	1
18	1	58	14	1	32	100	2
19	1	60	14	1	30	105	2
20	1	50	10	2	45	95	2

Here sex = 1 (male), sex = 0 (female), level of education is measured in completed years of schooling, occupation = 0 (housewife), occupation = 1 (service), occupation = 2 (business), level of blood pressure is diastolic, medicine used = 0 (not used), 1 = daily one time used, 2 = daily 2 times used.

Let us classify these observations according to sex and level of blood pressure. The classified results are shown in Table 2.

Table 2: Number of persons by sex and level of blood pressure

Sex	Level of blood pressure						Total
	80–85	85–90	90–95	95–100	100–105	105–110	
Male	1	2	1	4	2	3	13
Female	–	3	3	1	–	–	7
Total	1	5	4	5	2	3	20

The above classification is cross-classification of qualitative and quantitative characters. But, column total indicates the number of persons according to different levels of quantitative variable. Row total indicates the number of classified persons by qualitative character.

Another cross-classification can be made by both qualitative characters, viz., sex and occupation. The classified results are shown in Table 3. Here column total indicates number of persons of different occupations. In both row and column, levels of qualitative variables are shown. Sometimes the classification is done by only one character, qualitative or quantitative. For example, if the number of persons is shown by occupations only, then we can present this in a tabular form as below (Table 4):

Table 3: Number of persons by sex and level of occupation

Sex	Level of occupation			
	Housewife	Service	Business	Total
Male	–	7	6	13
Female	5	2		7
Total	5	9	6	20

Table 4: Number of persons by occupation

Occupation	Number of persons
Housewife	5
Service	9
Business	6
Total	20

Here levels of qualitative variable can be shown either in row or in column and number of observations are shown in column/row. This is a simple qualitative classification.

The number of observations can also be shown by levels of quantitative variable. As an example, we can mention the Table 5. This clasification is known as quantitative classification.

Table 5: Number of persons by levels of blood pressure

Levels of blood pressure	Number of persons
80–85	1
85–90	5
90–95	4
95–100	5
100–105	2
105–110	3
Total	20

It is also called frequency table. The process of preparation of frequency table is discussed below. Here the levels of quantitative variable is shown in rows of a column and number of observations of each level are shown in rows of another column. The number of observations of different levels are known as frequencies.

Classification can also be done by place and time. For example, let us consider the results of Table 6, where crude birth rate per 1000 are given for different years. These results can also

Table 6: The crude birth rate of Bangladesh in different years

Year	1991	1992	1993	1994	1995	1996	1997	1998
Crude birth rate per 1000	31.6	30.8	28.8	27.8	·26.5	25.6	21.0	19.9

be classified by place, viz., rural and urban. The results are shown in Table 7.

Table 7: The crude birth rate of Bangladesh in different years

Year	1991	1992	1993	1994	1995	1996	1997	1998
Urban CBR per 1000	23.9	23.7	21.0	20.2	19.4	19.0	16.2	14.0
Rural CBR per 1000	32.9	32.2	30.0	29.1	28.5	26.4	24.5	21.4
CBR for the country	31.6	30.8	28.8	27.8	26.5	25.6	21.0	19.9

[*Source*: Statistical Yearbook of Bangladesh, 2001.]

The data presented in Table 2, Table 3, Table 4 and Table 5 are shown from the data of Table 1, where the data of Table 1 are primary data. But data of Table 6 and Table 7 are secondary data.

Tabulation

The primary data shown in Table 1 are presented partly in Table 1, Table 2, Table 3, Table 4 and Table 5, where the tabulated values are classified data from raw form. Thus, the process of classification is the first step of tabulation. The tabulated information show statistical data having common characteristic in one place. This presentation of data simplifies complex data and facilitates comparison. The tabulated information make statistical analysis easier. According to A.M. Tutle, "A statistical table is the logical listing of related quantitative information in vertical columns and in horizontal rows with sufficient explanatory and qualifying words, phrases, and statements in the form of title, sub-titles, headings, and notes to make clear the full meaning of data and their origin".

2.4 Frequency Distribution

One of the important form of tabulation is the frequency distribution or frequency table [Table 5]. This table is prepared to present the number of observations in each class, where classes are made according to different values of quantitative variable. The Table 5 is an example of frequency distribution in which number of persons of different levels of blood pressure are shown. Sometimes number of observations of each value of the quantitative variable are also shown in a frequency table. The frequency table is prepared for both discrete and continuous variable. In its preparation the rules and regulations of preparation of table are followed. In the next section different steps of preparation of frequency table are discussed.

2.5 Steps in Preparing Frequency Table

The following steps are followed in preparing a frequency table.

(i) Each frequency table should be started with a short, concise and self-explanatory title.

(ii) There are rows and columns of a frequency table, where rows are used to present the individual values of the variable or intervals of the values of the variable so that the number of observatons of each interval or of each individual value can be shown. There are different columns in a frequency table. The first column is used to show individual values of the variables or intervals of the values of variables. This column is headed by "Class Interval". The second column headed by "Tally Marks" is used to accommodate the number of observations of each level/class interval of the values of variables. The third column headed by "Frequency, f" is used to show the number of observations of each level/class interval of the values of variable. The number of observations of each level/class is the total of tally marks of corresponding level/class. The fourth column headed by "Cumulative Frequency, cf" is used to present the number of observations of some lower value to highest value [if accumulation is done from bottom of the frequency table] or the number of observations of below an upper limit of a class [if accumulation is done from the top of the frequency table]. Sometimes fifth column headed by "Mid-value" of a class is used to present an approximate average value of a class.

The mid-value is denoted by X, where $X = \frac{1}{2}$ [lower limit of the class + upper limit of the class]. This mid-value is very important for further statistical analysis. If the lower limit of the first class and/or the upper limit of the last class are open, the mid-value for those classes are taken according to the change in mid-values of other classes [usually change by width of the class]. Thus, the mid-value of the first class is the mid-value of the second class less by width of the class and the mid-value of the last class is the mid-value of the last but one class plus width of the class.

(iii) *Decision regarding number of classes and class intervals*: Let $x_1 < x_2 < \cdots < x_N$, where x_1 is the lowest value of quantitative variable and x_N is the highest value of the variable. Then range of the values of the variable is

$$\text{Range, } R = x_N - x_1.$$

Let k be the number of classes of a frequency table. The value of k can be calculated mathematically, where

$$k = 1 + 3.3 \log_{10} N,$$

where N is the total number of observations.

However, mathematical calculation can be avoided. In that case the value of k should not be less than 5 and should not be more than 20.

After the decision of the value of k, the width of the class is calculated by

$$h = \frac{R}{k},$$

where h = width of the class [upper limit − lower limit of the class].

For large data set the value of h can be taken as 5 or multiple of 5.

The width of the class may not be constant for every class. Different classes may have different width. Sometimes the lower end of first class is kept open to accommodate very few observations of data in left end. In a similar way very few observations of data in the right end may be accommodated by keeping the upper end of last class open. For example, the classes may be of this type:

$$< (a+h), (a+h) - (a+2h), (a+2h) - (a+3h), \ldots, a + (k-1)h-$$

where lower end of first class and upper end of last class is open. If open end class is not needed the lower end of first class will be a and upper end of last class will be $(a + kh)$.

The classes discussed above are discontinuous in nature. This type of classes are suitable for both discrete and continuous variables. Discontinuous classes, such as 0–4, 5–9, 10–11, ⋯ may also be used if the variable is discrete. However, for continuous classes the number of observations in a class is accommodated by exclusion of the upper limit. For example, if a class is 5–10, the 10 is not included in this class. The observation 10 is included in the next class.

In case of discontinuous classes the lower limit and upper limit of classes are decided as follows:

$$\text{Lower limit} = \text{observed lower limit} - \tfrac{1}{2}x$$
$$\text{Upper limit} = \text{observed upper limit} + \tfrac{1}{2}x,$$

where x = observed lower limit of a class minus the observed upper limit of the preceding class.

For example, if classes are 0–4, 5–9, 10–14, ... then $x = 5 - 4 = 1$. Therefore, the real class will be $0 - 0.5$ to $4.0 + 4.5$, $5 - 0.5$ to $9 + 0.5$, ...

However, the classes are considered in such a way that each observation is accommodated once and only once in a class.

(iv) *Tally marks*: It is tally sign used to indicate that an observation is accommodated in a class. Tally mark is used for each observation. It facilitates the count of frequencies of a class.

(v) *Frequency*: Total tally marks of a class is the frequency of the class. The tally marks of i-th class is denoted by $f_i (i = 1, 2, \ldots, k)$.

(vi) *Cumulative frequency*: It is also a part of frequency table. It indicates the number of observations for less than a certain upper limit of a class if cumulation is done from top of the frequency table. The cumulated frequency calculated from bottom of the table indicates the number of observations for certain lower limit and above a class. It is usually denoted by $c.f.$

EXAMPLE 2.1 The following data represent the amount of s. urea (mg/dl) of some persons:

15, 18, 22, 27, 30, 35, 40, 18, 22, 24, 26, 45, 30, 17, 12, 32, 27,
41, 25, 15, 10, 12, 27, 26, 21, 20, 30, 28, 22, 25, 28, 42, 44, 48,
37, 25, 23, 24, 19, 20, 17, 13, 11, 15, 17, 32, 12, 14, 18, 11.

(i) Represent the above observations by a frequency table.
(ii) Find number of persons who have s. urea 25 and above.
(iii) Find percentage of persons who have s. urea less than 30.

Solution. (i) Here highest value and lowest value of observations are 48 and 10, respectively. There are $N = 50$ observations. $R = x_n - x_1 = 48 - 10 = 38$, where

x_n = highest value of observations

x_1 = lowest value of observations

k = number of classes $= 1 + 3.3 \log N = 1 + 3.3 \log 50 = 6.6 \approx 7$

h = width of the class $= \dfrac{R}{k} = \dfrac{38}{7} = 5.4 \approx 5.$

Title: *Frequency table of the amount of s. urea of some persons*

Class interval of s. urea	Tally marks	Frequency f	Cumulative frequency, $c.f.$	
10–15	𝍸 III	8	8	50
15–20	𝍸 𝍸	10	18	42
20–25	𝍸 IIII	9	27	32
25–30	𝍸 𝍸	10	37	23
30–35	𝍸	5	42	13
35–40	II	2	44	8
40+	𝍸 I	6	50	6
Total		$N = \Sigma f = 50$		

(ii) There are 23 persons who have s. urea 25 and above. [c.f. corresponding to class 25–30 calculated from bottom of the table]

(iii) There are 37 persons who have s. urea less than 30 [c.f. calculated from top of the table corresponding to class 25-30].

Percentage of persons who have s. urea less than 30 is

$$\frac{37}{50} \times 100 = 74\%.$$

EXAMPLE 2.2 The following are the amounts of diastolic blood pressure (mmHg) level of some persons:

Diastolic blood pressure (x): 70, 72, 78, 85, 82, 80, 65, 77, 75, 70, 72, 82, 74, 76, 85, 88, 90, 82, 66, 70, 72, 74, 75, 80, 75, 90, 88, 68, 70, 78.

(i) Represent the above observations by a frequency table.

(ii) Calculate the number of persons who have blood pressure levels less than 80.

(iii) Calculate the percentage of persons who have blood pressure levels 75 and above.

Solution. (i) Let us use the k = number of classes arbitrarily so that cumulative frequency provides the answer to question numbers (ii) and (iii). Here classes are used in such a way that 75 and 80 are in the lower and upper ends, respectively of classes.

Title: *Frequency distribution of the blood pressure levels of different persons*

Class interval of blood pressure	Tally marks	Frequency f	Cumulative frequency, $c.f.$	
65–70	|||	3	3	30
70–75	~~||||~~ ||||	9	12	27
75–80	~~||||~~ ||	7	19	18
80–85	~~||||~~	5	24	11
85–90	||||	4	28	6
90+	||	2	30	2
Total		$N = \Sigma f = 30$		

(ii) There are 19 persons [c.f. corresponding to the class 75–80, where c.f. is calculated from top of the frequency table.] who have blood pressures less than 80.

(iii) There are 11 persons [c.f. corresponding to the class 80–85, where c.f. is calculated from the bottom of the class] who have blood pressures 80 and above.

So, $\left(\frac{11}{30} \times 100\right) = 36.7\%.$

EXAMPLE 2.3 The following are the number of nurses working in different hospitals:

Number of nurses (x): 12, 18, 15, 22, 20, 16, 15, 18, 17, 20, 16, 15, 20, 22, 20.

Represent the data by a discrete type frequency table.

Solution. Discrete type frequency table is one where number of observations is shown against each value of the variable.

Title: *Frequency distribution of number of nurses working in hospitals*

<table>
<tr><th rowspan="2">Number of nurses</th><th rowspan="2">Tally marks</th><th rowspan="2">Frequency, f (Number of hospitals)</th><th colspan="2">Cumulative frequency, $c.f.$</th></tr>
<tr><th>From top</th><th>From bottom</th></tr>
<tr><td>12</td><td>|</td><td>1</td><td>1</td><td>15</td></tr>
<tr><td>15</td><td>|||</td><td>3</td><td>4</td><td>14</td></tr>
<tr><td>16</td><td>||</td><td>2</td><td>6</td><td>11</td></tr>
<tr><td>17</td><td>|</td><td>1</td><td>7</td><td>9</td></tr>
<tr><td>18</td><td>||</td><td>2</td><td>9</td><td>18</td></tr>
<tr><td>20</td><td>||||</td><td>4</td><td>13</td><td>6</td></tr>
<tr><td>22</td><td>||</td><td>2</td><td>15</td><td>2</td></tr>
<tr><td>Total</td><td></td><td>$N = \Sigma f = 15$</td><td></td><td></td></tr>
</table>

Bi-variate frequency table: The frequency tables shown in Example 2.1, Example 2.2 and in Example 2.3 are for presenting the values of one variable observed from units under investigation. These frequency tables are called uni-variate frequency tables. From each unit more than one variable can also be observed. If two variables are observed from each unit, the number of observations of certain level of one variable corresponding to the level of another can also be shown by representing the levels of one variable in column and levels of another variable in row. Such a table is called bi-variate frequency table.

EXAMPLE 2.4 The following are the gestation period (x in days) and birth-weight (y in kg) of some newborn babies of some mothers:

x : 260 265 250 255 260 262 270 268 250 245 240
y : 3.0 3.2 2.8 2.6 2.8 3.2 3.5 3.0 2.5 2.7 2.0
x : 252 255 258 250 255 260 265 270 262 260 255
y : 3.1 3.4 2.5 2.2 2.6 3.2 3.4 3.2 2.5 2.5 2.4
x : 245 248 250 255 265 270 264 258
y : 2.4 2.5 3.0 2.8 3.0 3.0 2.9 3.0

Represent the data by a bi-variate frequency table.

Solution.

Title: *The distribution of number of mothers according to their gestation period and birth-weight of baby*

<table>
<tr><th rowspan="2">Class interval of gestation period (x)</th><th rowspan="2"></th><th colspan="4">Class interval of birth-weight (y)</th><th rowspan="2">Frequency f_x</th></tr>
<tr><th>2.0–2.4</th><th>2.4–2.8</th><th>2.8–3.2</th><th>3.2–3.6</th></tr>
<tr><td rowspan="2">240–245</td><td>Tally Marks</td><td>|</td><td></td><td></td><td></td><td></td></tr>
<tr><td>Frequency</td><td>1</td><td></td><td></td><td></td><td>1</td></tr>
<tr><td rowspan="2">245–250</td><td>Tally Marks</td><td></td><td>|||</td><td></td><td></td><td></td></tr>
<tr><td>Frequency</td><td></td><td>3</td><td></td><td></td><td>3</td></tr>
<tr><td rowspan="2">250–255</td><td>Tally Marks</td><td>|</td><td>|</td><td>|||</td><td></td><td></td></tr>
<tr><td>Frequency</td><td>1</td><td>1</td><td>3</td><td></td><td>5</td></tr>
<tr><td rowspan="2">255–260</td><td>Tally Marks</td><td></td><td>||||</td><td>||</td><td>|</td><td></td></tr>
<tr><td>Frequency</td><td></td><td>4</td><td>2</td><td>1</td><td>7</td></tr>
<tr><td rowspan="2">260–265</td><td>Tally Marks</td><td></td><td>||</td><td>|||</td><td>||</td><td></td></tr>
<tr><td>Frequency</td><td></td><td>2</td><td>3</td><td>2</td><td>7</td></tr>
<tr><td rowspan="2">265–270</td><td>Tally Marks</td><td></td><td></td><td>||</td><td>||</td><td></td></tr>
<tr><td>Frequency</td><td></td><td></td><td>2</td><td>2</td><td>4</td></tr>
<tr><td rowspan="2">270–275</td><td>Tally Marks</td><td></td><td></td><td>|</td><td>||</td><td></td></tr>
<tr><td>Frequency</td><td></td><td></td><td>1</td><td>2</td><td>3</td></tr>
<tr><td>Frequency, f_y</td><td></td><td>2</td><td>10</td><td>11</td><td>7</td><td>30</td></tr>
</table>

2.6 Graphical and Diagrammatical Representation of Data

Raw data does not provide any comprehensive idea about the population. However, preliminary inference can be drawn from classified data when it is presented in tabular form. Further comprehensive idea about the data and the population can be obtained when data are presented in chart form or by graphs and diagrams. The graphical representation helps, even the layman, to have an idea about the population at a glance. In this section some important graphs and diagrams will be discussed with example.

2.6.1 General rules in constructing graphs and diagrams

The following are the general rules in constructing graphs and diagrams.

(i) *Title*: Each diagram should have a short, suitable and self-explanatory title. The title is placed in the middle either at the top or at the end of the diagram.

(ii) *Selection of the scale*: To present the values of the variable or the level of the variable, proper scale should be so selected that it is consistent with the size of the paper in which the diagram is to be drawn and also is consistent with the value of the variable to be presented. The scale is so chosen that the diagram is not looked too small or too big. Separate scale may be used for horizontal and vertical axes. Similar scale is to be used for each diagram when diagrams need comparison.

Let us consider the selection of scale by an example of the distribution of the number of persons having different blood pressure levels (mmHg) given below:

Class interval of blood pressure	< 65	65–70	70–75	75–80	80+
Number of persons, f	10	13	12	20	25

Here to present the values of blood pressure one can consider 1 cm = 5 in the X-axis in which the blood pressure is to be shown. Again, 1 cm = 2 can be taken to present the number of persons in the required direction.

(iii) *Axis to be used*: Information on variable, qualitative or quantitative, are collected from different units under study. Graphs and diagrams are used to show the number of units against different levels of qualitative or quantitative variable. The general principle to construct a diagram is to show the number of observations against the level of qualitative or quantitative variable. The values which are to be shown are plotted on Y-axis and against which these are to be shown are plotted on X-axis.

The other rules in constructing diagram are: (a) the diagram should be neat and clean, (b) An index to explain the symbols, colours, lines, etc. is to be given, (c) Explanatory notes to explains the important points should be given at the corner or at the bottom of the diagram.

2.6.2 Different graphs and diagrams

The important graphs and diagrams used in biostatistics are:

(a) Line diagram, (b) Bar diagram, (c) Pie diagram.

These 3 are used to present the values of qualitative variable.

(d) Stem-and-leaf plot, (e) Histogram, (f) Frequency curve,

(g) Ogive, (h) Scatter diagram.

These are used to present the values of quantitative variable.

Data can also be presented by hinges and 5 number summary with respect to measure of central tendency to be discussed in the next chapter.

2.6.3 Line diagram

This diagram is usually used to represent the time series data, where time is plotted in horizontal axis (X-axis) and the value of the variable against time is plotted in vertical axis (Y-axis) with appropriate scale. The plotted point in Y-axis is added to the X-axis by a straight line parallel to Y-axis. There will be several lines for all points of time. The resultant diagram is callled line diagram.

EXAMPLE 2.5 The following data represent the number of dead patients of a hospital in different years.

Year	2006	2007	2008	2009	2010	2011
Number of dead patients	725	680	550	650	540	500

Represent the data by a line diagram.

Solution.

Title: *Line diagram of the number of dead patients of a hospital in different years*

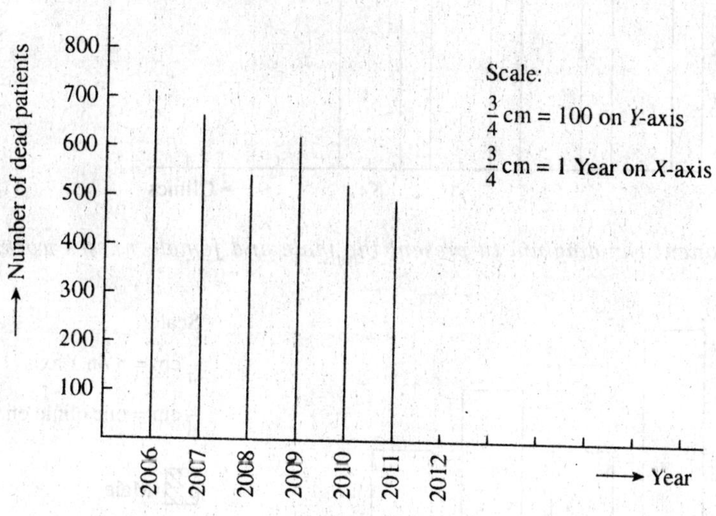

2.6.4 Bar diagram

This diagram is similar to line diagram except that the value of a variable is shown by a rectangle instead of a line. This is specially used to present the values of different levels of a qualitative variable, where levels of qualitative variable are plotted on X-axis and the values of different levels are plotted on Y-axis with appropriate scale. The value of each level is shown by a rectangle parallel to the Y-axis. The rectangle is also called a bar and hence, the name of the diagram is bar diagram. The distance from bar to bar should be half of the width of the bar.

There are different types of bar diagram, viz., (i) simple bar diagram, (ii) multiple bar diagram (it is used to show the values of a variable recorded from same level of the variable but on different occasions) (iii) component bar diagram (it is used to show the values of a variable recorded from same level of the variable when the total values are the sum of the components of the variable). Here example of simple bar diagram and component bar diagram are given.

EXAMPLE 2.6 The following are the number of nurses working in different clinics in a city.

Clinics	c_1	c_2	c_3	c_4	Total
Male nurse	12	8	5	6	31
Female nurse	18	12	20	12	62
Total	30	20	25	18	93

Represent the data by (i) bar diagram, (ii) component bar diagram.

Solution.

(i) **Title:** *Bar diagram representing the number of nurses working in clinics*

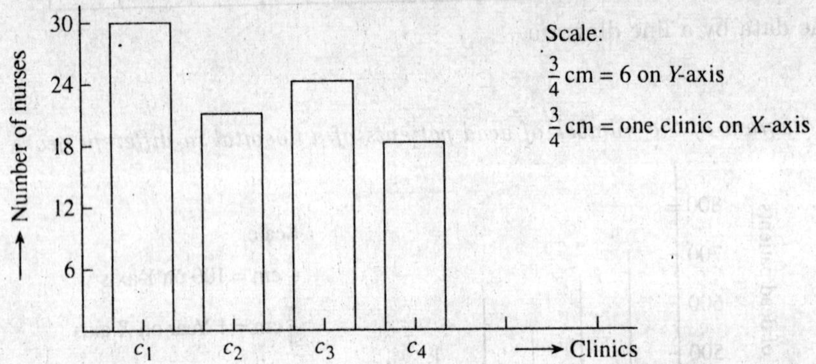

(ii) **Title:** *Component bar diagram to present the male and female nurses working in clinics*

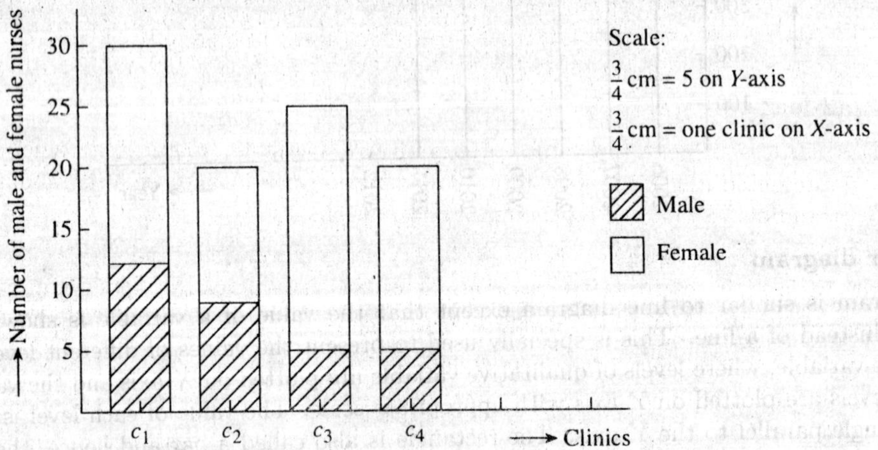

2.6.5 Pie diagram

This diagram is also used to present the values of different levels of a qualitative variable, where values are transformed to angle and the angles are drawn within a circle.

The total angle of a circle is 360°. Accordingly, the value of a particular level of the qualitative variable is transformed to angle, where the angle is proportional to total angle 360°. All the angles, except one, are drawn within the circle one by one.

EXAMPLE 2.7 The following are the number of heart patients who are under treatment in different hospitals.

Hospital	A	B	C	D	Total
Number of heart patients	105	40	120	60	325
Angle for patients	$\frac{105 \times 360}{325} = 116.3 \approx 116$	$\frac{40 \times 360}{325}$ $44.3 \approx 44$	$\frac{120 \times 360}{325} = 132.9 \approx 133$	67	360°

Last angle $= 360 - (116 + 44 + 133) = 67$.

Solution.

Title: *Pie diagram representing the number of heart patients of different hospitals.*

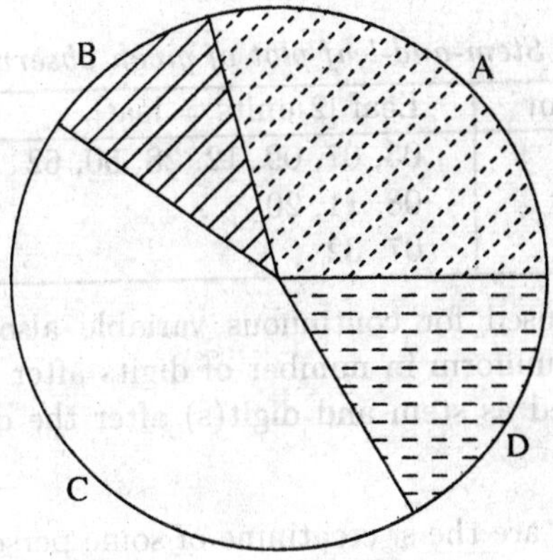

2.6.6 Stem-and-leaf plot

This diagram is used to present the values of quantitative variable, where large number of observations are presented into a compact form. This diagram provides information about the range of data set and also about the repetition of observations.

The quantitative variable is of two types, viz., (a) discrete variable, and (b) continuous variable. If the variable is discrete, the observations of the variable can be made uniform in number of digits by adding 0 (zero) in the left of the values.

For example, let us consider the ages of some persons as follows:

Age (x in years) : 42, 13, 8, 25, 32, 7, 15, 40, 28, 22.

These observations are made uniform in number of digits as follows:

Age (x in years) : 42, 13, 08, 25, 32, 07, 15, 40, 28, 22.

Here each observation is of two digits. In preparing stem-and-leaf plot the left side digit is used as stem and right side digit(s) is used as leaf. Both stem and leaf are arranged in ascending order. The resultant arrangement of the observations is known as stem-and-leaf plot. Let us arrange the observations of age in stem-and-leaf plot, where the plot is as follows:

Title: *Stem-and-leaf plot of the ages of some persons*

Stem	Leaf [1 unit = leaf]
0	7, 8
1	3, 5
2	2, 5, 8
3	2
4	0, 2

Let us consider another set of observations as follows:

$$\text{Observations } (x) : 50, 12, 108, 5, 62, 111, 207, 28, 9, 120, 232, 7.$$

These observations can be made uniform in number of digits as follows:

$$x : 050, 012, 108, 005, 111, 207, 028, 009, 120, 232, 007.$$

Now, all the observations are of 3 digits. In such a case, the stem-and-leaf plot is plotted using left side digit as stem and right side digits as leaf. Here two digits of right side are used as leaf.

Title: *Stem-and-leaf plot of given observations*

Stem	Leaf [2 units = leaf]
0	05, 07, 09, 12, 28, 50, 62
1	08, 11, 20
2	07, 32

The stem-and-leaf plot is used for continuous variable also. The given observatons of continuous variable are made uniform in number of digits after decimal by adding zero. The digit(s) before decimal are used as stem and digit(s) after the decimal are used as leaf. The decimal is overlooked.

EXAMPLE 2.8 The following are the s. creatinine of some persons

s. creatinine (x) : 0.62, 0.6, 1, 1.2, 0.67, 0.65, 0.8, 0.85, 1, 1.1, 1.05, 0.7, 0.95, 1.1, 0.78, 0.88, 0.73, 1.06, 0.72

Represent the data by a stem-and-leaf plot.

Solution. The observations can be rewritten by adding a zero in the right side to make the observations uniform in digits after the decimal. The rewritten observations are

x : **0.62, 0.60, 1.00, 1.20, 0.67, 0.65, 0.80, 0.85, 1.00, 1.10, 1.05, 0.70, 0.95, 1.10, 0.78, 0.88, 0.73, 1.06, 0.72.**

Title: ***Stem-and-leaf plot of the amount of s. creatinine of persons***

Stem	**Leaf [2 units = leaf]**
0	**60, 62, 65, 67, 70, 72, 73, 78, 80, 85, 88, 95**
1	**00, 00, 05, 06, 10, 10, 20**

2.6.7 Histogram

This diagram is used to represent the frequencies of continuous classes of a frequency distribution of quantitative variable. In this diagram class intervals are plotted in X-axis and frequencies are plotted in Y-axis with appropriate scales. Since each class has a width, the

frequency of a class is shown by erecting a bar with width equal to class width of the frequency distribution. As classes are continuous, the diagram shows consecutive rectangles. The height of the rectangle is proportional to the width of the classes, if classes are of unequal width.

Frequencies of discontinuous classes can also be depicted by drawing histogram. This is possible by changing the class limits to class boundaries. The frequencies of classes are then shown against classes by bars.

Histogram can also be drawn for open-end classes, even the magnitude of first class is not the same as that of the following class or the magnitude of the last class is not the same as that of the preceding class.

In this diagram there are adjacent bars looking like a bar diagram. Still there are some differences in bar diagram and histogram. These are:

(a) Bar diagram is used to present the value of a qualitative variable or value against time period, while histogram is used to present the values of a quantitative variable.

(b) The bars of bar diagram are at equal distance, while the bars of histogram are adjacent.

(c) The height of the bar of histogram is proportional to the width of a class, where the height of a bar diagram depends on the value of a level of qualitative variable.

(d) Histogram is a two-dimensional diagram, where length and breadth are under consideration, while in bar diagram only length of bar is under consideration and breadth of the bar depends on the space available for the diagram.

EXAMPLE 2.9 The following is the frequency distribution of S. GGT of some persons:

Class interval of S. GGT (U/L)	15–20	20–25	25–30	30–35	35–40	Total
Number of persons, f	4	6	8	5	2	25

Draw histogram of the distribution of S. GGT.

Solution.

Title: *Histogram of the distribution of S. GGT of some persons.*

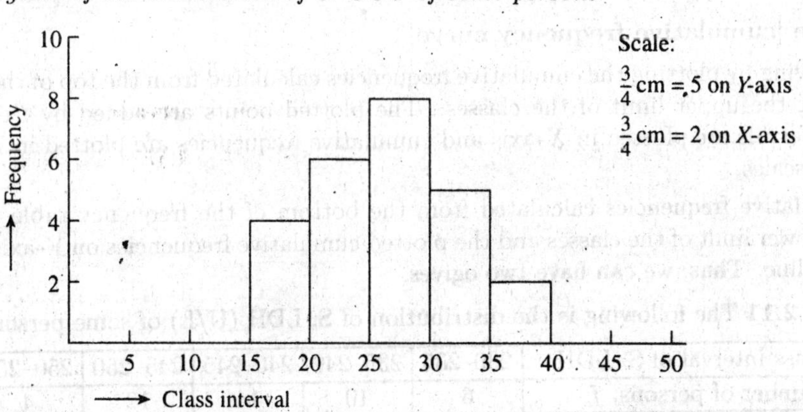

In drawing histogram the class interval can be plotted in X-axis by using zigzag in the origin of X-axis, when class limit is of higher value. In such case, origin of the graph may not be considered as $(0, 0)$. This is true in drawing frequency curve, ogive and scatter diagram also.

2.6.8 Frequency curve

This graph is also used to represent the frequency of a continuous class of the quantitative variable, where class intervals are plotted on X-axis and frequencies are plotted on Y-axis with appropriate scales. The frequency of a class interval is plotted against the mid-point of the class. The frequency points of different classes are added by freehand line or smooth line. The area under the curve is enclosed by extending the line to the X-axis against the mid-point of the assumed preceding and following classes of the frequency table.

The frequency curve can also be drawn by drawing histogram first. Then the mid-points of the upper ends of bars of different classes are added by smooth line. The line is extended to the X-axis as mentioned above.

EXAMPLE 2.10 The following is the frequency distribution of s. sodium of some persons:

Class interval of s. sodium (mmol/l)	136–138	138–140	140–142	142–144
Number of persons, f	5	12	6	2

Draw frequency curve of the distribution of s. sodium.

Solution.

Title: *Frequency curve of the distribution of s. sodium.*

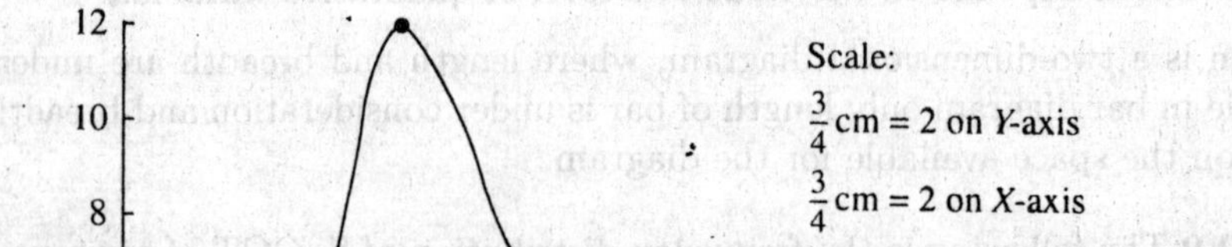

2.6.9 Ogive [cumulative frequency curve]

Ogive is drawing by plotting the cumulative frequencies calculated from the top of the frequency table against the upper limit of the classes. The plotted points are added by freehand line. The class intervals are plotted in X-axis and cumulative frequencies are plotted in Y-axis with appropriate scales.

The cumulative frequencies calculated from the bottom of the frequency table are plotted against the lower limit of the classes and the plotted cumulative frequencies on Y-axis are added by freehand line. Thus, we can have two ogives.

EXAMPLE 2.11 The following is the distribution of S. LDH (U/L) of some persons:

Class interval of S. LDH	230–235	235–240	240–245	245–250	250–255
Number of persons, f	6	10	14	12	4
Cumulative frequency, $c.f.$	6	16	30	42	46
	46	40	30	16	4

Draw ogive(s) of the distribution of S. LDH.

Solution.

Title: *Ogive(s) of the distribution of S. LDH of some persons.*

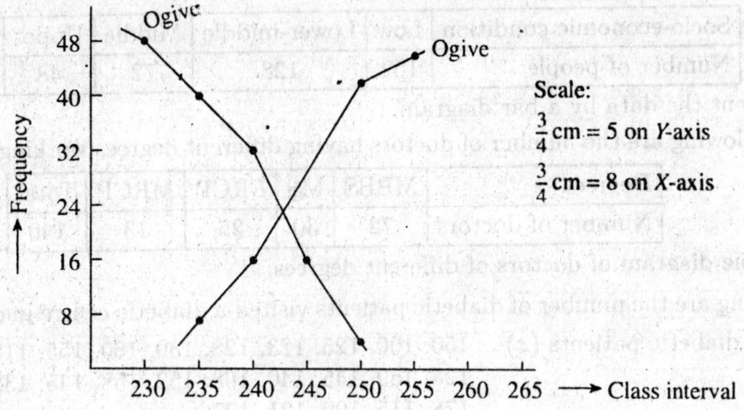

2.6.10 Scatter diagram

If two variables are observed from same unit, the variables are assumed to be correlated. For example, age (x) and blood pressure (y) of a person are assumed to be correlated. Here y is dependent on x. Similarly, weight (y) of a person depends on height (x) of him. The scatter diagram is drawn to show the pairs of values of x and y, where the values of independent variable are plotted on X-axis and values of dependent variable are plotted on Y-axis with scales. The pair of values are shown by dots and there are several dots to show the pairs of values observed from different units. The dots are scattered over the diagram and hence, the diagram is called scatter diagram. It is also called dot diagramme as there are dots to present the coordinate of two values of two variables.

EXAMPLE 2.12 The following are the ages (x in years) and systolic blood pressure (y mmHg) of some persons:

x	30	40	35	55	52	50	45	50	55	55
y	95	80	85	90	90	85	95	100	100	95

Represent the data by a scatter diagram.

Solution.

Title: *Scatter diagram of the age and systolic blood pressure of some persons.*

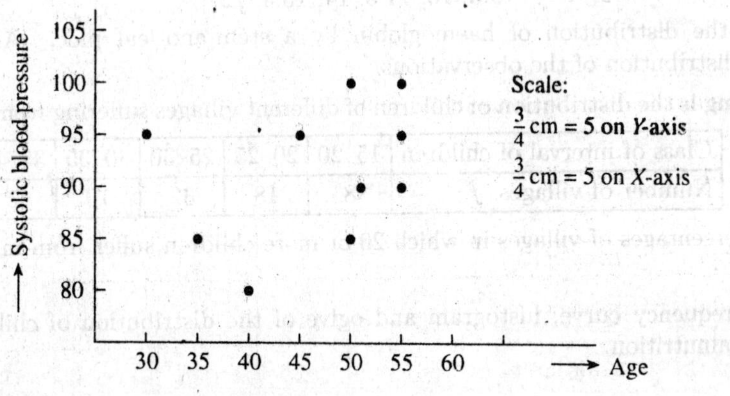

EXERCISE – 2

1. (a) The following are the number of people of different socio-economic conditions who suffer from malnutrition:

Socio-economic condition	Low	Lower-middle	Middle	Upper	Total
Number of people	152	128	72	48	400

Represent the data by a bar diagram.

(b) The following are the number of doctors having different degrees working in a hospital:

Degrees	MBBS	MS	FRCP	MRCP	Total
Number of doctors	72	30	25	13	140

Draw pie diagram of doctors of different degrees.

2. The following are the number of diabetic patients visited a diabetic centre in different days:

Number of diabetic patients (x) : 150, 100, 125, 123, 128, 130, 160, 155, 113, 118, 128, 159, 145, 140, 168, 152, 158, 145, 132, 130, 178, 115, 109, 121, 127.

(a) Prepare a frequency table of the number of diabetic patients of different days.

(b) Find number of days in which 140 or more diabetic patients visited.

(c) Find percentage of days in which less than 130 patients visited.

(d) Draw frequency curve, histogram and ogive of the distribution of a number of diabetic patients.

3. The following are the amount of S. TCO_2 of some persons:

Amount of S. TCO_2 (x) : 27, 23, 25, 20, 28, 32, 18, 30, 19, 22, 27, 26, 25, 18, 33, 30, 29, 26, 24, 25, 20, 30.

Represent the distribution of amount of S. TCO_2 by a stem-and-leaf plot. Also, find median of S. TCO_2.

4. The following is the distribution of infected patients of different hospitals in a month.

Class interval of patients	2–4	4–6	6–8	8–10	10–12	Total
Number of hospitals, f	8	6	3	3	2	22

Draw histogram of the distribution of infected patients.

5. The following are the amount of haemoglobin of some persons:

Amount of haemoglobin (x) : 12.6, 10.8, 8.2, 9.0, 10.0, 11, 12, 14, 16, 17.3, 11.6, 9.3, 10, 11.3, 14, 13.4, 12.

Represent the distribution of haemoglobin by a stem-and-leaf plot. Also, prepare a frequency distribution of the observations.

6. The following is the distribution of children of different villages suffering from malnutrition:

Class of interval of children	15–20	20–25	25–30	30–35	35–40
Number of villages, f	8	18	32	10	7

(a) Find percentages of villages in which 20 or more children suffer from malnutrition.

[Ans. 89.33%]

(b) Draw frequency curve, histogram and ogive of the distribution of children suffering from malnutrition.

7. The following are the age (x years) and blood pressure levels (y mmHg) of some persons:

x	58	42	62	67	55	50	60
y	75	82	85	90	70	78	75

Draw scatter diagram of the distribution of ages and blood pressure levels.

8. (a) Write down the necessity of graphical representation of biostatistical data.
   (b) Explain classification and tabulation.
   Write down the uses of classification and tabulation.
   (c) Explain the method of data condensation.

9. (a) What are the different graphs and diagrams we use to represent data of bioscience?
   (b) Write down the objectives of constructing a frequency table.
   (c) The following are the number of outdoor patients of a hospital in different days:
   Patients of different days (x) : 150, 121, 105, 127, 140, 145, 150, 120, 116, 100, 128, 142, 147, 137, 120, 115, 118, 129, 130, 135, 120, 110, 105, 138, 147, 125, 120, 117, 124, 110.
   Prepare a frequency table of the distribution and present the data by a frequency curve.

Chapter 3

Measure of Central Tendency and Location

3.1 Introduction

The statistical data collected either by census or sample survey are used to study the characteristics of population under investigation. A comprehensive idea about the population characteristic can be obtained when collected data are presented by graphs and diagrams. Further comprehensive idea is obtained by estimating or calculating the unknown characteristic of the population mean. This mean can be calculated if data are collected from all population units or can be estimated if data are collected from randomly selected units of the population. Similar to mean there are other measures of the characteristics. These measures are known as '*measure of central tendency*' and '*measures of location*'. In this chapter these measures are discussed with examples.

3.2 Measures of Central Tendency

Let us consider the data of pulse rate (x bits/minute) of some persons as follows: The data are collected from $N = 5$ persons.

Pulse rate (x bits/minute) : 72, 76, 70, 65, 82.

These observations can be arranged in ascending or descending order as follows:

Ascending order (x)	65	70	72	76	82
Descending order (x)	82	76	72	70	65

These arrangements of data in ascending order or in descending order is called 'array'. These observations can be presented by a simple mathematical measure, called *arithmetic mean*, where the mean is

$$\bar{X} = \frac{1}{N}\Sigma x = \frac{1}{5}(72 + 76 + 70 + 65 + 82) = 73.$$

This measure falls near the centre of the array. Again, the value 72 exactly falls in the centre of the array and this value divides the array into two equal parts. Here we can say that the average pulse rate per person is 73 bits/minute. This measure provides some idea about the data set and falls near the centre of the array. As it has a tendency to fall near the centre of the array, it is called *measure of central tendency*.

There are different measures of central tendency.

These are: (i) Mean
(ii) Median
(iii) Mode.

Again, mean is of three types, viz., (a) arithmetic mean ($\bar{x}, \bar{X}$), (b) geometric mean (G), and (c) harmonic mean (H).

3.3 Arithmetic Mean

Let $x_1, x_2, \ldots, x_N$ be N observations recorded from N units in the population. Then arithmetic mean is measured by

$$\bar{X} = \frac{1}{N}\sum_{i=1}^{N} x_i.$$

For example, let us consider the gestation period (x in months) of some mothers as follows:

Gestation period (x) : 8.5, 9.2, 8.8, 8.4, 9.0, 8.0 ($N = 6$)

Then the average gestation period per mother is

$$\bar{X} = \frac{1}{N}\sum_{i=1}^{N} x_i = \frac{1}{6}[8.5 + 9.2 + 8.8 + 8.4 + 9.0 + 8.0] = 8.65 \text{ months}$$

Let us consider another set of observations on amount of s. triglyceride (x mg/dl) collected from $N = 25$ persons. The observations are:

x : 65, 120, 60, 66, 100, 120, 125, 75, 65, 60, 66, 140, 120, 125,
65, 75, 75, 90, 100, 120, 60, 75, 140, 125, 66

These observations can be presented in a discrete frequency table as follows:

S. triglyceride, X	60	65	66	75	90	100	120	125	140	Total
Number of persons, f	3	3	3	4	1	2	4	3	2	25

The average amount of s. triglyceride per person is measured by

$$\bar{X} = \frac{1}{N}\sum_{i=1}^{k} f_i X_i = \frac{1}{25}[3 \times 60 + 3 \times 65 + 3 \times 66 + 4 \times 75 + 1 \times 90 + 2 \times 100 + 4 \times 120 + 3 \times 125 + 2 \times 140]$$

$$= \frac{2298}{25} = 91.92.$$

Here $N = \sum_{i=1}^{k} f_i = 25.$

The formula of $\bar{X}$ is also used to calculate mean from continuous type frequency table also.

EXAMPLE 3.1 The following is the distribution of s. chloride (mmol/l) of a group of persons:

Class interval of s. chloride	98–100	100–102	102–104	104–106	106–108	Total
Number of persons, f	12	20	11	5	2	50

Calculate mean amount of s. chloride per person.

Solution.

Class interval of s. chloride	98–100	100–102	102–104	104–106	106–108	Total
Number of persons, f	12	20	11	5	2	50
Mid-value, X	99	101	103	105	107	–
fX	1188	2020	1133	525	214	

$$\text{Mean, } \bar{X} = \frac{1}{N}\sum_{i=1}^{k} f_i X_i = \frac{5080}{50} = 101.6\,\text{mmol/l.}$$

This arithmetic mean is affected by extreme value in the data set. For example, let us consider the ages (x in years) of some persons as below:

Ages (x in years) : 12, 18, 10, 15, 95 ($N = 5$).

The arithmetic mean is $\bar{X} = \frac{1}{5}(12 + 18 + 10 + 15 + 95) = 30$ years.

This average is such a value which does not represent any of the observations in the data. It is inflated by the extreme value 95.

3.4 Geometric Mean

Let $x_1, x_2, \ldots, x_N$ be N non-zero and non-negative observations recorded from N units. Then geometric mean is measured by

$$G = (x_1 x_2 \cdots x_N)^{1/N}$$

$$\text{or, } \log G = \frac{1}{N}\log(x_1 x_2 \cdots x_N) = \frac{1}{N}\sum_{i=1}^{N}\log x_i.$$

$$\therefore G = \text{antilog } \frac{1}{N}\sum_{i=1}^{N}\log x_i$$

Let us calculate the geometric mean of the gestation period of mothers given in section 3.3, where the gestation periods are:

x (in months) : 8.5, 9.2, 8.8, 8.4, 9.0, 8.0.

The geometric mean of these data is

$$G = (8.5 \times 9.2 \times 8.8 \times 8.4 \times 9.0 \times 8.0)^{\frac{1}{6}} = (416199.168)^{\frac{1}{6}} = 8.64$$

$$\text{or, } G = \text{antilog } \frac{1}{6}[\log 8.5 + \log 9.2 + \log 8.8 + \log 8.4 + \log 9.0 + \log 8.0]$$

$$= \text{antilog } \frac{1}{6} \times 5.6193 = 8.64.$$

The arithmetic mean of these observations is $\bar{X} = 8.65$ (section 3.3)

It is observed that $\bar{X} > G$.

Let us consider another set of observations as follows:

Observations (x) : 12, 13, 18, 12, 13

The geometric mean of these observations is

$$G = (12 \times 13 \times 18 \times 12 \times 13)^{\frac{1}{5}} = \left[12^2 \times 13^2 \times 18^1\right]^{\frac{1}{5}}$$

$$= (438048)^{\frac{1}{5}} = 13.44.$$

The observations can be presented in a discrete frequency distribution as follows:

X	12	13	18
Frequency, f	2	2	1

Then $G = \left(X_1^{f_1} X_2^{f_2} X_3^{f_3}\right)^{\frac{1}{N}}, N = \sum_{i=1}^{k} f_i = 5.$

This formula can be written for frequency distribution of k classes as follows:

$$G = \left(X_1^{f_1} X_2^{f_2} \cdots X_k^{f_k}\right)^{\frac{1}{N}}, N = \sum_{i=1}^{k} f_i$$

$$\text{or, } \log G = \frac{1}{N}[f_1 \log X_1 + f_2 \log X_2 + \cdots + f_k \log X_k] = \frac{1}{N}\sum_{i=1}^{k} f_i \log X_i$$

$$G = \text{antilog } \frac{1}{N}\sum_{i=1}^{k} f_i \log X_i.$$

Let us calculate the geometric mean of the data of Example 3.1

Class interval	Frequency f	Mid-value X	$f \log X$
98–100	12	99	23.94762
100–102	20	101	40.08543
102–104	11	103	22.14121
104–106	5	105	10.10595
106–108	2	107	4.05877
Total	$N = \Sigma f = 50$		100.33898

$$G = \text{antilog } \frac{1}{N}\Sigma f_i \log X_i = \text{antilog } \frac{100.33898}{50} = 101.57.$$

This result also indicates that $\bar{X} > G$.

Let us consider a set of observations as follows:

Observations (x) : $-2, -4, 8$.

The geometric mean of these observations is $G = [(-2)(-4)(8)]^{\frac{1}{3}} = 4$.

The arithmetic mean of these observations is $\bar{X} = \frac{1}{3}(-2-4+8) = \frac{2}{3}$.

It implies that $G > \bar{X}$. It contradicts with the previous conclusion $[\bar{X} > G]$.

Again, let us consider another set of observations as follows:

Observations (x) : 5, 8, 2, 0.

The geometric mean of these observations is $G = (5 \times 8 \times 2 \times 0)^{\frac{1}{4}} = 0$.

But, measure of central tendency of these observations should not be zero. This implies that geometric mean is not calculated if any observation is zero or negative. Finally, let us consider another set of observations as follows:

Observations (x) : 4, 4, 4.

The geometric mean of these observations is $G = (4 \times 4 \times 4)^{\frac{1}{3}} = 4$.

The arithmetic mean is $\bar{X} = \frac{1}{3}(4+4+4) = 4$. Therefore, we can conclude that $\bar{X} \geq G$.

As geometric mean is suitable for non-negative values, it is calculated if observations are of rates or ratios.

EXAMPLE 3.2 The following are the maternal mortality rates of Bangladesh in different years.

Year	1994	1995	1996	1997	1998
Maternal mortality rate per 1000	4.5	4.4	4.4	3.5	3.0

Calculate average maternal mortality rate per year.

Solution. As observations are rates, we can use geometric mean to calculate average maternal mortality rate per year. Here

$$G = [4.5 \times 4.4 \times 4.4 \times 3.5 \times 3.0]^{\frac{1}{5}} = (914.76)^{\frac{1}{5}} = 3.91.$$

3.5 Harmonic Mean

Let $x_1, x_2, \ldots, x_N$ be N non-negative and non-zero observations. Then harmonic mean is measured by

$$H = \frac{N}{\frac{1}{x_1} + \frac{1}{x_2} + \cdots + \frac{1}{x_N}} = \frac{N}{\Sigma\frac{1}{x}}.$$

Let us calculate the harmonic mean of gestation period of mothers discussed in section 3.4. We have

$$H = \frac{N}{\Sigma\frac{1}{x}} = \frac{6}{\frac{1}{8.5} + \frac{1}{9.2} + \frac{1}{8.8} + \frac{1}{8.4} + \frac{1}{9.0} + \frac{1}{8.0}} = \frac{6}{0.6951} = 8.63.$$

For the same set of observations, we have $\bar{X} = 8.65, G = 8.64$ and $H = 8.63$. This implies that $\bar{X} > G > H$. Again, when $x = 4, 4, 4$

$$H = \frac{3}{\frac{1}{4} + \frac{1}{4} + \frac{1}{4}} = 4$$

Finally, we conclude that $\bar{X} \geq G \geq H$. This is an important feature of three means.

Let us consider the observations of s. uric acid (mg/dl) of some persons as follows:

S. uric acid (x) : 3.4, 3.5, 4.5, 4.0, 3.4, 4.5, 4.5, 4.5, 3.5, 4.0, 4.0

The harmonic mean of these observations is

$$H = \frac{N}{\Sigma\frac{1}{x}} = \frac{10}{\frac{1}{3.4} + \frac{1}{3.5} + \frac{1}{4.5} + \frac{1}{4.0} + \frac{1}{3.4} + \frac{1}{4.5} + \frac{1}{4.5} + \frac{1}{3.5} + \frac{1}{4.0} + \frac{1}{4.0}}$$

$$= \frac{10}{\frac{2}{3.4} + \frac{2}{3.5} + \frac{3}{4.0} + \frac{3}{4.5}} = \frac{10}{2.5763} = 3.88.$$

The observations can be shown in a frequency table of discrete type as follows:

X	3.4	3.5	4.0	4.5
f	2	2	3	3

Accordingly, the formula for harmonic mean becomes $H = \frac{N}{\Sigma\frac{f_i}{X_i}}$.

EXAMPLE 3.3 The following distribution is for the S. CPK (U/L) of some persons.

Class interval of S. CPK	25–35	35–45	45–55	55–65	65–75	Total
Number of persons, f	7	13	20	6	4	50

Calculate harmonic mean of the amount of S. CPK.

Solution.

Class interval	Frequency f	Mid-value X	f/X
25–35	7	30	0.2333
35–45	13	40	0.3250
45–55	20	50	0.4000
55–65	6	60	0.1000
65–75	4	70	0.0571
Total	$N = 50$		1.1154

The harmonic mean is $H = \dfrac{N}{\Sigma \frac{f}{X}} = \dfrac{50}{1.1154} = 44.83.$

3.6 Median

Median is a value in the data set which divides the array into two equal parts such that half of the observations are below it and another half of the observations are above it. For example, let us consider the observations on birth-weight (x kg) of some new born babies as follows:

Birth-weight (x kg) : 3.2, 2.5, 2.8, 2.2, 3.0 $(N = 5)$.

The array of the observations is

x : 2.2, 2.5, 2.8, 3.0, 3.2.

It is observed that the observation 2.8 divides the array into two equal parts. So, by definition median of the given data set is Me = 2.8.

Let us consider another array of age (x years) of some persons as follows:

Age (x years) : 12, 17, 20, 22, 42, 70 $(N = 6)$.

By definition median is located between the observations 20 and 22. So, median is measured by an average of these two values, where

$$\text{Me} = \frac{1}{2}(20 + 22) = 21.$$

From the results of median the formula to find median from raw data can be written as

$$\text{Me} = \text{value of } \left(\frac{N+1}{2}\right)\text{th observation in the array, if } N \text{ is odd}$$

$$= \text{value of } \frac{1}{2}\left[\frac{N}{2}\text{th} + \left(\frac{N}{2}+1\right)\text{th}\right] \text{ observation in the array, if } N \text{ is even.}$$

EXAMPLE 3.4 The following is the distribution of s. cholesterol (mg/dl) of some persons

S. cholesterol, x	125	132	142	150	165	170	175	176	200	Total
Number of persons, f	8	15	10	25	28	12	10	8	5	121
$c.f.$	8	23	33	58	86	98	108	116	121	

Find median amount of s. cholesterol of the persons.

Solution. Here $N = 121$ (odd number of observations). Therefore,

$$\text{Median, Me} = \text{value of } \left(\frac{N+1}{2}\right)\text{th observation in the array as } N \text{ is odd}$$

$$= \text{value of } \left(\frac{121+1}{2}\right)\text{th observation in the array}$$

$$= \text{value of 61st observation.}$$

Here cumulative frequency indicates the number of observations for a certain value of x or less. The $c.f. \geq 61$ is 86 and this $c.f.$ is against 165. So, Me = 165.

Median is also calculated from frequency distribution of continuous type. To find it we need to identify the median class of the frequency table. The median class is decided as follows:

> Median class is that class of a frequency table for which $c.f. \geq \frac{N}{2}$

where N = total frequency. The value of the median will be minimum of the lower limit of the median class. The formula for median is

$$\text{Me} = l + \frac{h}{f}\left(\frac{N}{2} - c\right), \text{ where}$$

l = lower limit of the median class
h = width of median class
f = frequency of median class
c = $c.f.$ of the class preceding the median class
N = total frequency.

EXAMPLE 3.5 The following is the distribution of s. amylase of some persons:

Class interval of s. amylase	50–55	55–60	60–65	65–70	70–75	Total
Number of persons, f	13	17	25	18	7	80
$c.f.$	13	30	55	73	80	

Calculate median amount of s. amylase.

Solution. Here $c.f. \geq \frac{N}{2} = 40$ is 55.

Therefore, median is falling in the class 60–65 and $[l = 60, h = 5, f = 25, c = 30]$ median is as follows:

$$\text{Me} = l + \frac{h}{f}\left(\frac{N}{2} - c\right) = 60 + \frac{5}{25}\left(\frac{80}{2} - 30\right) = 62.$$

As median falls in the centre of the array, it is not affected by the extreme value or negative value or zero value in the data set. It is a good measure of central tendency for a markedly skewed distribution such as income distribution. It is also a good measure of central tendency if observations are given in ranks or scores.

3.7 Mode

Mode is a value in the data set which occurs most frequently. For example, let us consider the ages (x years) of some children, where ages are

Age (x years) : 8, 5, 10, 9, 5, 8, 10, 8.

Here 8 occurs maximum time. Therefore, Mo = 8 as 8 occurs most frequently in the given data set. Again, consider another set of data of number of ever born children in different families as follows:

Number of ever born children (x) : 2, 3, 1, 2, 4, 4, 2, 4, 2, 2, 3, 3, 2, 4, 4, 4.

Here 2 and 4 occur 6 times. So Mo = 2, 4. There are two modes in the given data set. It is called bi-modal distribution.

In the previous example there is one mode and it is a uni-modal distribution. Similarly, there may be multi-modal distribution having more than two modes.

Mode can also be found out from continuous frequency distribution, where modal class is that class for which frequency is maximum. The value of mode is minimum of lower limit of modal class. Thus, the formula for mode is

$$\text{Mo} = l + \frac{h(f_1 - f_0)}{2f_1 - f_0 - f_2}, \text{ where}$$

l = lower limit of modal class

f_1 = frequency of modal class

f_0 = frequency of the class before modal class

f_2 = frequency of the class after modal class

h = width of modal class.

EXAMPLE 3.6 The following is the distribution of family planning adopter couples in different villages:

Class interval of adopter couples	40–50	50–60	60–70	70–80	80–90	90–100	Total
Number of villages, f	5	17	33	40	15	8	118

Find mode of the distribution.

Solution. Here maximum frequency in a class is 40 and it is for the class 70–80. So, 70–80 is the modal class.

Here $l = 70, f_1 = 40, f_0 = 33, f_2 = 15$ and $h = 10$.

$$\text{Therefore, Mo} = l + \frac{h(f_1 - f_0)}{2f_1 - f_0 f_2} = 70 + \frac{10(40 - 33)}{2 \times 40 - 33 - 15} = 72.19.$$

From the above discussion it is clear that mode may not fall in the centre of the array. If the extreme value occurs maximum time, mode will fall in either left end or right end of the array. Moreover, there may be many modes. Mode is not well defined if modal class is the first class or last class of a frequency table.

As mode is the value of a data set having maximum frequency, it is used to study the maximum number of patients admitted in a hospital in a day, maximum number patients admitted having a particular disease, maximum number of people coming to the hospital for treatment against snake bite in a day, maximum number medicine strips damaged in a day during production period, number of days having maximum temperature or having maximum rainfall, etc.

3.8 Measures of Location

It is observed that median is a value in the data set which falls in the centre of the array or it is located in the centre of the array. As it is located in the centre of the array, it can also be called measure of location. It fall in a location which divides the array into two equal parts, In a similar way, array can be divided into 4 equal parts, 10 equal parts and 100 equal parts. The division of array into 4 equal parts can be indicated by 3 points located in 3 positions in the array and from each position we can have a measure of data set. The measure is located in different places in the array and hence, such a measure is called measure of location.

There are different measures of location.

These are: (i) Quartile, (ii) Decile, and (iii) Percentile.

3.9 Quartile

It is a measure which divides the array into 4 equal parts. It is denoted by $Q_i (i = 1, 2, 3)$, where Q_i is the i-th quartile of the data set. For example, let us consider the set of observations of number of dead patients of a hospital in some days:

Number of dead patients (x) : 5, 8, 3, 7, 10, 4, 7, 5, 9, 12, 6, 4 $(N = 12)$.

The array of these observations is

$$\text{Array } (x) : 3, 4, 4 \Updownarrow 5, 5, 6 \Updownarrow 7, 7, 8 \Updownarrow 9, 10, 12.$$

The array is divided into 4 equal parts showing 3 positions in the array. We get 3 quartiles, where

$$Q_1 = \frac{4+5}{2} = 4.5,\ Q_2 = \frac{6+7}{2} = 6.5 \text{ and } Q_3 = \frac{8+9}{2} = 8.5.$$

The position between 6 and 7 in the array divides the array into two equal parts and a measure from this position is 6.5. But this positional measurement is median also. It implies that $Q_2 = \text{Me}$. Thus, we can generalise the formula of Q_i from the formula of median as follows:

$$\text{Me} = Q_2 = \text{value of } \frac{2(N+1)}{4}\text{th observation in the array, if } N \text{ is odd.}$$

$$= \text{value of } \frac{1}{2}\left[\frac{2N}{4}\text{th} + \left(\frac{2N}{4}+1\right)\text{th}\right] \text{ observation in the array, if } N \text{ is even.}$$

Thus, for Q_i, we have

$$Q_i = \text{value of } \frac{i(N+1)}{4}\text{th observation in the array, if } N \text{ is odd.}$$

$$= \text{value of } \frac{1}{2}\left[\frac{iN}{4}\text{th} + \left(\frac{iN}{4}+1\right)\text{th}\right] \text{ observation in the array, if } N \text{ is even.}$$

In the above example of number of dead patients, we have

$$Q_1 = \text{value of } \frac{1}{2}\left[\frac{1\times 12}{4}\text{th} + \left(\frac{1\times 12}{4}+1\right)\text{th}\right] \text{ observation in the array,}$$

$$= \text{value of } \frac{1}{2}[\text{3rd} + \text{4th}] \text{ observation}$$

$$= \frac{1}{2}(4+5) = 4.5.$$

Similarly, Q_2 and Q_3 are as follows:

$$Q_2 = \text{value of } \frac{1}{2}\left[\frac{2N}{4}\text{th} + \left(\frac{2N}{4}+1\right)\text{th}\right] \text{ observation in the array}$$

$$= \text{value of } \frac{1}{2}\left[\frac{2\times 12}{4}\text{th} + \left(\frac{2\times 12}{4}+1\right)\text{th}\right] \text{ observation in the array}$$

$$= \text{value of } \frac{1}{2}[\text{6th} + \text{7th}] \text{ observation in the array}$$

$$= \frac{1}{2}(6+7) = 6.5.$$

$$Q_3 = \text{value of } \frac{1}{2}\left[\frac{3N}{4}\text{th} + \left(\frac{3N}{4}+1\right)\text{th}\right] \text{ observation in the array}$$

$$= \text{value of } \frac{1}{2}\left[\frac{3\times 12}{4}\text{th} + \left(\frac{3\times 12}{4}+1\right)\text{th}\right] \text{ observation in the array}$$

$$= \text{value of } \frac{1}{2}[\text{9th} + \text{10th}] \text{ observation in the array}$$

$$= \frac{1}{2}(8+9) = 8.5.$$

Here Q_1 indicates the maximum number of dead patients in 25% days in which death is lower. On the other hand, Q_3 indicates minimum number of dead patients in 25% days in which death is higher.

Quartile can also be calculated from frequency distribution. As median $= Q_2$, the formula for median can be generalized for Q_i by dividing N by 4, where

$$Q_i = l + \frac{h}{f}\left(\frac{iN}{4} - c\right), \quad i = 1, 2, 3.$$

Here l = lower limit of Q_i class
h = width of Q_i class
f = frequency of Q_i class
c = $c.f.$ of the class before Q_i class
N = total frequency.

Q_i class is decided as follows:

Q_i class is that class for which $c.f. \geq \frac{iN}{4}$

EXAMPLE 3.7 The following are the number of pregnant mothers visiting a primary health care unit in an union in different days:

Number of mothers visited (x) : 12, 9, 10, 15, 11, 7, 6, 8, 10, 12, 7.

Calculate (i) Q_1, (ii) Q_2, and (iii) Q_3 of the distribution of visiting pregnant mothers.

Solution. The array of the observations is

Array (x) : 6, 7, 7, 8, 9, 10, 10, 11, 12, 12, 15, $N = 11$.

$$Q_1 = \text{value of } \frac{1(N+1)}{4}\text{th observation in the array as } N \text{ is odd}$$
$$= \text{value of } \frac{1(11+1)}{4}\text{th observation in the array}$$
$$= \text{value of 3rd observation in the array}$$
$$= 7.$$

$$Q_2 = \text{value of } \frac{2(N+1)}{4}\text{th observation in the array}$$
$$= \text{value of } \frac{2(11+1)}{4}\text{th observation in the array}$$
$$= \text{value of 6th observation in the array}$$
$$= 10.$$

$$Q_3 = \text{value of } \frac{3(N+1)}{4}\text{th observation in the array}$$
$$= \text{value of } \frac{3(11+1)}{4}\text{th observation in the array}$$
$$= \text{value of 9th observation in the array}$$
$$= 12.$$

EXAMPLE 3.8 The following is the distribution of ever married women according to their ever born children:

Class interval of ever born children	0–2	2–4	4–6	6–8	8^+	Total
Number of ever married women, f	15	38	12	10	5	80
$c.f.$	15	53	65	75	80	

(i) Calculate Q_1, Q_2 and Q_3 of the distribution of ever born children.

(ii) Represent the data by a Box-and-Whisker plot.

Solution.

(i) $Q_1 = l + \frac{h}{f}\left(\frac{1N}{4} - c\right)$, **where Q_1 class is that class for which $c.f. \geq 20$, $\left[\frac{N}{4}\right]$**

$$= 2 + \frac{2}{38}\left(\frac{1 \times 80}{4} - 15\right) = 2.26.$$

$Q_2 = l + \frac{h}{f}\left(\frac{2N}{4} - c\right)$, **where Q_2 class is that class for which $c.f. \geq 40$, $\left[\frac{2N}{4}\right]$**

$$= 2 + \frac{2}{38}\left(\frac{2 \times 80}{4} - 15\right) = 3.32.$$

$Q_3 = l + \frac{h}{f}\left(\frac{3N}{4} - c\right)$, **where Q_3 class is that class for which $c.f. \geq 60$, $\left[\frac{3N}{4}\right]$**

$$= 4 + \frac{2}{12}\left(\frac{3 \times 80}{4} - 53\right) = 5.17.$$

(ii) Box-and-Whisker plot is drawn to show the position of quartiles in the limit of the values of variable by drawing box parallel to the X-axis. The lower limit and upper limit of values

of variable are indicated by drawing lines parallel to X-axis from the middle of the box. The lines extended to lower limit and upper limit of the values of variable are known as Whisker.

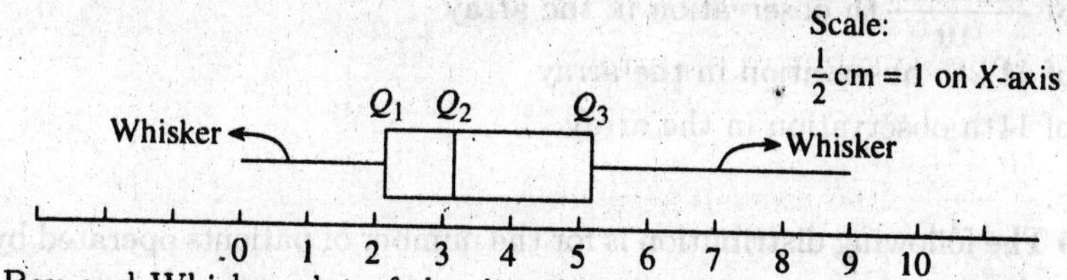

Fig. Box-and-Whisker plot of the distribution of number of ever born children

3.10 Decile

It is also a measure of location which divides the array into 10 equal parts. We have 9 deciles. These are $D_1, D_2, \cdots, D_9$.

Here D_1 measures the maximum value of the observations of first 10% lowest observations and D_9 measures the minimum value of the last 10% highest observations in the data set.

D_i values are calculated similar to Q_i values except that N is divided by 10 instead of 4. Thus,

$$D_i = \text{value of } \frac{i(N+1)}{10}\text{th observation in the array, if } N \text{ is odd.}$$

$$= \text{value of } \frac{1}{2}\left[\frac{iN}{10}\text{th} + \left(\frac{iN}{10}+1\right)\text{th}\right] \text{observation in the array, if } N \text{ is even.}$$

EXAMPLE 3.9 The following are the number of mothers recorded from different hospitals, who are aware of child health care and nutrition.

Number of mothers (x) : 11, 15, 8, 7, 12, 20, 6, 10, 13, 12, 8, 15, 10 [$N = 13$].

Calculate (i) D_3, (ii) D_7, and (iii) D_8 of the number of mothers.

Solution. Here $N = 13$ is odd.

The array of the observations is

Array (x) : 6, 7, 8, 8, 10, 10, 11, 12, 12, 13, 15.15, 20.

(i) $D_3 = \text{value of } \frac{3(N+1)}{10}\text{th observation in the array}$

$= \text{value of } \frac{3(13+1)}{10}\text{th observation in the array}$

= value of 4.2th observation in the array

= value of 4th observation in the array

= 8.

(ii) $D_7 = \text{value of } \frac{7(N+1)}{10}\text{th observation in the array}$

$= \text{value of } \frac{7(13+1)}{10}\text{th observation in the array}$

= value of 9.8th observation in the array

= value of 10th observation in the array

= 13.

(iii) D_8 = value of $\frac{8(N+1)}{10}$th observation in the array

= value of $\frac{8(13+1)}{10}$th observation in the array

= value of 11.2th observation in the array

= value of 11th observation in the array

= 15.

EXAMPLE 3.10 The following distribution is for the number of patients operated by a surgeon in different days:

Number of patients operated	3	4	5	6	7	Total
Number of days, f	12	28	30	5	3	78
$c.f.$	12	40	70	75	78	

Calculate (i) D_4, (ii) D_6, and (iii) D_9 of the operated patients.

Solution. Here $N = 78$ (even).

(i) D_4 = value of $\frac{1}{2}\left[\frac{4N}{10}\text{th} + \left(\frac{4N}{10}+1\right)\text{th}\right]$ observation in the array

= value of $\frac{1}{2}\left[\frac{4\times 78}{10}\text{th} + \left(\frac{4\times 78}{10}+1\right)\text{th}\right]$ observation in the array

= value of $\frac{1}{2}$[31.2th + 32.2th] observation in the array

= value of $\frac{1}{2}$[31st + 32nd] observation in the array

= 4 [From $c.f.$ it is seen that 31st and 32nd observations is 4].

(ii) D_6 = value of $\frac{1}{2}\left[\frac{6N}{10}\text{th} + \left(\frac{6N}{10}+1\right)\text{th}\right]$ observation in the array

= value of $\frac{1}{2}\left[\frac{6\times 78}{10}\text{th} + \left(\frac{6\times 78}{10}+1\right)\text{th}\right]$ observation in the array

= value of $\frac{1}{2}$[46.8th + (47.8)th] observation in the array

= value of $\frac{1}{2}$[47th + 48th] observation in the array

= 5 [From $c.f.$ it is observed that the value of 47th and 48th observations is 5].

(iii) D_9 = value of $\frac{1}{2}\left[\frac{9N}{10}\text{th} + \left(\frac{9N}{10}+1\right)\text{th}\right]$ observation in the array

= value of $\frac{1}{2}\left[\frac{9\times 78}{10}\text{th} + \left(\frac{9\times 78}{10}+1\right)\text{th}\right]$ observation in the array

= value of $\frac{1}{2}$[70.2th + 71.2th] observation in the array

= value of $\frac{1}{2}$[70th + 71st] observation in the array

= $\frac{5+6}{2}$ = 5.5 [From $c.f.$ it is observed that 70th observation is 5 and 71st observation is 6].

Decile can also be calculated from frequency table in a similar way as Q_i is calculated except that N is to be divided by 10 instead of 4.

Thus, $D_i = l + \frac{h}{f}\left(\frac{iN}{10} - c\right)$, where $[i = 1, 2, \ldots, 9]$

l = lower limit of D_i class

h = width of D_i class

f = frequency of D_i class

c = $c.f.$ of the class before D_i class

N = total frequency.

> D_i class is that class for which $c.f. \geq \frac{iN}{10}$

EXAMPLE 3.11 The following is the distribution of birth-weight (lb) of some new born babies:

Class interval of birth-weight	< 5.0	5.0–5.5	5.5–6.0	6.0–6.5	6.5–7.0	7.0+	Total
Number of babies, f	15	18	50	70	47	25	225
$c.f.$	15	33	83	153	200	225	–

Calculate (i) D_7, and (ii) D_8 of the **birth-weight.**

Solution.

(i) D_7 class is that class for which $c.f. \geq \frac{7 \times 225}{10} = 157.5$.

The class is 6.5–7.0.

$$D_7 = l + \frac{h}{f}\left(\frac{7N}{10} - c\right) = 6.5 + \frac{0.5}{47}\left(\frac{7 \times 225}{10} - 153\right) = 6.55.$$

(ii) D_8 class is that class for which $c.f. \geq \frac{8 \times 225}{10} = 180$.

The class is 6.5–7.0.

$$D_8 = l + \frac{h}{f}\left(\frac{8N}{10} - c\right) = 6.5 + \frac{0.5}{47}\left(\frac{8 \times 225}{10} - 153\right) = 6.79.$$

3.11 Percentile

Percentile is a measure of data set which divides the array into 100 equal parts. We have 99 percentiles. These are $P_1, P_2, \cdots, P_{99}$, where P_i is the i-th percentile.

Here P_{10} measures the maximum value of the first 10% lowest observations and P_{90} measures the minimum value of the last 10% highest observations of the data set.

The P_i values are calculated in a similar way as it is done in calculating D_i except that N is divided by 100 instead of 10. Thus,

$$P_i = \text{value of } \frac{i(N+1)}{100}\text{th observaton in the array, if } N \text{ is odd.}$$

$$= \text{value of } \frac{1}{2}\left[\frac{iN}{100}\text{th} + \left(\frac{iN}{100} + 1\right)\text{th}\right] \text{ observation in the array, if } N \text{ is even.}$$

EXAMPLE 3.12 The following are the number of outdoor patients of a hospital in different days:

Number of patients (x) : 32, 40, 28, 50, 35, 44, 40, 52, 45, 38, 42, 35, 30, 30, 37, 44, 45, 48, 40, 37, 35.

Find (i) P_{30}, and (ii) P_{80} of the number of patients.

Solution. The array of the observations is

Array (x) : 28, 30, 30, 32, 35, 35, 35, 37, 37, 38, 40, 40, 40, 42, 44, 44, 45, 45, 48, 50, 52 $(N = 21)$

(i) P_{30} = value of $\frac{30(N+1)}{100}$th observation in the array as N is odd

= value of $\frac{30(21+1)}{100}$th observation in the array

= value of 6.6th observation in the array

= value of 7th observation in the array

= 35.

(ii) P_{80} = value of $\frac{80(N+1)}{100}$th observation in the array

= value of $\frac{80(21+1)}{100}$th observation in the array

= value of 17.6th observation in the array

= value of 18th observation in the array

= 45.

EXAMPLE 3.13 The following distribution represents the number of operated patients in a hospital in different days:

Number of operated patients	8	10	12	13	14	15	20	Total
Number of days, f	12	16	25	30	30	25	12	150
$c.f.$	12	28	53	83	113	138	150	

Calculate (i) P_{40}, and (ii) P_{70} of the number of operated patients.

Solution. Here observations are arranged in array of number of observations where $N = 150$.

(i) P_{40} = value of $\frac{1}{2}\left[\frac{40N}{100}\text{th} + \left(\frac{40N}{100}+1\right)\text{th}\right]$ observation in the array

= value of $\frac{1}{2}\left[\frac{40 \times 150}{100}\text{th} + \left(\frac{40 \times 150}{100}+1\right)\text{th}\right]$ observation in the array

= value of $\frac{1}{2}$[60th + 61st] observation in the array

= 13 [$c.f.$ indicates that the value of 60th and 61st observations is 13.]

(ii) P_{70} = value of $\frac{1}{2}\left[\frac{70N}{100}\text{th} + \left(\frac{70N}{100}+1\right)\text{th}\right]$ observation in the array

= value of $\frac{1}{2}\left[\frac{70 \times 150}{100}\text{th} + \left(\frac{70 \times 150}{100}+1\right)\text{th}\right]$ observation in the array

= value of $\frac{1}{2}$[105th + 106th] observation in the array

= 14 [$c.f.$ indicates that the value of 60th and 61st observations is 14.]

Percentile can also be calculated from frequency distribution in a similar way as it is done in calculating D_i except that N is to be divided by 100 instead of 10. Thus,

$$P_i = l + \frac{h}{f}\left(\frac{iN}{100} - c\right), \quad i = 1, 2, 3, \ldots, 99.$$

Here l = lower limit of P_i class
h = width of P_i class
f = frequency of P_i class
c = $c.f.$ of the class before P_i class
N = total frequency

P_i class is that class for which $c.f. \geq \dfrac{iN}{100}$

EXAMPLE 3.14 The following is the distribution of s. sodium (mmol/l) of some persons:

Class interval of s. sodium	136–138	138–140	140–142	142–146	146–148	Total
Number of persons, f	15	28	35	22	10	110
$c.f.$	15	43	78	100	110	

Calculate (i) P_{60}, and (ii) P_{40} of s. sodium.

Solution.

(i) $P_{60} = l + \frac{h}{f}\left(\frac{60N}{100} - c\right)$, P_{60} class is the class for which $c.f. \geq \frac{60 \times 110}{100} = 66$. Here P_{60} class is 140–142.

$$= 140 + \frac{2}{35}\left(\frac{60 \times 110}{100} - 43\right)$$
$$= 141.31.$$

(ii) $P_{40} = l + \frac{h}{f}\left(\frac{40N}{100} - c\right)$, P_{40} class is that class for which $c.f. \geq \frac{40 \times 110}{100} = 44$. Here P_{40} class is 140–142.

$$= 140 + \frac{2}{35}\left(\frac{40 \times 110}{100} - 43\right)$$
$$= 140.06.$$

The measure of central tendency and measure of location is also used to study the shape characteristic of frequency curve. The shape of the curve is changed due to the changes in tail of the curve. For a distribution, if

$$\text{Mean} = \text{Median} = \text{Mode},$$

the tails of the curve is uniform in both right and left sides of the curve. There is symmetry in the tails in both sides. Such a curve is called *symmetric curve.* If mean, median and mode are not equal, the tails of the curve will not be uniform in both left and right sides of the curve. If

$$\text{Mean} > \text{Median} > \text{Mode},$$

there will be a long tail in the right side of the curve. The curve is called *asymmetric curve* or *skewed curve* and the curve is skewed to the right or positively skewed. If

$$\text{Mean} < \text{Median} < \text{Mode},$$

there will be a long tail in the left side of the curve. The curve is called negatively skewed curve.

The asymmetric property of frequency curve is called skewness.

The skewness is measured by Sk $= \bar{X} -$ Me or Sk $= \bar{X} -$ Mo.

Another measure of skewness is Sk $= Q_3 + Q_1 - 2$Me.

These are all absolute measure of skewness. The value of Sk may be zero, positive or negative. Accordingly, one can comment about the shape characteristic of the curve as symmetric, positively skewed or negatively skewed, respectively.

EXAMPLE 3.15 The following is the distribution of number of nurses working in different clinics of a city

Class interval of nurses	20–22	22–24	24–26	26–28	28–30	Total
Number of clinics, f	5	12	18	25	10	70
$c.f.$	5	17	35	60	70	
Mid-value, X	21	23	25	27	29	
fX	105	276	450	675	290	1796

Calculate skewness and comment.

Solution. Mean, $\bar{X} = \frac{1}{N}\Sigma fX = \frac{1796}{70} = 25.66.$

$$\text{Median, Me} = l + \frac{h}{f}\left(\frac{N}{2} - c\right) = 24 + \frac{2}{18}\left(\frac{70}{2} - 17\right) = 26 = Q_2.$$

$$\text{Mode, Mo} = l + \frac{h(f_1 - f_0)}{2f_1 - f_0 - f_2} = 26 + \frac{2(25-18)}{2 \times 25 - 18 - 10} = 26.64$$

$$Q_1 = l + \frac{h}{f}\left(\frac{1N}{4} - c\right) = 24 + \frac{2}{18}\left(\frac{1 \times 70}{4} - 17\right) = 24.06$$

$$Q_3 = l + \frac{h}{f}\left(\frac{3N}{4} - c\right) = 26 + \frac{2}{25}\left(\frac{3 \times 70}{4} - 35\right) = 27.40$$

Skewness, Sk $= \bar{X} -$ Me $= 25.66 - 26 = -0.34$

or, Sk $= \bar{X} -$ Mo $= 25.66 - 26.64 = -0.98.$

The frequency curve of the distribution is negatively skewed.

Again, Sk $= Q_3 + Q_1 - 2\text{Me} = 27.40 + 24.06 - 2 \times 26 = -0.54.$

The curve is negatively skewed. There is a long tail in the left side of the curve.

EXERCISE – 3

1. The following are the post-operation infected patients of a hospital in different months. Infected patients (x) : 4, 3, 6, 2, 4, 5, 7, 4, 3, 5, 5, 4.

   (a) Calculate mean, median and mode of the number of infected patients.
   [**Ans.** $\bar{X} = 4.33$, Me $= 4$, Mo $= 4$]

   (b) Calculate Q_1, Q_2, Q_3, D_7 and P_{40} of the number of infected patients.
   [**Ans.** $Q_1 = 3.5, Q_2 = 4, Q_3 = 5, D_7 = 5, P_{40} = 4$]

   (c) Draw Box-and-Whisker plot of the number of infected patients.

2. The following are the amount of F.B.S. (x mmol/l) of some persons:
Amount of F.B.S. (x) : 4.8, 6.7, 7.0, 6.3, 5.8, 8.6, 10.2, 6.0, 6.3.

(a) Calculate mean, median and mode of F.B.S. [**Ans.** $\bar{X} = 6.86$, Me $= 6.3$, Mo $= 6.3$]

(b) Calculate Q_1, Q_2, Q_3, D_4 and P_{70} of the amount of F.B.S.
[**Ans.** $Q_1 = 6.0, Q_2 = 6.3, Q_3 = 8.6, D_4 = 6.3, P_{70} = 7.0$]

(c) Draw Box-and-Whisker plot of the amount of F.B.S.

3. The following is the distribution of smokers among school students of different high schools:

Class interval of smokers	50–75	75–100	100–125	125–150	150–175
Number of schools, f	18	32	28	12	10

(a) Calculate mean, median and mode of the number of smoker students.
[**Ans.** $\bar{X} = 103.5$, Me $= 100.0$, Mo $= 94.44$]

(b) Calculate Q_1, Q_2, Q_3, D_8 and P_{30} of the distribution of smoker students.
[**Ans.** $Q_1 = 80.47, Q_2 = 100.0, Q_3 = 122.32, D_8 = 129.17, P_{30} = 84.375$]

(c) Draw Box-and-Whisker plot of the distribution of smoker students.

4. The following is the distribution of body mass index (kg/m^2) of some persons:

Class interval of body mass index	15–20	20–25	25–30	30–35	35–40
Number of persons, f	7	18	10	8	7

(a) Calculate mean, median and mode of body mass index.
[**Ans.** $\bar{X} = 26.5$, Me $= 25$, Mo $= 22.84$]

(b) Calculate Q_1, Q_2, Q_3, D_6 and P_{70} of the body mass index.
[**Ans.** $Q_1 = 21.53, Q_2 = 25, Q_3 = 31.56, D_6 = 27.5, P_{70} = 30$]

(c) Draw Box-and-Whisker plot of the distribution of body mass index.

5. (a) What are the different measures of central tendency? Why these are called measure of central tendency? Which measure do you think best?

(b) The following are the s. lipase of some persons:
S. lipase (x) : 15, 12, 18, 28, 27, 32, 16, 20, 18, 21.
Calculate mean, median and mode of s. lipase.

6. (a) Write down the difference between measure of central tendency and measure of location. What are the different measures of location?

(b) The following are s. triglyceride (x mg/dl) of some persons:
x : 58, 95, 110, 165, 150, 50, 70, 78, 110, 120, 114, 120, 90, 88, 40, 66.
Calculate Q_1, Q_3, D_4 and P_{70} of the amount of s. triglyceride.

7. (a) When geometric mean is preferable to arithmetic mean?

(b) The following is the frequency distribution of s. amylase (U/L) of some persons:

Class interval of s. amylase	50–60	60–70	70–80	80–90	90–100	Total
Number of persons, f	12	28	32	18	10	100

Calculate geometric mean of the distribution. Also, calculate median and mode of s. amylase.

8. (a) The following is the distribution of s. sodium (mmol/l) of some persons:

Class interval of s. sodium	130–134	134–138	138–142	142–146	146–150	Total
Number of person, f	10	18	22	20	5	75

Calculate Q_1, Q_2, Q_3, D_9, and P_{70} of the distribution.

(b) Represent the distribution by Box-and-Whisker plot.

9. (a) Two bottles of saline are pushed to a patients at a rate of 20 drops per minute, 3 bottles are pushed at a rate of 10 drops per minute and one bottle is pushed at a rate of 5 drops per minute. Find the average rate of drops per minute.

(b) The following is the distribution of s. urea (mg/dl) of some persons:

Class interval of s. urea	12–16	16–20	20–24	24–28	28–32	Total
Number of persons, f	5	12	16	10	7	50

Calculate arithmetic mean, median and mode of s. urea.

Chapter 4

Measures of Dispersion

4.1 Dispersion

The deviation of observations from some central value is called *dispersion.* For example, let us consider the observations of ages (x years) of some children in a family, where $x : 2, 5, 8, 11, 14$. The average age of these children is

$$\bar{X} = \frac{1}{N}\Sigma x = \frac{1}{5}(2 + 5 + 8 + 11 + 14) = 8 \text{ years.}$$

This is a central value of the array of ages which indicates that on an average, the ages of children of this family is 8 years. This central value represents the whole of the data set. But there is deviations of this value from all observations, where the deviations are

$$x - \bar{X} : -6, -3, 0, 3, 6.$$

This deviation is known as *dispersion.*

This deviation helps us to study the formation of data set. We can explain it by another example. Consider that a person walks 8 kilometres everyday to attend his office. In five working days of a week he walks

$$y \text{ kilometres} : 8, 8, 8, 8, 8.$$

The person walks, on an average,

$$\bar{Y} = \frac{1}{N}\Sigma y = \frac{1}{5}(8 + 8 + 8 + 8 + 8) = 8 \text{ kilometres per day.}$$

But there is no deviation of observations from mean. It is seen that the formation of data set of x and the formation of data set of y are not same. This formation of data set can be studied by the *topic dispersion.*

4.2 Measures of Dispersion

In section 4.1, it is seen that the first observation of data set x is deviated from mean $\bar{X}$ by an amount -6. The deviation of other observations are $-3, 0, 3$ and 6. These individual deviations do not depict the clear formation of data set. Thus, we need such a value of deviations by which clear picture of formation of data set can be guessed at a glance. This leads us to study measure of dispersion. Instead of individual deviations of observations we can have an average of deviations of observations from some central value.

The average of deviations of observations from some central value is called measure of dispersion. The mean square of deviations is also used as a measure of variability among the observations with respect to some central value or with respect to some arbitrary value. This latter type of measure is called *second order average*. Similarly, the *r-th order average* of deviations can also be calculated.

4.3 Different Measures of Dispersion

Let us consider that the systolic blood pressure (x mmHg) of a person is recorded for seven days and the recorded values are as follows:

$$x : 85, 90, 92, 87, 88, 100, 102 \quad (N = 7).$$

The average blood pressure per day is $\bar{X} = \frac{1}{N}\Sigma x = \frac{644}{7} = 92$ mmHg.

The deviations of x from $\bar{X}$ is $x - \bar{X} : -7, -2, 0, -5, -4, 8, 10$.

The average of deviatons is 0.

$$\frac{1}{N}\Sigma(x - \bar{X}) = \frac{1}{7}(-7 - 2 + 0 - 5 - 4 + 8 + 10) = 0.$$

Here mean deviation from mean is zero. It is an important characteristic of arithmetic mean. This measure does not depict anything about the formation of observations. In reality we need to measure the distance of the observations from some central value. In measuring distance we do not usually use negative distance. Distance is measured in absolute form. So, we have

$$|x - \bar{X}| : 7, 2, 0, 5, 4, 8, 10.$$

The average of this absolute deviation is called *mean deviation from mean* [*M.D.*] and it is given by

$$\text{M.D.}(\bar{X}) = \frac{1}{N}\Sigma|x - \bar{X}| = \frac{1}{7}(7 + 2 + 0 + 5 + 4 + 8 + 10) = 5.14\,\text{mmHg}.$$

This measure indicates that each value of blood pressure is deviated, on an average, from mean blood pressure by an amount 5.14 mmHg. This measure provides a preliminary information about the formation of data set of blood pressure level. It is used as a measure of dispersion. In a similar way, we can measure the average deviation from median [M.D. (Me)] and from mode [M.D. (Mo)]. So, there are different measures of dispersion.

Again, let us consider another set of observations of seven persons on s. triglyceride (y mg/dl), where

$$y : 85, 90, 92, 87, 88, 102, 100 \quad (N = 7).$$

Here $\bar{Y} = \frac{1}{N}\Sigma y = 92$ mg/dl, and $|y - \bar{Y}| : 7, 2, 0, 5, 4, 10, 8$

$$\text{M.D.}(\bar{Y}) = \frac{1}{N}\Sigma|y - \bar{Y}| = 5.14\,\text{mg/dl}.$$

Here M.D. $(\bar{X})$ and M.D. $(\bar{Y})$ are the measures of dispersion of two distributions of X and Y, respectively. These two measures are similar in value. But these measures cannot be used to compare the dispersion of the distributions of X and Y as these are dependent on unit of variable. In such a case, if we need to compare the dispersion of two or more distributions, we need a measure which is free of unit of variable.

The measure can be made unit-free as follows:

$$\frac{\text{M.D.}(\bar{X})}{\bar{X}} = \frac{5.14\,\text{mmHg}}{92\,\text{mmHg}} = 0.056 \text{ or } 5.6\%.$$

$$\frac{\text{M.D.}(\bar{Y})}{\bar{Y}} = \frac{5.14\,\text{mmHg}}{92\,\text{mmHg}} = 0.056 \text{ or } 5.6\%.$$

Now, on the basis of these two measures, we can say that both the distributions of X and Y are uniformly dispersed. The formation of data set in the values of X and Y are similar.

Therefore, we have two different types of measure of dispersion.

These are:

(i) Absolute Measure of Dispersion

(ii) Relative Measure of Dispersion.

The absolute measures of dispersion are:

(a) Range (R_N), where $R_N = x_N - x_1$; $x_1 < x_2 < \cdots < x_N$

(b) Mean Deviation [M.D.]

Again mean deviation is of three types. These are:

(i) $\text{M.D.}(\bar{X}) = \frac{1}{N}\Sigma|x - \bar{X}|$ for primary (raw) data.

$= \frac{1}{N}\Sigma f|X - \bar{X}|$ for frequency table.

(ii) $\text{M.D.}(\text{Me}) = \frac{1}{N}\Sigma|x - \text{Me}|$ for primary data.

$= \frac{1}{N}\Sigma f|X - \text{Me}|$ for frequency table.

(iii) $\text{M.D.}(\text{Mo}) = \frac{1}{N}\Sigma|x - \text{Mo}|$ for primary data.

$= \frac{1}{N}\Sigma f|X - \text{Mo}|$ for frequency table.

(c) Standard Deviation (σ)

(d) Semi-interquartile range or Quartile Deviation [Q.D.]

$$\text{Q.D.} = \frac{1}{2}[Q_3 - Q_1].$$

Here Q_1 and Q_3 are first and third quartile respectively.

The relative measures of dispersion are:

(a) Coefficient of Range, Coeff. $R_N = \frac{x_N - x_1}{x_N + x_1}$, where $x_1 < x_2 < \cdots < x_N$

(b) Coefficient of Mean Deviation

(i) Coeff. $\text{M.D.}(\bar{X}) = \frac{\text{M.D.}(\bar{X})}{\bar{X}}$

(ii) Coeff. $\text{M.D.}(\text{Me}) = \frac{\text{M.D.}(\text{Me})}{\text{Me}}$

(iii) Coeff. $\text{M.D.}(\text{Mo}) = \frac{\text{M.D.}(\text{Mo})}{\text{Mo}}$

(c) Coefficient of variation [C.V.], where $C.V. = \frac{\sigma}{\bar{X}} \times 100$.

(d) Coefficient of quartile deviation, Coeff. Q.D. $= \frac{Q_3 - Q_1}{Q_3 + Q_1}$.

EXAMPLE 4.1 The following are the s. sodium (mmol/l) of a person in different days when he is under treatment.

$$\text{S. sodium } (x) : 110, 118, 115, 125, 134, 140, 142.$$

Calculate (i) range, (ii) mean deviation, and (iii) quartile deviation of s. sodium. Also, calculate their relative measures of dispersion..

Solution. We have $N = 7$.

(i) $x_N = 142,\ x_1 = 110$.

Therefore, range is $R_N = x_N - x_1 = 142 - 110 = 32$ mmol/l

Its relative measure is Coeff. $R_N = \frac{x_N - x_1}{x_N + x_1} = \frac{142 - 110}{142 + 110} = 0.127$.

(ii) Mean Deviation [M.D.]:

$\text{M.D.}(\bar{X}) = \frac{1}{N}\Sigma|x - \bar{X}|$, where

$\bar{X} = \frac{1}{N}\Sigma x = \frac{1}{7}[110 + 118 + 115 + 125 + 134 + 140 + 142]$

$= 126.3$ mmol/l.

$|x - \bar{X}| : 16.3, 8.3, 11.3, 1.3, 7.7, 13.7, 15.7$.

$\therefore \text{M.D.}(\bar{X}) = \frac{1}{7}[16.3 + 8.3 + 11.3 + 1.3 + 7.7 + 13.7 + 15.7]$

$= 10.61$ mmol/l.

Its relative measure is

$$\text{Coeff. M.D.}(\bar{X}) = \frac{\text{M.D.}(\bar{X})}{\bar{X}} = \frac{10.61}{126.3} = 0.084.$$

This measure is free of unit of variable but not free of algebraic mistake.

Mean deviation from median [M.D. (Me)]

The array of the observations (x) : 110, 115, 118, 125, 134, 140, 142.

Median, Me = value of $\frac{N+1}{2}$th observation, as N is odd (in the array)

= value of 4th observaton = 125 mmol/l.

$|x - \text{Me}| : 15, 7, 10, 0, 9, 15, 17$

$\text{M.D.}(\text{Me}) = \frac{1}{N}\Sigma|x - \text{Me}| = \frac{1}{7}[15 + 7 + 10 + 0 + 9 + 15 + 17]$

$= 10.43$ mmol/l.

Its relative measure is

$$\text{Coeff. M.D.}(\text{Me}) = \frac{\text{M.D.}(\text{Me})}{\text{Me}} = \frac{10.43}{125} = 0.083.$$

Mean deviation from mode is not calculated as there is no mode in the data.

Quartile deviation is Q.D. $= \frac{1}{2}[Q_3 - Q_1]$, where

$$Q_1 = \text{value of } \frac{N+1}{4}\text{th observation, as } N \text{ is odd (in the array).}$$
$$= \text{value of 2nd observation} = 115.$$
$$Q_3 = \text{value of } \frac{3(N+1)}{4}\text{th observation}$$
$$= \text{value of 6th observation} = 140.$$
$$\therefore \text{Q.D.} = \frac{1}{2}[140 - 115] = 12.5 \text{ mmol/l}$$

Its relative measure is

$$\text{Coeff. Q.D.} = \frac{Q_3 - Q_1}{Q_3 + Q_1} = \frac{140 - 115}{140 + 115} = 0.098.$$

Standard Deviation (σ): The mean deviation what is used as a measure of dispersion is the average deviation of first order, where deviation is measured from mean, or median or mode. But in its calculation we do an algebraic mistake as absolute deviation is used instead of real deviation of observation from central value. This is the disadvantage of mean deviation. This disadvantage can be overcome if we take deviation of second order. For this, let us consider five ($N = 5$) observations of s. urea (x mg/dl) of five persons as follows:

$$x: \ 18, 15, 20, 28, 29.$$

Here $\bar{X} = \frac{1}{N}\Sigma x = \frac{1}{5}(18 + 15 + 20 + 28 + 29) = 22$ mg/dl.

Deviations of first order of observations from mean are $x - \bar{X}: -4, -7, -2, 6, 7$.

Deviations of second order are $(x - \bar{X})^2: \ 16, 49, 4, 36, 49$.

Mean square deviations of observations from mean is

$$\text{M.S.}(x) = \frac{1}{N}\Sigma(x - \bar{X})^2 = \frac{1}{5}(16 + 49 + 4 + 36 + 49) = 30.8\ (\text{mg/dl})^2.$$

This mean square deviations [M.S.] is free of algebraic mistake. This is also used as a measure of dispersion and is called variance (σ^2), where

$$V(x) = \sigma^2 = \frac{1}{N}\Sigma(x - \bar{X})^2 = \frac{1}{N}\left[\Sigma x^2 - \frac{(\Sigma x)^2}{N}\right].$$

As we measure the distance of the observations from mean, we take positive square root of variance (σ^2) as a measure of standard deviation, i.e.,

$$\text{Standard deviation, } \sigma = \sqrt{V(x)}.$$

For given example of s. urea standard deviation is

$$\sigma = \sqrt{V(x)} = \sqrt{30.8\ (\text{mg/dl})^2} = 5.55 \text{ mg/dl}.$$

This measure is free of algebraic mistake but not free of unit of variable. So, it is also not used to compare the dispersion of two or more distributions. The formula to calculate variance (σ^2) from frequency table is

$$V(x) = \sigma^2 = \frac{1}{N}\Sigma f\,(X - \bar{X})^2 = \frac{1}{N}\left[\Sigma f X^2 - \frac{(\Sigma f X)^2}{N}\right].$$

The coefficient of variation [C.V.] of the data of s. urea is

$$\text{C.V.} = \frac{\sigma}{\bar{X}} \times 100 = \frac{5.55}{22} \times 100 = 25.23.$$

This value of C.V. indicates that with respect to mean the observations vary around 25%.

This coefficient of variation is free of unit of variable as well as free of algebraic mistake. Hence, it can be used to compare the dispersion of two or more distributions.

Though standard deviation is not used to compare the dispersion of two or more distributions, yet it is a good measure of dispersion to study the formation of observations in a distribution. Moreover, it is used to estimate the lower limit and upper limit ($\bar{X} \pm \sigma$) of the observations in a distribtion. For normal distribution 95% observations fall in the limit $\bar{X} \pm 2\sigma$. Standard deviation is also used to estimate the precision of an estimate.

EXAMPLE 4.2 The following distributions are for the S. GGT(U/L) of some male and female population.

Class interval of S. GGT	15–19	19–23	23–27	27–31	31–35	Total
Number of males, f	7	15	18	6	4	50
Number of females, f	15	20	9	4	2	50

Calculate (i) Mean deviation from mean, median and mode of the observations of two distribution, (ii) Coefficient of mean deviations of two distributions, (iii) Coefficient of quartile deviations of two distributions, (iv) Variances of the distributions, (v) Standard deviations of the distribtions, (vi) C.V. of the distributions, (vii) Compare the dispersion of two distributions and comment.

Solution. For the distribution of males:

Class interval	Frequency f	$c.f.$	Mid-value X	fX	fX^2	$f\|X-\bar{X}\|$	$f\|X-\text{Me}\|$	$f\|X-\text{Mo}\|$
15–19	7	7	17	119	2023	47.6	46.69	47.6
19–23	15	22	21	315	6615	42.0	40.05	42.0
23–27	18	40	25	450	11250	21.6	23.94	21.6
27–31	6	46	29	174	5046	31.2	31.98	31.2
31–35	4	50	33	132	4224	36.8	37.32	36.8
Total	$N = \Sigma f = 50$			1190	29158	179.2	179.98	179.2

$$\text{Mean, } \bar{X} = \frac{1}{N}\Sigma fX = \frac{1190}{50} = 23.8\,\text{U/L}.$$

$$\text{Median, Me} = l + \frac{h}{f}\left(\frac{N}{2} - c\right) = 23 + \frac{4}{18}\left(\frac{50}{2} - 22\right) = 23.67\,\text{U/L}.$$

$$\text{Mode, Mo} = l + \frac{h(f_1 - f_0)}{2f_1 - f_0 - f_2} = 23 + \frac{4(18-15)}{2 \times 18 - 15 - 6} = 23.8\,\text{U/L}.$$

(i) Mean deviation from mean

$$\text{M.D.}\,(\bar{X}) = \frac{1}{N}\Sigma f|X - \bar{X}| = \frac{179.2}{50} = 3.584\,\text{U/L}.$$

Mean deviation from median

$$\text{M.D. (Me)} = \frac{1}{N}\Sigma f|X - \text{Me}| = \frac{179.98}{50} = 3.5996\,\text{U/L.}$$

Mean deviation from mode

$$\text{M.D. (Mo)} = \frac{1}{N}\Sigma f|X - \text{Mo}| = \frac{179.2}{50} = 3.584\,\text{U/L.}$$

(ii) $\text{Coeff. M.D.}\,(\bar{X}) = \dfrac{\text{M.D.}\,(\bar{X})}{\bar{X}} = \dfrac{3.584}{23.8} = 0.151.$

$\text{Coeff. M.D. (Me)} = \dfrac{\text{M.D. (Me)}}{\text{Me}} = \dfrac{3.5996}{23.67} = 0.152.$

$\text{Coeff. M.D. (Mo)} = \dfrac{\text{M.D. (Mo)}}{\text{Mo}} = \dfrac{3.584}{23.8} = 0.151.$

These results indicate that the observations are similarly dispersed with respect to mean, median and mode.

(iii) $Q_1 = l + \dfrac{h}{f}\left(\dfrac{N}{4} - c\right) = 19 + \dfrac{4}{15}\left(\dfrac{50}{4} - 7\right) = 20.47.$

$Q_3 = l + \dfrac{h}{f}\left(\dfrac{3N}{4} - c\right) = 23 + \dfrac{4}{18}\left(\dfrac{3 \times 50}{4} - 22\right) = 26.44.$

Coefficient of quartile deviation

$$\text{Coeff. Q.D.} = \frac{Q_3 - Q_1}{Q_3 + Q_1} = \frac{26.44 - 20.47}{26.44 + 20.47} = 0.127.$$

(iv) Variance $\sigma^2 = \dfrac{1}{N}\left[\Sigma f X^2 - \dfrac{(\Sigma f X)^2}{N}\right] = \dfrac{1}{50}\left[29158 - \dfrac{(1190)^2}{50}\right] = 16.72\,(\text{U/L})^2.$

(v) Standard deviation, $\sigma = \sqrt{\sigma^2} = \sqrt{16.42} = 4.09\,\text{U/L.}$

(vi) Coefficient of variation C.V. $= \dfrac{\sigma}{\bar{X}} \times 100 = \dfrac{4.09}{23.8} \times 100 = 17.18.$

For the distribution of females:

Class interval	Frequency f	$c.f.$	Mid-value X	fX	fX^2	$f\lvert X - \bar{X}\rvert$	$f\lvert X - \text{Me}\rvert$	$f\lvert X - \text{Mo}\rvert$
15–19	15	15	17	255	4335	69.6	60.0	48.75
19–23	20	35	21	420	8820	12.8	0.0	15.00
23–27	9	44	25	225	5625	30.24	36.0	42.75
27–31	4	48	29	116	3364	29.44	32.0	35.00
31–35	2	50	33	66	2178	22.72	24.0	25.50
Total	$N = 50$			1082	24322	164.8	152.0	167.0

$$\text{Mean, } \bar{X} = \frac{1}{N}\Sigma fX = \frac{1082}{50} = 21.64\,\text{U/L.}$$

$$\text{Median, Me} = l + \frac{h}{f}\left(\frac{N}{2} - c\right) = 19 + \frac{4}{20}\left(\frac{50}{2} - 15\right) = 21\,\text{U/L.}$$

$$\text{Mode, Mo} = l + \frac{h(f_1 - f_0)}{2f_1 - f_0 - f_2} = 19 + \frac{4(20 - 15)}{2 \times 20 - 15 - 9} = 20.25\,\text{U/L.}$$

(i) $\text{M.D.}(\bar{X}) = \frac{1}{N}\Sigma f|X - \bar{X}| = \frac{164.8}{50} = 3.296\,\text{U/L}.$

$\text{M.D.}(\text{Me}) = \frac{1}{N}\Sigma f|X - \text{Me}| = \frac{152}{50} = 3.04\,\text{U/L}.$

$\text{M.D.}(\text{Mo}) = \frac{1}{N}\Sigma f|X - \text{Mo}| = \frac{167}{50} = 3.34\,\text{U/L}.$

(ii) $\text{Coeff. M.D.}(\bar{X}) = \frac{\text{M.D.}(\bar{X})}{\bar{X}} = \frac{3.296}{21.64} = 0.152.$

$\text{Coeff. M.D.}(\text{Me}) = \frac{\text{M.D.}(\text{Me})}{\text{Me}} = \frac{3.04}{21} = 0.145.$

$\text{Coeff. M.D.}(\text{Mo}) = \frac{\text{M.D.}(\text{Mo})}{\text{Mo}} = \frac{3.34}{20.25} = 0.165.$

It is seen that the observations are more dispersed with respect to mode than their dispersion with respect to mean and median. From the result of coefficient of mean deviation from median it can be concluded that the observations are concentrated near the centre of the array.

(iii) $Q_1 = l + \frac{h}{f}\left(\frac{N}{4} - c\right) = 15 + \frac{4}{15}\left(\frac{50}{4} - 0\right) = 18.33\,\text{U/L}.$

Here c is taken as 0 as the $c.f.$ before Q_1 class is not there [First class of the frequency table is the class for Q_1].

$Q_3 = l + \frac{h}{f}\left(\frac{3N}{4} - c\right) = 23 + \frac{4}{9}\left(\frac{3 \times 50}{4} - 35\right) = 24.1\,\text{U/L}$

$\text{Coeff. of Q.D.} = \frac{Q_3 - Q_1}{Q_3 + Q_1} = \frac{24.1 - 18.33}{24.1 + 18.33} = 0.136.$

(iv) $\sigma^2 = \frac{1}{N}\left[\Sigma f X^2 - \frac{(\Sigma f X)^2}{N}\right] = \frac{1}{50}\left[24322 - \frac{(1082)^2}{50}\right] = 18.150\,(\text{U/L})^2.$

(v) $\sigma = \sqrt{\sigma^2} = \sqrt{18.1504} = 4.26\,\text{U/L}.$

(vi) $\text{C.V.} = \frac{\sigma}{\bar{X}} \times 100 = \frac{4.26}{21.64} \times 100 = 19.69.$

(vii) From the results of C.V. it is seen that the amount of S. GGT of males are less dispersed than that of females with respect to mean.

It has already been discussed that the deviations of observations from some central value provide a guess about the formation of distribution of observations. The formation of data set can also be studied by frequency curve, if the variable under study is quantitative. Let us discuss the frequency curve of the following 3 distributions.

The following are the distributions of ever born children in different families:

Class interval of children	2–4	4–6	6–8	8–10	10–12	Total
Number of first group of families, f	2	4	8	4	2	20
Number of second group of families, f	2	4	6	7	1	20
Number of third group of families, f	1	7	6	4	2	20

The calculation of mean, median and mode of these 3 distributions are shown below:

Class interval	Frequency of Group-1	Frequency of Group-2	Frequency of Group-3	C.f. of Group-1	C.f. of Group-2	C.f. of Group-3	Mid-value X	fx 1st	fx 2nd	fx 3rd
2–4	2	2	1	2	2	1	3	6	6	3
4–6	4	4	7	6	6	8	5	20	20	35
6–8	8	6	6	14	12	14	7	56	42	42
8–10	4	7	4	18	19	18	9	36	63	36
10–12	2	1	2	20	20	20	11	22	11	22
Total	$N = 20$	$N = 20$	$N = 20$					140	142	138

For the first group of families the mean, median and mode are as follows:

$$\text{Mean, } \bar{X} = \frac{1}{N}\Sigma fX = \frac{140}{20} = 7.$$

$$\text{Median, Me} = l + \frac{h}{f}\left(\frac{N}{2} - c\right) = 6 + \frac{2}{8}\left(\frac{20}{2} - 6\right) = 7.$$

$$\text{Mode, Mo} = l + \frac{h(f_1 - f_0)}{2f_1 - f_0 - f_2} = 6 + \frac{2(8-4)}{2 \times 8 - 4 - 4} = 7.$$

These measures for second group of families are:

$$\bar{X} = \frac{1}{N}\Sigma fX = \frac{142}{20} = 7.1.$$

$$\text{Me} = l + \frac{h}{f}\left(\frac{N}{2} - c\right) = 6 + \frac{2}{6}\left(\frac{20}{2} - 6\right) = 7.3.$$

$$\text{Mo} = l + \frac{h(f_1 - f_0)}{2f_1 - f_0 - f_2} = 8 + \frac{2(7-6)}{2 \times 7 - 6 - 1} = 8.3.$$

The measures for the third group of families are:

$$\bar{X} = \frac{1}{N}\Sigma fX = \frac{138}{20} = 6.9.$$

$$\text{Me} = l + \frac{h}{f}\left(\frac{N}{2} - c\right) = 6 + \frac{2}{6}\left(\frac{20}{2} - 8\right) = 6.7.$$

$$\text{Mo} = l + \frac{h(f_1 - f_0)}{2f_1 - f_0 - f_2} = 4 + \frac{2(7-1)}{2 \times 7 - 1 - 6} = 5.7.$$

Let us now draw the frequency curve of these 3 distributions:

Title: *Frequency curve of the distribution of number of ever born children in different families.*

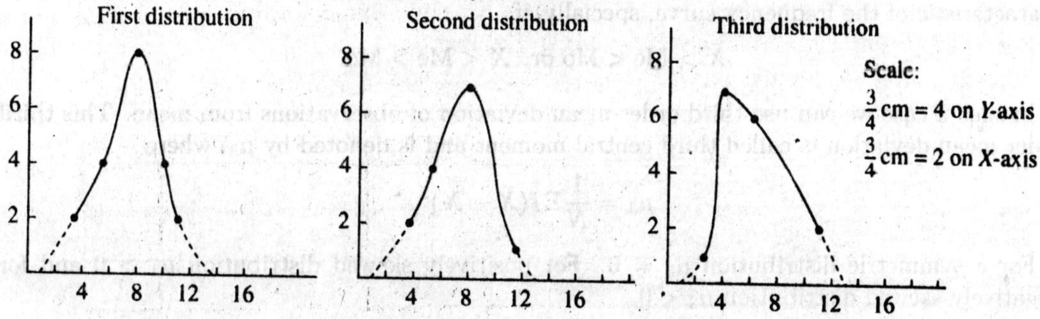

For the first distribution $\bar{X}$ = Me = Mo and the frequency curve is symmetric with uniform tails in both sides with respect to mean. For the second distribution $\bar{X}$ < Me < Mo and there is a long tail in the left side of the curve with respect to mean. This curve is skewed to the left or negatively skewed. For the third distribution $\bar{X}$ > Me > Mo and there is a long tail in the right side of the curve. This curve is skewed to the left or positively skewed.

These frequency curves also provide some information about the formation of data set. The first curve is symmetric and other two curves are asymmetric. *This asymmetric character of frequency curve is called skewness.*

For the symmetric distribution there is no difference between mean and median or between mean and mode. For asymmetric distribution there are differences between mean and median and between mean and mode. Hence, these differences can be used as a measure of skewness. The measure of skewness (Sk) is

$$\text{Sk} = \bar{X} - \text{Me}$$
$$\text{or, Sk} = \bar{X} - \text{Mo}.$$

For the first distribution,

$$\text{Sk} = \bar{X} - \text{Me} = 7 - 7 = 0$$
$$\text{or, Sk} = \bar{X} - \text{Mo} = 7 - 7 = 0.$$

Any one of these measures leads us to conclude that the distribution is symmetric.

For the second distribution,

$$\text{Sk} = \bar{X} - \text{Me} = 7.1 - 7.3 = -0.2$$
$$\text{or, Sk} = \bar{X} - \text{Mo} = 7.1 - 8.3 = -1.2.$$

We conclude that the distribution is negatively skewed.

For the third distribution,

$$\text{Sk} = \bar{X} - \text{Me} = 6.9 - 6.7 = 0.2$$
$$\text{or, Sk} = \bar{X} - \text{Mo} = 6.9 - 5.7 = 1.2.$$

We conclude that the distribution is positively skewed.

From the above measures, it may be concluded that in a symmetric distribution the observations are uniformly dispersed in the left and right sides of the mean. In a negatively skewed distribution more of the observations are in the left side of the mean. In a positively skewed distribution more of the observations are in the right side of the mean.

The above measures of skewness may not always depict the clear idea about the shape characteristic of the frequency curve, specially, if

$$\bar{X} > \text{Me} < \text{Mo or, } \bar{X} < \text{Me} > \text{Mo}.$$

In such a case we can use third order mean deviation of observations from mean. This third order mean deviation is called third central moment and is denoted by μ_3, where

$$\mu_3 = \frac{1}{N}\Sigma f(X - \bar{X})^3.$$

For a symmetric distribution $\mu_3 = 0$. For positively skewed distribution $\mu_3 > 0$ and for negatively skewed distribution $\mu_3 < 0$.

Let us investigate the value of μ_3 for the three distributions discussed above.

Class interval	Frequency for			$f(X-\bar{X})^3$ for			X
	1st	2nd	3rd	1st	2nd	3rd	
2–4	2	2	1	−128	−137.842	−59.319	3
4–6	4	4	7	−32	−37.044	−48.013	5
6–8	8	6	6	0	−0.006	0.006	7
8–10	4	7	4	32	48.013	37.044	9
10–12	2	1	2	128	59.319	137.842	11
Total	$N = 20$	$N = 20$	$N = 20$	0	−67.56	67.56	

For first, second and third distributions μ_3 values are respectively as below:

$$\mu_3 = \frac{1}{N}\Sigma f(X-\bar{X})^3 = 0.$$

$$\mu_3 = \frac{1}{N}\Sigma f(X-\bar{X})^3 = \frac{-67.56}{20} = -3.378.$$

$$\text{and } \mu_3 = \frac{1}{N}\Sigma f(X-\bar{X})^3 = \frac{67.56}{20} = 3.378.$$

EXERCISE – 4

1. The following are the number of **cancer patients** (x) **who are under** treatment in different hospitals.

   Number of cancer patients (x): **32, 23, 18, 16, 30.**

   (a) Calculate mean deviation from **mean and from** median of the number of cancer patients.
   [**Ans.** M.D. ($\bar{X}$) = 5.76, M.D. (Me) = 5.6]

   (b) Calculate coefficient of mean deviation from mean and from median.
   [**Ans.** Coeff. M.D. ($\bar{X}$) = 0.242, Coeff. M.D. (Me) = 0.243]

   (c) Calculate variance of the number of cancer patients. [**Ans.** $\sigma^2 = 40.16$]

   (d) Calculate standard deviation of the number of cancer **patients.** [**Ans.** $\sigma = 6.34$]

   (e) Calculate coefficient of variation [C.V.] of the number **of cancer** patients.
   [**Ans.** C.V. = 26.63]

   (f) Find range of the number of cancer patients. [**Ans.** $R_N = 14$]

2. The following is the distribution of ages of drug addicted **people coming** to the hsopital for treatment:

Class interval of ages	18–22	22–26	26–30	30–34	34–38	Total
Number of people, f	12	17	28	22	21	100

   (a) Calculate mean deviation from mean, median and mode of age.
   [**Ans.** M.D. ($\bar{X}$ = 4.3289, M.D. (Me) = 4.34, M.D. (Mo) = 4.2826]

   (b) Calculate coefficient of mean deviation from mean, median and mode of age.
   [**Ans.** Coeff. M.D. ($\bar{X}$) = 0.1497, Coeff. M.D. (Me) = 0.1497. Coeff M.D. (Mo) = 0.1498]

(c) Calculate variance of the distribution of age. [**Ans.** $\sigma^2 = 25.5136$]

(d) Calculate standard deviation of age. [**Ans.** $\sigma = 5.15$]

(e) Calculate coefficient of variation of age. [**Ans.** C.V. = 17.81]

(f) Find range of age. [**Ans.** $R_N = 20$]

3. The following are the gestation period (x months) of some mothers:

Gestation period (x months) : 9.0, 8.5, 8.4, 7.5, 8.8, 8.5.

(a) Find coefficient of mean deviation from mean, median and mode of the gestation period.
[**Ans.** Coeff. M.D. ($\bar{X} = 0.041$, Coeff. M.D. (Me) = 0.037, Coeff. M.D. (Mo) = 0.037]

(b) Calculate coefficient of variation [C.V.] of age. [**Ans.** C.V. = 5.58]

(c) Find range of age. [**Ans.** $R_N = 1.5$]

4. (a) Define dispersion and measure of dispersion. What are the different measures of dispersion? Why do we need measure of dispersion?

(b) Which measure of dispersion do you think best? Why?

Chapter 5.

Probability and Probability Distribution

5.1 Introduction

Probability can be considered as a measure of uncertainty of an event that can be defined from a random experiment. Let us explain it by an example. Suppose, a doctor operates patients and 50 per cent of his patients suffer due to post-operation infection. A researcher may be interested to measure the chance of a selected patient to be infected after operation and he can select two patients one by one for his investigation. These two patients may be infected (I) or may not be infected (N) or one of them may be infected. The probable results of his investigation may be of the following type:

$$\{II,\ IN,\ NI,\ NN\}.$$

One cannot claim which of these results will be observed unless the investigation is completed. These results are uncertain. We need to measure this uncertainty.

The results of this investigation of two patients, one by one, are not pre-determined. So, this work of investigation of patients is known as *Random Experiment.*

In the above investigation of two patients, there are 4 possible outcomes. Each of the outcomes is known as *Sample Point* and the totality of the sample points constitutes a *sample space* (S), where

$$S: \{II,\ IN,\ NI,\ NN\}.$$

These 4 outcomes are exhaustive as we cannot expect any other results of this random experiment. Again, during investigation one can find any one of these 4 results. No two results can be observed at the same time and hence, the results are called *Mutually Exclusive* outcomes. As there is no prior information about the occurrence of these 4 outcomes, any of the outcomes can occur with equal chance. Hence, these outcomes are *Equally Likely* outcomes.

However, due to prior information that 50% patients are infected after operation, the chance of I is 0.5 and chance of N is also 0.5. This prior information tells that I and N are equally likely.

In the above experiment (investigation), the investigator may be interested to measure the chance of observing one infected patient. He is making a statement about 2 points $\{IN,\ NI\}$ in the sample space. This statement is known as 'Event'. Event is generally denoted by A or B or C or ....

In this particular case, let us write:

$$A: \text{one patient is infected.}$$

This A can be shown by a diagram as follows:

$$A \boxed{IN,\ NI}$$

This diagram is known as *Venn Diagram*. A is constituted with 2 points in favour of it. These 2 points are known as *Favourable Outcomes* to the event. The number of favourable outcomes is denoted by $m(\leq n)$, where n is the number of equally likely outcomes.

Probability: If a random experiment shows n equally likely, exhaustive and mutually exclusive outcomes and if $m(\leq n)$ of these outcomes are in favour of an event A, then the probability of A is measured by

$$P(A) = \frac{m}{n}.$$

The example of random experiment cited above provides $m = 2$ cases in favour of A and there are $n = 4$ equally likely outcomes. So

$$P(A) = \frac{m}{n} = \frac{2}{4} = \frac{1}{2}.$$

From the above definition of $P(A)$, it is seen that $0 \leq P(A) \leq 1$. If it is assumed that 80 per cent patients are not infected (N), then $P(I) = 0.2$ and $P(N) = 0.8$. The outcomes I and N are not equally likely. In such a case the investigation of two patients, one by one, will not provide equally likely outcomes. The sample space will be constituted with 4 mutually exclusive and exhaustive outcomes. In that case, we can apply this definition of probability.

Here $A \boxed{IN,\ NI}$

We need $P(A) = P(IN) + P(NI)$

$$= 0.2 \times 0.8 + 0.8 \times 0.2$$

$$= 0.32.$$

One may be interested to find the probability that, out of 2 patients investigated, none is infected. That is he needs

$$P(C) = P(NN) = 0.8 \times 0.8 = 0.64, \text{ where } C \boxed{NN}.$$

Again, one may be interested to find the probability that at least one is infected. Let us write this event as B, where

$$B \boxed{II,\ IN,\ NI}.$$

$$\therefore P(B) = P(II) + P(IN) + P(NI)$$

$$= 0.2 \times 0.2 + 0.2 \times 0.8 + 0.8 \times 0.2$$

$$= 0.36.$$

It is seen that $P(C) + P(B) = 0.64 + 0.36 = 1.00$. This B is defined with the sample points of S except the points of C. This latter event B is known as *Complementary Event* to C. Usually, if A is any event, the complementary event is denoted by $\bar{A}$. Thus,

$$P(A) + P(\bar{A}) = 1$$

$$\Rightarrow P(A) = 1 - P(\bar{A}) \text{ or, } P(\bar{A}) = 1 - P(A).$$

Independent events: Two events A and B are said to be independent, if and only if

$$P(AB) = P(A)P(B).$$

For example, let us consider that in a family there are two children. The birth-order of boy (B) or girl (G) is not known. However, assume that the birth of a boy or a girl is equally likely.

$$\therefore \left[P(B) = \frac{1}{2}, P(G) = \frac{1}{2}\right].$$

The sample space of the birth of children is

$$S: \{BB, BG, GB, GG\}.$$

Let A : the first child is a boy, B : the second child is a girl.

Then A [BB [BG] GG] B

Here AB : [BG]

The events A and B are not mutually exclusive as they can occur together [both A and B have common point BG].

Here equally likely cases are $n = 4$; favourable cases to A are $m = 2$ and favourable cases to B are $m = 2$.

Hence, $P(A) = \frac{m}{n} = \frac{2}{4} = \frac{1}{2}$ and $P(B) = \frac{m}{n} = \frac{2}{4} = \frac{1}{2}$.

Again, $P(AB) = \frac{m}{n} = \frac{1}{4}$, m = favourable case to $AB = 1$.

Now $P(A)P(B) = \frac{1}{2} \times \frac{1}{2} = \frac{1}{4} = P(AB)$.

So, A and B are independent events.

Factorial: Let $1, 2, 3, \ldots, n$ be n natural numbers. Using these natural numbers factorial $n(n!)$ is measured as the product of all n natural numbers. That is

$$n! = 1 \times 2 \times 3 \times \cdots \times n.$$

For example, $1! = 1$

$2! = 1 \times 2 = 2$

$3! = 1 \times 2 \times 3 = 6$

$4! = 1 \times 2 \times 3 \times 4 = 24.$

Again, $0! = 1$.

Permutation: Let us consider that in a clinic three doctors D_1, D_2 and D_3 work. One of them works in the morning and one works in the evening. The duty of the doctors can be assigned as follows:

Morning	D_1	D_2	D_1	D_3	D_2	D_3
Evening	D_2	D_1	D_3	D_1	D_3	D_2

These arrangement of doctors for duties is known as permutation. Here there are 6 permutations. Mathematically, it is measured by

$$^3P_2 = \frac{3!}{(3-2)!} = 6.$$

Here $^3P_2$ is called permutation of 2 doctors from 3 doctors. Generally, permutation of r objects taken together from n objects is written as

$$^nP_r = \frac{n!}{(n-r)!}.$$

But the number of different permutations of n different objects taken r at a time is

$$^nP_r = n^r.$$

Combination: From three doctors D_1, D_2 and D_3 if two are assigned duty to work at a time, then the arrangement will be D_1D_2 or D_1D_3 or D_2D_3. There are 3 probable cases of selection of 2 doctors at a time from 3 doctors. This grouping of 3 doctors taking 2 at a time is called combination of 3 doctors taken 2 at a time and it is written as

$$^3C_2 = \frac{3!}{2!(3-2)!} = 3.$$

In general, combination of n objects taking r at a time is measured by

$$^nC_r = \frac{n!}{r!(n-r)!}.$$

5.2 Example on Probability

EXAMPLE 5.1 In a clinic 50% are male nurses and another 50% are female nurses. Once 3 nurses are selected to supervise the work of nurses in 3 duty hours. Find the probability that (i) all selected nurses are males, (ii) all selected nurses are females, (iii) 2 selected nurses are males, (iv) at least 2 selected nurses are males, (v) at best one selected nurse is male.

Solution. Let M = male nurse, F = female nurse.

The sample space of this experiment is

$$S: \{MMM, MMF, MFF, MFM, FMM, FFM, FMF, FFF\}$$

Given $P(M) = 0.5, P(F) = 0.5$; male and female are equally likely.

There are $n = 8$ equally likely outcomes in the sample space.

(i) Let A: all selected nurses are males;

A $\boxed{MMM}$, where m = favourable outcomes = 1.

$\therefore P(A) = \frac{m}{n} = \frac{1}{8}.$

(ii) Let B: all selected nurses are females;

B $\boxed{FFF}$, where $m = 1$.

$\therefore P(B) = \frac{m}{n} = \frac{1}{8}.$

(iii) Let C: 2 selected nurses are males;

C $\boxed{MMF, MFM, FMM}$, where $m = 3$.

$\therefore P(C) = \frac{m}{n} = \frac{3}{8}.$

(iv) Let D: at least 2 selected nurses are males;

D $\boxed{MMM, MMF, MFM, FMM}$, where $m = 4$.

$\therefore P(D) = \frac{m}{n} = \frac{4}{8} = \frac{1}{2}.$

(v) Let E: at best one selected nurse is male;

E $\boxed{MFF, FFM, FMF, FFF}$, where $m = 4$.

$\therefore P(E) = \dfrac{m}{n} = \dfrac{4}{8} = \dfrac{1}{2}$.

EXAMPLE 5.2 In a town there are 3 hospitals. Three patients are admitted in hospitals. Find the probability that (i) all 3 are admitted in the same hospital, (ii) they are admitted in different hospitals, (iii) 2 are admitted in one hospital.

Solution. Let H_1, H_2 and H_3 be the 3 hospitals and P_1, P_2 P_3 be the 3 patients. The arrangements of patients in hospitals can be shown as follows:

H_1	P_1	P_1	P_2	P_2	P_3	P_3	$P_1P_2P_3$	P_1P_2	P_1P_2	P_1P_3	P_1P_3	P_2P_3	P_2P_3	P_3	P_3
H_2	P_2	P_3	P_1	P_3	P_1	P_2	—	P_3	—	P_2	—	P_1	—	P_1P_2	—
H_3	P_3	P_2	P_3	P_1	P_2	P_1	—	—	P_3	—	P_2	—	P_1	—	P_1P_2
H_1	—	—	—	—	—	P_2	—	P_1	—	—	—	P_2	—	P_1	—
H_2	$P_1P_2P_3$	—	—	—	P_1P_2	P_1P_3	P_1P_3	P_2P_3	P_2P_3	—	P_3	—	P_2	—	P_1
H_3	—	$P_1P_2P_3$	—	—	P_3	—	P_2	—	P_1	—	P_1P_2	P_1P_3	P_1P_3	P_2P_3	P_2P_3

There are 27 arrangements of patients in hospitals. This arrangement is different permutations of 3 different objects taken 3 at a time and is given by $3^3 = 27$.

(i) Let A: All 3 are admitted in the same hospital.

There are 3 arrangements in favour of $A (m = 3)$.

$\therefore P(A) = \dfrac{m}{n} = \dfrac{3}{3^3} = \dfrac{1}{9}$.

(ii) Let B: They are admitted in different hospitals.

There are $3! = 6$ arrangements $(m = 6)$.

$\therefore P(B) = \dfrac{m}{n} = \dfrac{6}{3^3} = \dfrac{2}{9}$.

(iii) Let C: Two are admitted in one hospital. From the arrangements it is seen that there are 18 such arrangements $(m = 27 - 3 - 6 = 18)$

$\therefore P(C) = \dfrac{m}{n} = \dfrac{18}{3^3} = \dfrac{2}{3}$.

EXAMPLE 5.3 In a clinic there are 6 MBBS and 4 FRCP doctors. A medical board of 4 doctors is to be formed. Find the probability that the board will consist of (i) 4 FRCP doctors, (ii) 4 MBBS doctors, (iii) 2 FRCP and 2 MBBS doctors.

Solution. A board of 4 doctors from $6 + 4 = 10$ doctors can be formed in

$$n = {}^{10}C_4 = \frac{10!}{4!(10-4)!} = 210 \text{ ways.}$$

(i) Let A: Board will consist of 4 FRCP doctors.

A can occur in $m = {}^4C_4 = 1$ way.

$\therefore P(A) = \dfrac{m}{n} = \dfrac{1}{210}$.

(ii) Let B: Board will consist of 4 MBBS doctors.
B can occur in $m = {}^6C_4 = \frac{6!}{4!(6-4)!} = 15$ ways.
$\therefore P(B) = \frac{m}{n} = \frac{15}{210} = \frac{1}{14}$.

(iii) Let C: Board will consist of 2 MBBS doctors and 2 FRCP doctors.
C can occur in $m = {}^6C_2 \times {}^4C_2 = 90$ ways.
$\therefore P(C) = \frac{m}{n} = \frac{90}{210} = \frac{3}{7}$.

EXAMPLE 5.4 In a hospital there are 15 male and 20 female nurses. Every day 10 nurses are selected for duties in a shift. Find the probability that, out of 10 nurses, (i) 6 are females and 4 are males, (ii) all selected nurses are females.

Solution. Total number of nurses are $15 + 20 = 35$. Ten from 35 can be selected in

$$n = {}^{35}C_{10} = \frac{35!}{10!(35-10)!} = 183579396 \text{ ways.}$$

(i) Let A: 6 are female and 4 are male nurses.
A can occur in $m = {}^{15}C_4 \times {}^{20}C_6 = 52907400$ ways.
$\therefore P(A) = \frac{m}{n} = \frac{{}^{15}C_4 \times {}^{20}C_6}{{}^{35}C_{10}} = 0.2882.$

(ii) Let B: All selected nurses are female.
B can occur in $m = {}^{20}C_{10} = 184756$ ways.
$\therefore P(B) = \frac{m}{n} = \frac{{}^{20}C_{10}}{{}^{35}C_{10}} = 0.001006.$

EXAMPLE 5.5 In a hospital there are 20 male and 30 female nurses. Among male nurses 12 are graduate and 8 have nursing diploma degree. The corresponding figures for female are 20 and 10 respectively. One nurse is selected at random. Find the probability that the selected one is (i) either a male or a graduate, (ii) female under the condition that she has nursing diploma degree, (iii) male and graduate.

Solution. The distribution of nurses can be shown as follows:

Qualification	Sex		Total
	Male (B)	Female ($\bar{B}$)	
Graduate (A)	12 AB	20 $A\bar{B}$	32
Diploma ($\bar{A}$)	8 $\bar{A}B$	10 $\bar{A}\bar{B}$	18
Total	20	30	50

Let A: A nurse is graduate, $\bar{A}$: (not A) A nurse has diploma degree
B: A nurse is male, $\bar{B}$: (not B) A nurse is female.

From $(20 + 30) = 50$ nurses one can be selected in $n = {}^{50}C_1 = 50$ ways.

(i) We need

$$P(A \text{ or } B) = P(A \cup B) = P(A) + P(B) - P(AB)$$
$$= \frac{32}{50} + \frac{20}{50} - \frac{12}{50} = \frac{4}{5}.$$

This rule $P(A \cup B) = P(A) + P(B) - P(AB)$ is called addition rule of probability for two not mutually exclusive events. Here A and B have common points ($AB = 12$) and A and B are not mutually exclusive events.

(ii) $P(\bar{B}/\bar{A}) = \dfrac{P(\bar{A}\bar{B})}{P(\bar{A})} = \dfrac{10/50}{18/50} = \dfrac{5}{9}.$

This $P(\bar{B}/\bar{A})$ is called *conditional probability* of $\bar{B}$ under the condition that $\bar{A}$ occurs first. In general, we write:

$$P(A/B) = \frac{P(AB)}{P(B)} \text{ or, } P(B/A) = \frac{P(AB)}{P(A)}$$

$$\text{or, } P(AB) = P(B)P(A/B) \text{ or, } P(AB) = P(A)P(B/A).$$

These latter two rules are called *Multiplication Rule of probability.*

(iii) $P(AB) = \dfrac{12}{50} = \dfrac{6}{25}.$

EXAMPLE 5.6 In a rural clinic there are 5 beds for patients. One day all beds are found empty and two patients are to be admitted in the clinic. Beds are identified by serial number $1, 2, 3, 4, 5$. Find the arrangements of beds for the patients. Find the probability that (i) sum of the numbers of selected beds is even, (ii) sum of the numbers of selected beds is more than 8 or the bed for first incoming patient bears number 2, (iii) sum of the numbers of selected beds is 8 under the condition that the serial number of bed for first incoming patient is 3.

Assume that one patient is male (M) and one patient is female (F), and male and female can be accommodated in any of two beds.

Solution. The arrangements of beds for male (M) and female (F) patients are as follows:

$$S : \left\{\begin{matrix} MF & FM & MF & FM & MF & FM & MF & MF & MF & FM \\ 1\,2 & 1\,2 & 2\,1 & 2\,1 & 3\,1 & 3\,1 & 4\,1 & 4\,1 & 5\,1 & 5\,1 \\ 1\,3 & 1\,3 & 2\,3 & 2\,3 & 3\,2 & 3\,2 & 4\,2 & 4\,2 & 5\,2 & 5\,2 \\ 1\,4 & 1\,4 & 2\,4 & 2\,4 & 3\,4 & 3\,4 & 4\,3 & 4\,3 & 5\,3 & 5\,3 \\ 1\,5 & 1\,5 & 2\,5 & 2\,5 & 3\,5 & 3\,5 & 4\,5 & 4\,5 & 5\,4 & 5\,4 \end{matrix}\right\}$$

There are $n = 40$ arrangements of beds for male and female and all 40 cases are mutually exclusive, exhaustive and equally likely.

(i) Let A: Sum of the numbers of selected beds is even.

The cases in favour of A are as follows:

$$A : \left\{\begin{matrix} MF & FM & MF & FM \\ 1\,3 & 1\,3 & 4\,2 & 4\,2 \\ 1\,5 & 1\,5 & 5\,1 & 5\,1 \\ 2\,4 & 2\,4 & 5\,3 & 5\,3 \\ 3\,1 & 3\,1 & & \\ 3\,5 & 3\,5 & & \end{matrix}\right\}$$

Favourable cases to A are $m = 16$.

$$\therefore P(A) = \frac{m}{n} = \frac{16}{40} = 0.4.$$

(ii) Let B: Sum of the numbers of selected bed is more than 8.

The cases in favour of B are

$$B : \left\{\begin{matrix} MF & FM \\ 4\,5 & 4\,5 \\ 5\,4 & 5\,4 \end{matrix}\right\}$$

Favourable cases to B are $m = 4$

Let C: Bed number 2 for first incoming patient

$$C:\begin{Bmatrix} MF & FM \\ 2\ 4 & 2\ 4 \end{Bmatrix}$$

Favourable cases to C are $m = 2$.

B and C have no common points. They are mutually exclusive events. We need

$$P(B \cup C) = P(B) + P(C) = \frac{4}{40} + \frac{2}{40} = \frac{3}{20}.$$

Here $P(B \cup C) = P(B) + P(C)$ is known as *addition rule of probability* for two mutually exclusive events.

(iii) Let D: Sum of the numbers of beds is 8.

The cases in favour of D are

$$C:\begin{Bmatrix} MF & FM \\ 3\ 5 & 3\ 5 \\ 5\ 3 & 5\ 3 \end{Bmatrix}$$

Let E: The number of bed for first incoming patient is 3.

The cases in favour of E are

$$E:\begin{Bmatrix} MF & FM \\ 3\ 1 & 3\ 1 \\ 3\ 2 & 3\ 2 \\ 3\ 4 & 3\ 4 \\ 3\ 5 & 3\ 5 \end{Bmatrix}$$

Here D and E are not mutually exclusive events. They have common points and these are

$$DE \quad \boxed{\begin{matrix} MF & FM \\ 3\ 5 & 3\ 5 \end{matrix}}$$

Favourable cases to DE are $m = 2$. We need

$$P(D/E) = \frac{P(DE)}{P(E)} = \frac{2/40}{8/40} = \frac{1}{4}.$$

EXAMPLE 5.7 In a clinic 6 male patients and 4 female patients have come for treatment. Two patients are asked (i) at random to the doctors, (ii) one by one for first 15 minutes of working time. Find the probability that (a) both male patients are asked, (b) one male and one female patients are asked.

Solution.

(i) There are $6 + 4 = 10$ patients. Two can be asked for two doctors in $n = {}^{10}C_2 = 45$ ways.

(a) Let A: Both patients are male.

A can occur in $m = {}^6C_2 = 15$ ways.

$$\therefore P(A) = \frac{m}{n} = \frac{15}{45} = \frac{1}{3}.$$

(b) Let B: One male patient and one female patient are asked.

B can occur in $m = {}^6C_1 \times {}^4C_1 = 24$ ways.

$$\therefore P(B) = \frac{m}{n} = \frac{24}{45} = \frac{8}{15}.$$

(ii) Patients, one by one, can be asked in the following order:

$$\{MM,\ MF,\ FM\ FF\}$$

One male patient can be asked in ${}^6C_1$ ways and one female patient can be asked in ${}^4C_1$ ways. One patient from 10 patients can be asked in ${}^{10}C_1 = 10$ ways.

(a) We need $P(A) = P(MM) = \dfrac{{}^6C_1}{{}^{10}C_1} \times \dfrac{{}^5C_1}{{}^9C_1} = \dfrac{1}{3}$.

If one male is asked first, the next male will be asked from 5 males and second one will be asked from total 9 patients.

(b) $P(B) = P(MF) + P(FM) = \dfrac{{}^6C_1}{{}^{10}C_1} \times \dfrac{{}^4C_1}{{}^9C_1} + \dfrac{{}^4C_1}{{}^{10}C_1} \times \dfrac{{}^6C_1}{{}^9C_1} = \dfrac{8}{15}$.

EXAMPLE 5.8 A distributor has sent a box containing 5 bottles of cough syrup (C) and 5 bottles of dextrose (D) to a shop. The owner of the shop picks up two bottles from the box (a) at random, (b) one by one without replacement. Find the probability that the selected bottles are (i) of cough syrup, (ii) of cough syrup and dextrose.

Solution.

(a) There are $5 + 5 = 10$ bottles.

Two bottles can be taken from 10 bottles in $n = {}^{10}C_2 = 45$ ways.

(i) Let A: Both bottles are of cough syrup.

A can occur in $m = {}^5C_2 = 10$ ways.

$\therefore P(A) = \dfrac{m}{n} = \dfrac{10}{45} = \dfrac{2}{9}$.

(ii) Let B: One bottle of cough syrup and one bottle of dextrose.

B can occur in $m = {}^5C_1 \times {}^5C_1 = 25$ ways.

$\therefore P(B) = \dfrac{m}{n} = \dfrac{25}{45} = \dfrac{5}{9}$.

(b) Bottles can be taken in the following order:

$$\{CC,\ CD,\ DC,\ DD\}.$$

First bottle can be taken in ${}^{10}C_1 = 10$ ways and second bottle can be taken in ${}^9C_1 = 9$ ways. First cough syrup bottle can be taken in ${}^5C_1 = 5$ ways and second cough syrup bottle can be taken in ${}^4C_1 = 4$ ways. Similar are the cases of D also.

(i) $P(A) = P(CC) = \dfrac{5}{10} \times \dfrac{4}{9} = \dfrac{2}{9}$.

(ii) $P(B) = P(CD) + P(DC) = \dfrac{5}{10} \times \dfrac{5}{9} + \dfrac{5}{10} \times \dfrac{5}{9} = \dfrac{5}{9}$.

EXAMPLE 5.9 A company supplies vitamin tablets in plastic container. It is known that 80% of the containers are in good condition. Once 3 containers are selected one after another during production process. Find the probability that (i) all 3 selected containers are defective, (ii) two selected containers are good, (iii) at least one is good, (iv) at best two are good.

Solution. Let G = good container, D = defective container.

The selection of container one by one will be of the form:

$$S: \{GGG,\ GGD,\ GDD,\ GDG,\ DDG,\ DGG,\ DGD,\ DDD\}.$$

Given $P(G) = 0.8$, $P(D) = 0.2$.

Here good and defective containers are not equally likely.

(i) Let A: All 3 are defective containers.

We need $P(A) = P(DDD) = 0.2 \times 0.2 \times 0.2 = 0.008$.

(ii) Let B: Two selected containers are good.

We need $P(B) = P(GGD) + P(GDG) + P(DGG)$

$$= 0.8 \times 0.8 \times 0.2 + 0.8 \times 0.2 \times 0.8 + 0.2 \times 0.8 \times 0.8$$

$$= 0.384.$$

(iii) Let C: At least one is good.

We have A indicating all 3 are defectives.

Here C is complementary to A. That is $C = \bar{A}$.

Therefore, $P(C) = P(\bar{A}) = 1 - P(A) = 1 - 0.008 = 0.992$.

(iv) Let D: At best 2 are good.

Let E: All are good.

Here D is the complementary event to E.

That is $D = \bar{E}$.

We have $P(E) = P(GGG) = 0.8 \times 0.8 \times 0.8 = 0.512$.

$\therefore P(D) = P(\bar{E}) = 1 - P(E) = 1 - 0.512 = 0.488$.

EXAMPLE 5.10 Out of 100 patients attended an outdoor of a hospital 60 came from slum area and 40 from normal residential area. Among 60 patients of slum area 35 were found to suffer from stomach problem and 25 were suffering from other diseases. The corresponding figures among patients of normal residential area are 18 and 22. A patient is randomly selected. Find the probability that the selected one is (i) either from slum area or is a patient of stomach, (ii) a patient of stomach under the condition that he is from slum area.

Solution. Let A : The patient is from slum area,

B : The patient is of stomach problem.

(i) We need $P(A \cup B) = P(A) + P(B) - P(AB)$.

Residential status	Problems		Total
	B Stomach	Others, $\bar{B}$	
A	35 $A \cap B$	25 $A \cap \bar{B}$	60
$\bar{A}$	18 $\bar{A} \cap B$	22 $\bar{A} \cap \bar{B}$	40
Total	53	47	100

$$\therefore P(A \cup B) = \frac{60}{100} + \frac{53}{100} - \frac{35}{100} = 0.78.$$

(ii) We need $P(B/A) = \dfrac{P(A \cap B)}{P(A)} = \dfrac{35/100}{60/100} = \dfrac{7}{12}$.

EXAMPLE 5.11 A company produces medicine using two plants. Plant-1 produces 600 packets per hour and Plant-2 produces 400 packets per hour. It is known that 10% packets of Plant-1 and 15% packets of Plant-2 are usually found defective. Once a defective packet is identified. Find the probability that it is a product of Plant-1.

Solution. Let H_1 : A product is of Plant-1,

H_2 : A product is of Plant-2,

E : A product is defective.

The events can be shown by Venn diagram as follows:

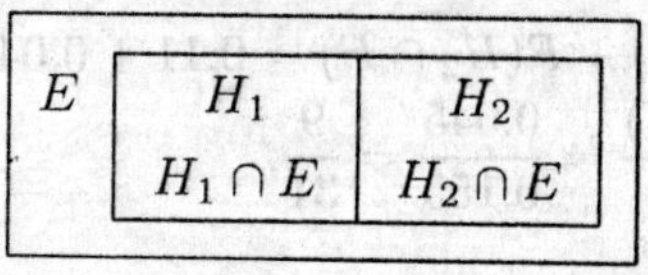

It is seen that $E = H_1 \cap E + H_2 \cap E$.

$$\therefore P(E) = PH_1 \cap E) + P(H_2 \cap E).$$

From the definition of conditional probability, we can write:

$$P(E/H_i) = \frac{P(H_i \cap E)}{P(H_i)} (i = 1, 2).$$

$$\therefore P(H_i \cap E) = P(H_i)P(E/H_i).$$

We need $P(H_1/E) = \dfrac{P(H_1 \cap E)}{P(E)}$.

This rule of probability is known as *Bayes Theorem*.

From the given information, we have

$$P(H_1) = \frac{600}{1000} = 0.6, \ P(H_2) = \frac{400}{1000} = 0.4.$$

Total production is $600 + 400 = 1000$ per hour.

$$P(E/H_1) = 0.1, \ P(E/H_2) = 0.15.$$

Now, $P(H_1 \cap E) = P(H_1)P(E/H_1) = 0.6 \times 0.1 = 0.06.$

$$P(H_2 \cap E) = P(H_2)P(E/H_2) = 0.4 \times 0.15 = 0.06.$$

$$\therefore P(E) = P(H_1 \cap E) + P(H_2 \cap E) = 0.06 + 0.06 = 0.12.$$

$$\therefore P(H_1/E) = \frac{P(H_1 \cap E)}{P(E)} = \frac{0.06}{0.12} = 0.5.$$

EXAMPLE 5.12 In a hospital there are two surgeons A and B. A operates 55% patients and B operates 45% patients. Twenty per cent patients of A and 10% patients of B are suffering from heart disease. Once a patient of heart disease is identified. Find the probability that he is a patient of B.

Solution. Let H_1: Patient of A, H_2: Patient of B, E: Patient of heart disease.

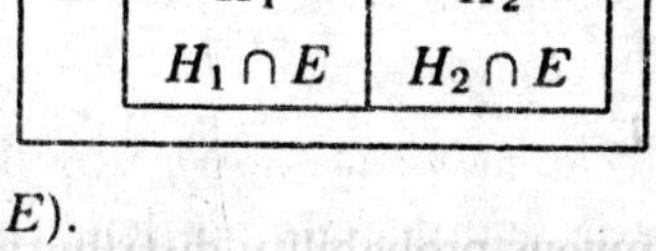

$P(E) = P(H_1 \cap E) + P(H_2 \cap E)$.

We need $P(H_2/E) = \dfrac{P(H_2 \cap E)}{P(E)}$.

Given $P(H_1) = 0.55,\ P(H_2) = 0.45$

$$P(E/H_1) = 0.20,\ P(E/H_2) = 0.10$$

$$P(H_1 \cap E) = P(H_1)P(E/H_1) = 0.55 \times 0.20 = 0.11$$

$$P(H_2 \cap E) = P(H_2)P(E/H_2) = 0.45 \times 0.10 = 0.045$$

$$P(E) = P(H_1 \cap E) + P(H_2 \cap E) = 0.11 + 0.045 = 0.155$$

$$\therefore P(H_2/E) = \frac{P(H_2 \cap E)}{P(E)} = \frac{0.045}{0.155} = \frac{9}{31}.$$

5.3 Random Variable and Its Distribution

Let us consider that 70% doctors of a hospital are males. Once two doctors are assigned duties for two working hours. The arrangements of doctors will be of the form given below:

$$S : \{MM,\ MF,\ FM,\ FF\},\ \text{where } M = \text{Male},\ F = \text{Female}.$$

This arrangement is a random one. From this random experiment one may be interested in finding the probabilities of selecting no male doctors or one male doctor or two male doctors. In general, we can write:

Let X = Number of male doctors, where $x = 0, 1, 2$.

From the given condition, we have $P(M) = 0.7$ and $P(F) = 0.3$.

Therefore, $P(X = 0) = P(FF) = 0.3 \times 0.3 = 0.09$

$$P(X = 1) = P(MF) + P(FM)$$

$$= 0.7 \times 0.3 + 0.3 \times 0.7 = 0.42$$

$$P(X = 2) = P(MM) = 0.7 \times 0.7 = 0.49.$$

These results of probabilities can be written as follows:

X	0	1	2
$P(x)$	0.09	0.42	0.49

It is seen that the variable X takes different values with associated probabilities. So, the variable X is called *Random Variable* and $P(x)$ is called *Probability Function* of X, where $P(x)$ satisfies the conditions:

(i) $P(x) \geq 0$, (ii) $\sum_{x=0}^{\infty} P(x) = 1$.

Here X takes only integral values and hence, it is known as *Discrete Random Variable*, where $x = 0, 1, 2, \ldots, \infty$.

The set of values $\{x, P(x)\}$ is known as *probability distribution*.

The random variable X can take values from $-\infty$ to ∞. In that case, X is called *Continuous Random Variable*. It has a probability density function denoted by $f(x)$, where $f(x)$ satisfies the conditions:

(i) $f(x) \geq 0$, (ii) $\int_{-\infty}^{\infty} f(x)dx = 1$.

Here $\{x, f(x)\}$ is known as continuous probability distribution.

For example, let $f(x) = 2x,\ 0 < x < 1$.

This $f(x)$ satisfies both the conditions discussed above and hence, $f(x)$ is a probability density function.

Important probability distributions are: (i) Binomial distribution, (ii) Poisson distribution, (iii) Normal distribution, and (iv) Exponential distribution. Application of these distributions will be discussed in the following sections.

5.4 Binomial Distribution

It is a discrete probability distribution, where the random variable X takes the values $x = 0, 1, 2, \ldots, n$; if the random experiment is repeated n times. This random variable is observed from a random experiment, where experimental results can be classified into two classes, namely 'Success' S and 'Failure' F. For example, from a hospital 10 nurses are selected. A selected nurse will be either male or female. The experimental results are of two types. In such an experiment if we are interested in finding the probability of selection of x number of male nurses, then x will be the value of random variable X and observing a particular value of x will be our success 'S'. If in the hospital there are 70% female nurses and 30% male nurses, then the probability of success is

$$p = P[\text{a male nurse}] = 0.3, \text{ and}$$

probability of female nurse (Failure) is

$$q = P(\text{a female nurse}) = 0.7 = 1 - 0.3 = 0.7 = 1 - p$$

such that $p + q = 1$. The probability of selection of x male nurses is given by

$$P(x) = {}^{10}C_x(0.3)^x(0.7)^{10-x}, x = 0, 1, \ldots, 10.$$

This $P(x)$ is called binomial probability function of the random variable X. In general, the binomial probability function is given by

$$P(x) = {}^nC_x p^x q^{n-x}, \; x = 0, 1, 2, \ldots, n.$$

The mean of the random variable X is np and variance of it is npq.

EXAMPLE 5.13 A medical representative supplies medicines in packets of 10 bottles of a syrup and ensures that 10% bottles will be damaged. Once a medicine shop-keeper checks the bottles at random. Find the probability that he will identify (i) 2 damaged bottles, (ii) at least one damaged bottle, (iii) at best 2 damaged bottles.

Solution. Let X be the number of damaged bottles.

Given $p = P[\text{damaged bottles}] = 0.1$, $q = 1 - p = 1 - 0.1 = 0.9$.

The probability function of X is

$$P(x) = {}^nC_x p^x q^{n-x} = {}^{10}C_x(0.1)^x(0.9)^{10-x}, x = 0, 1, \ldots, 10.$$

(i) $P(X = 2) = {}^{10}C_2(0.1)^2(0.9)^{10-2} = 0.1937.$

(ii) $P(X \geq 1) = 1 - P(X < 1) = 1 - P(X = 0)$
$= 1 - {}^{10}C_0(0.1)^0(0.9)^{10-0}$
$= 0.6513$

(iii) $P(X \leq 2) = P(X = 0) + P(X = 1) + P(X = 2)$
$= {}^{10}C_0(0.1)^0(0.9)^{10-0} + {}^{10}C_1(0.1)(0.9)^{10-1} + {}^{10}C_2(0.1)^2(0.9)^{10-2}$
$= 0.9298.$

EXAMPLE 5.14 In a hospital every day babies are born. The birth of a boy (B) or a girl (G) are equally likely. Once records of 6 births are examined at random. Find the probability that, out of 6 births, (i) there are 3 boys, (ii) at best 1 is a boy, (iii) at least 2 are boys.

Solution. Let X be the number of boys.

Given $P(B) = \frac{1}{2} = p$ and $P(G) = \frac{1}{2} = q$, $n = 6$. Then

$$P(x) = {}^nC_x p^x q^{n-x} = {}^6C_x \left(\frac{1}{2}\right)^x \left(\frac{1}{2}\right)^{6-x} = {}^6C_x \left(\frac{1}{2}\right)^6, \quad x = 0, 1, \ldots, 6.$$

(i) $P(X = 3) = {}^6C_3 \left(\frac{1}{2}\right)^6 = \frac{20}{64} = \frac{5}{16}$

(ii) $P(X \leq 1) = P(X = 0) + P(X = 1)$

$$= {}^6C_0 \left(\frac{1}{2}\right)^6 + {}^6C_1 \left(\frac{1}{2}\right)^6 = \frac{7}{64}$$

(iii) $P(X \geq 2) = 1 - P(X < 2) = 1 - P(X = 0) - P(X = 1)$

$$= 1 - {}^6C_0 \left(\frac{1}{2}\right)^6 + {}^6C_1 \left(\frac{1}{2}\right)^6 = 1 - \frac{7}{64} = \frac{57}{64}.$$

EXAMPLE 5.15 A company produces paracetamol tablets in strips. Ten per cent of the strips are usually found defective. Once 5 strips are observed at random. Find the probability that, out of 5 strips, (i) one is defective, (ii) at best one is defective, (iii) at least one is defective.

Solution. Let X be the number of defective strips.

Given $n = 5$, $p = P[\text{Defective}] = 0.1, q = 1 - p = 1 - 0.1 = 0.9$.

$$P(x) = {}^nC_x p^x q^{n-x} = {}^5C_x (0.1)^x (0.9)^{5-x}, \quad x = 0, 1, 2, 3, 4, 5.$$

(i) $P(X = 1) = {}^5C_1 (0.1)(0.9)^{5-1} = 0.32805.$

(ii) $P(X \leq 1) = P(X = 0) + P(X = 1)$

$$= {}^5C_0 (0.1)^0 (0.9)^{5-0} + {}^5C_1 (0.1)(0.9)^{5-1} = 0.91854.$$

(iii) $P(X \geq 1) = 1 - P(X < 1) = 1 - P(X = 0)$

$$= 1 - {}^5C_0 (0.1)^0 (0.9)^{5-0} = 0.40951.$$

From the above discussion we can mention examples of binomial variable as follows:

(1) Number of male/female doctors working in a hospital

(2) Number of live birth/still birth of baby,

(3) Number of male/female nurses working in a clinic,

(4) Number of cured/dead patients admitted in a hospital,

(5) Number of post-operation infected/non-infected patients of a hospital,

(6) Number of defective/good medicine strips produced by a manufacturing company.

5.5 Poisson Distribution

Let us discuss the last example of the binomial variable mentioned above, where the variable is defective/good medicine strips produced by a company. The company produces a large number of strips per hour/day and if the production system is good, a very few defective strips are expected to be produced. In that case, if any one is interested to investigate the defective strips, he needs to investigate a large number of strips ($n \to \infty$), where $p = P[\text{Defective}] \to 0$. This random variable ($X$ number of defective strips) is transformed to Poisson variable under the condition that

$$n \to \infty,\ p \to 0 \text{ and } np = m \text{ is finite.}$$

The probability function of X is given by

$$P(x) = \frac{e^{-m} m^x}{x!}, \quad x = 0, 1, 2, \ldots.$$

The mean and variance of X is m.

Examples of poisson variable

(1) Number of defective items produced by a manufacturer
(2) Number of suicide cases in a society
(3) Number of infected patients after operation in a well-managed hospital
(4) Number of α-particles radiated from an atomic installation
(5) Number of telephone calls received in an emergency section of a hospital in a particular period of time.

EXAMPLE 5.16 A manufacturing company produces tablets in strips. Two per cent of their strips are usually found defective. The company produces 200 strips per 15 minutes. Find the probability that in a randomly selected period of 15 minutes the number of defective strips will be (i) 2, (ii) more than 2, (iii) less than 2.

Solution. Let X be the number of defective strips.
Given $n = 200$, $p = P[\text{defective}] = 0.02; m = np = 200 \times 0.02 = 4$.

$$\therefore P(x) = \frac{e^{-m} m^x}{x!} = \frac{e^{-4} 4^x}{x!}, \quad x = 0, 1, 2, \ldots$$

(i) $P(X = 2) = \dfrac{e^{-4} 4^2}{2!} = 0.1465.$

(ii)
$$\begin{aligned} P(X > 2) &= 1 - P(X \le 2) \\ &= 1 - P(X = 0) - P(X = 1) - P(X = 2) \\ &= 1 - e^{-4}\left[\frac{4^0}{0!} + \frac{4}{1!} + \frac{4^2}{2!}\right] \\ &= 1 - 0.2381 = 0.7619. \end{aligned}$$

(iii)
$$\begin{aligned} P(X < 2) &= P(X = 0) + P(X = 1) \\ &= e^{-4}\left[\frac{4^0}{0!} + \frac{4}{1!}\right] = 0.0916. \end{aligned}$$

EXAMPLE 5.17 The average number of post-operation infected patients per month in a hospital is 3. Find the probability that in a randomly selected month there will be (i) 2 infected patients, (ii) at least two infected patients, (iii) at best 2 infected patients.

Solution. Let X be the number of infected patients.

Given mean of X is $m = 3$.

Therefore, $P(x) = \dfrac{e^{-m}m^x}{x!} = \dfrac{e^{-3}3^x}{x!}$, $x = 0, 1, 2, \ldots$

(i) $P(X = 2) = \dfrac{e^{-3}3^2}{2!} = 0.22404.$

(ii) $P(X \geq 2) = 1 - P(X < 2) = 1 - P(X = 0) - P(X = 1)$

$$= 1 - e^{-3}\left[\frac{3^0}{0!} + \frac{3}{1!}\right] = 0.80085.$$

(iii) $P(X \leq 2) = P(X = 0) + P(X = 1) + P(X = 2)$

$$= e^{-3}\left[\frac{3^0}{0!} + \frac{3}{1!} + \frac{3^2}{2!}\right] = 0.4232.$$

5.6 Normal Distribution

This is a continuous distribution, where the random variable X ranges from $-\infty$ to ∞ and the probability density function of X is

$$f(x) = \frac{1}{\sigma\sqrt{2n}} e^{-\frac{1}{2}\left(\frac{x-M}{\sigma}\right)^2}, \quad -\infty < x < \infty.$$

Here mean of X is M and variance of X is σ^2. The shape of the normal probability curve is as follows:

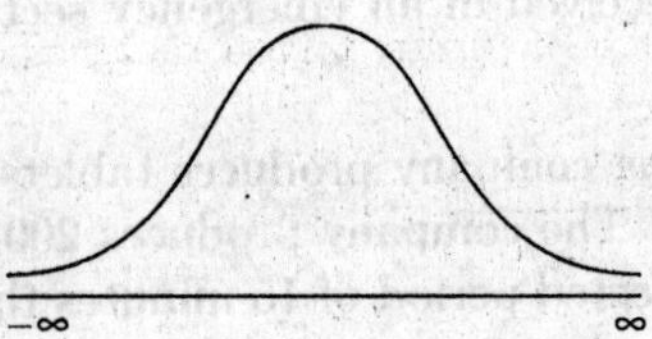

Total area under the curve is 1 and the area (probability) from $-\infty$ to certain value of x is available from the normal probability table. The tabulated value is available for the random variable z, where

$$z = \frac{X - M}{\sigma}$$

This z is known as standard normal variable and its mean is zero and variance is 1. From the normal probability table it is seen that

$$P(z < -1.96) = 0.025 \text{ and } P(z > 1.96) = 0.025.$$

If 1.96 is taken approximately 2, then for normal distribution 95% of its values fall in the limit $\bar{x} \pm 2\sigma$.

From this limit it may be concluded that if 95% values of a continuous random variable X fall in the limit $\bar{x} \pm 2\sigma$, then X can be assumed as normal variable.

Example of normal variable

If age, height, weight, blood pressure, blood sugar, body temperature, etc. are observed from random experiment, then these variables may follow normal distribution. Normal distribution is very important in studying statistical tests since most of the statistical tests are based on sampling distributions and sampling distributions are based on samples selected from normal population.

EXAMPLE 5.18 The blood sugar (mmol/l) level is assumed to follow normal distribution with mean, $M = 6.2$ and variance, $\sigma^2 = 4.5$. Find the probability that the blood sugar level of a randomly selected person (i) exceeds 6.8, (ii) is less than 7.5, (iii) is between 5.5 to 7.5.

Solution. Let X be the amount of blood sugar. Given mean of X, $M = 6.2$ and variance of X, $\sigma^2 = 4.5$. We have $\sigma = \sqrt{4.5} = 2.12$. Now

(i) $P(X > 6.8)$. We have $z = \dfrac{X - M}{\sigma} = \dfrac{6.8 - 6.2}{2.12} = 0.28$.

$\therefore P(X > 6.8) = P(z > 0.28) = 1 - P(z \leq 0.28) = 1 - 0.6103 = 0.3897.$

(ii) $P(X < 7.5)$. We have $z = \dfrac{X - M}{\sigma} = \dfrac{7.5 - 6.2}{2.12} = 0.61$.

$P(X < 7.5) = P(Z < 0.61) = 0.7291.$

(iii) $P(5.5 < X < 7.5)$. We have

$$z_1 = \frac{X_1 - M}{\sigma} = \frac{5.5 - 6.2}{2.12} = -0.33, \quad z_2 = \frac{X_2 - M}{\sigma} = \frac{7.5 - 6.2}{2.12} = 0.61.$$

$$P(5.5 < X < 7.5) = P(-0.33 < z < 0.61) = P(z < 0.61) - P(z < -0.33)$$
$$= 0.7291 - 0.3707 = 0.3584.$$

EXAMPLE 5.19 The amount of s. sodium of human being is assumed to follow normal distribution with mean, $M = 142$ mmol/l and variance, $\sigma^2 = 146\,(\text{mmol/l})^2$. Find the probability that the amount of s. sodium of a randomly selected person is (i) more than 145, (ii) less than 136, (iii) between 140 to 150.

Solution. Let X be the amount of s. sodium. Given mean of X, $M = 142$ and variance of X, $\sigma^2 = 146$; $\sigma = \sqrt{146} = 12.08$. Now

(i) $P(X > 145)$, $z = \dfrac{X - M}{\sigma} = \dfrac{145 - 142}{12.08} = 0.25$

$P(X > 145) = P(z > 0.25) = 1 - P(z < 0.25) = 1 - 0.5987 = 0.4013.$

(ii) $P(X < 136)$, $z = \dfrac{X - M}{\sigma} = \dfrac{136 - 142}{12.08} = -0.50$

$P(X < 136) = P(z < -0.50) = 0.3085.$

(iii) $P(140 < X < 150)$

$$z_1 = \frac{X_1 - M}{\sigma} = \frac{140 - 142}{12.08} = -0.17, \quad z_2 = \frac{X_2 - M}{\sigma} = \frac{150 - 142}{12.08} = 0.66$$

$$P(140 < X < 150) = P(-0.17 < z < 0.66)$$
$$= P(z < 0.66) - P(z < -0.17)$$
$$= 0.7454 - 0.4325 = 0.3129.$$

5.7 Exponential Distribution

Let X be a random variable and $f(x)$ be its probability density function, where

$$f(x) = \frac{1}{\lambda} e^{-x/\lambda}, \quad \lambda > 0, x > 0$$

Here $\{x, f(x)\}$ is exponential distribution. Mean of X is λ and variance of X is λ^2. Randomly observed 'waiting time' is the example of exponential variable.

We need to calculate (i) $P(X > x) = e^{-x/\lambda}$

(ii) $P(X < x) = 1 - e^{-x/\lambda}$

(iii) $P(x_1 < X < x_2) = e^{-x_1/\lambda} - e^{-x_2/\lambda}$.

EXAMPLE 5.20 The average time needed to check a patient by a doctor is 10 minutes. Find the probability that a patient will be checked (i) after 10 minutes, (ii) before 10 minutes, (iii) between 6 to 12 minutes.

If a man is in queue for one hour, what is the probability that he will be checked after one hour and fifteen minutes?

Solution. Let X be the time needed to check a patient. Given mean of X, $\lambda = 10$. We need

(i) $P(X > x) = P(X > 10) = e^{-10/10} = 0.36788.$

(ii) $P(X < x) = P(X < 10) = 1 - e^{-10/10} = 1 - 0.36788 = 0.63212.$

(iii) $P(X_1 < X < X_2) = P(6 < X < 12)$

$$= e^{-6/10} - e^{-12/10} = 0.5488 - 0.3012 = 0.2476.$$

Last part. As the man is in queue for 1 hour or 60 minutes, he will be checked after $(75 - 60) = 15$ minutes. Here queue time is subtracted because queue time is not the time to check a patient. Therefore, we have

$$P(X > x) = P(X > 15) = e^{-15/10} = 0.2231.$$

EXAMPLE 5.21 The average time needed to give a blood sample in a diagnostic centre is 15 seconds. Find the probability that a sample will be given (i) after 20 seconds, (ii) before 25 seconds, (iii) between 10 to 20 seconds.

If a man is in queue for 10 minutes, what is the probability that he will be able to give blood before 10 minutes 10 seconds?

Solution. Let X be the time to give blood sample. Given mean of X, $\lambda = 15$. We have

(i) $P(X > x) = P(X > 20) = e^{-20/15} = 0.2636.$

(ii) $P(X < x) = P(X < 25) = 1 - e^{-25/15} = 1 - 0.1889 = 0.811.$

(iii) $P(x_1 < X < x_2) = P(10 < X < 20) = e^{-10/15} - e^{-20/25}$

$$= 0.5134 - 0.2636 = 0.2498.$$

Last part. As the man is in queue for 10 minutes = 600 seconds, he will be able to give blood before $(610 - 600) = 10$ seconds. We need

$$P(X < 10) = 1 - e^{-10/15} = 1 - 0.5134 = 0.4866.$$

EXERCISE – 5

1. (a) From a locality 3 persons are selected one by one. It is assumed that 50% people of the locality suffer from malnutrition. Find the probability that (i) all 3 selected people suffer from malnutrition, (ii) at best 2 of the selected people suffer from malnutrition, (iii) one selected people suffers from malnutrition, (iv) none of the selected people suffers from malnutrition, (v) at least one selected people suffers from malnutrition.

[**Ans.** (i) $\frac{1}{8}$, (ii) $\frac{7}{8}$, (iii) $\frac{3}{8}$, (iv) $\frac{1}{8}$, (v) $\frac{7}{8}$]

(b) Sixty per cent of the people of another locality also suffer from malnutrition. Three people are selected one by one. Find the probabilities of the events discussed in (a).

[**Ans.** (i) 0.216, (ii) 0.784, (iii) 0.288, (iv) 0.064, (v) 0.936]

2. In a family there are 3 children. The birth order of the children are not known. Find the probability that (i) all 3 are boys, (ii) at best 2 are boys, (iii) one is a boy, (iv) all 3 are girls, (v) at least one is a boy. Boy and girl are equally probable.

[**Ans.** (i) $\frac{1}{8}$, (ii) $\frac{7}{8}$, (iii) $\frac{7}{8}$, (iv) $\frac{1}{8}$, (v) $\frac{7}{8}$]

3. The quality control department of a pharmaceutical company investigates the industrial product to suggest ways and means for all non-defective products. Once the department checks 3 items one by one during production process. It is known that 90% of the products are non-defective. Find the probability that out of the investigated items (i) all 3 are non-detective, (ii) at best two are non-detective, (iii) one is detective, (iv) all are defectives, (v) at least one is defective. [**Ans.** (i) 0.729, (ii) 0.271, (iii) 0.027, (iv) 0.001, (v) 0.999]

4. In a hospital 60% are male doctors. One day 3 doctors are selected for duties in 3 different periods. Find the probability that (i) all 3 selected doctors are male, (ii) at best 2 selected doctors are male, (iii) two selected doctors are female, (iv) all 3 selected doctors are females, (v) at least one selected doctor is male.

[**Ans.** (i) 0.216, (ii) 0.784, (iii) 0.288, (iv) 0.064, (v) 0.936]

5. In a hospital there are 20 male and 15 female doctors. Among male doctors 12 have FRCP degree. Among female doctors 10 have FRCP degree.

(a) A medical board of 5 doctors is to be formed. Find the probability that in the board (i) all are male doctors, (ii) 2 are female doctors, (iii) all have FRCP degree, (iv) none has FRCP degree.

[**Ans.** (i) $\frac{^{20}C_5}{^{35}C_5}$, (ii) $\frac{^{20}C_3 \times ^{15}C_2}{^{35}C_5}$, (iii) $\frac{^{22}C_5}{^{35}C_5}$, (iv) $\frac{^{13}C_5}{^{35}C_5}$]

(b) One doctor is selected at random. Find the probability that the selected doctor is (i) either male or having FRCP degree, (ii) female having FRCP degree, (iii) female under the condition that she has FRCP degree. [**Ans.** (i) $\frac{6}{7}$, (ii) $\frac{2}{7}$, (iii) $\frac{5}{11}$]

6. In a clinic there are 20 beds identified by serial number 1, 2, 3, . . . , 20. In the beds there are patients. Patient of one of the beds is to be selected for special care. Find the probability that the number of selected bed is (i) even, (ii) multiple of 4, (iii) multiple of 5 or an even number, (iv) multiple of 3 under the condition that it is even.

[**Ans.** (i) $\frac{1}{2}$, (ii) $\frac{1}{4}$, (iii) $\frac{3}{5}$, (iv) $\frac{3}{10}$]

7. A pharmaceutical company produces medicines using two machines A and B. Machine A produces 60% of the medicines and machine B produces 40% of the medicines. During production process 10% of A and 20% of B are damaged. Once a damaged medicine is identified. Find the probability that it is the product of machine A. [**Ans.** $\frac{3}{7}$]

8. A new medicine to control blood pressure is discovered. It is given to two hypertension patients. Find the probability that the medicine is effective to (i) one patient, (ii) at least one patient. [**Ans.** (i) $\frac{1}{4}$, (ii) $\frac{3}{4}$]

9. One hundred patients are classified by ages and diseases. The classified results are shown below:

Diseases	Ages (in years)			Total
	< 40	40–50	50+	
Diabetes	5	12	20	37
Heart problem	13	10	25	48
Kidney problem	2	3	10	15
Total	20	25	55	100

One patient is selected at random. Find the probability that the selected one is a (i) diabetic patient, (ii) patient of kidney problem or of ages 40–50, (iii) diabetic patient under the condition that he/she is of ages 40–50 years. [**Ans.** (i) $\frac{37}{100}$, (ii) $\frac{37}{100}$, (iii) $\frac{12}{25}$]

10. Two medical scientists have started independently two research works under the condition that they will finish the work within a stipulated period of time. After some time the researcher A has understood that the probability of finishing the work within the stipulated period is 0.8. The same probability for B is 0.7. Find the probability that (i) either of one will finish the work within the stipulated period of time, (ii) both of them will not be able to finish the work within the stipulated period of time. [**Ans.** (i) 0.94, (ii) 0.06]

11. In a blood bank there are 70 bottles of blood of males and 30 bottles of blood of females. 60% blood of males are A^+. The corresponding figure in case of females is 50%. One bottle of A^+ is found out. Find the probability that it is of male. [**Ans.** $\frac{14}{19}$]

12. In a clinic there are 6 male and 4 female doctors. Among male doctors 3 are heart specialist. The corresponding figure for female is 2. Everyday in the morning one doctor works in the outdoor of the clinic. Find the probability that one day in the morning treatment will be done by female heart specialist. [**Ans.** $\frac{2}{5}$]

13. In a ward of a hospital there are 10 beds identified by serial number $1, 2, \ldots, 10$. The patient of a bed is selected for special care. Find the probability that the selected bed number is multiple of (i) 2 or 4, (ii) 3 or 4, (iii) 3 under the condition that it is even number. [**Ans.** (i) $\frac{1}{2}$, (ii) $\frac{1}{2}$, (iii) 1]

14. In a box there are 6 packets of nidipro and 4 packets of mitosun 40. Two packets are taken (a) at random, (b) one by one with replacement, (c) one by one without replacement. Find the probability that (1) both packets are of nidipro, (ii) one packet is of nidipro.

[**Ans.** (a) (i) $\frac{1}{3}$, (ii) $\frac{8}{15}$; (b) (i) $\frac{9}{25}$, (ii) $\frac{12}{25}$; (c) (i) $\frac{1}{3}$, (ii) $\frac{8}{15}$]

15. In a hospital there are 20 male nurses and 30 female nurses. Among male nurses 12 have BPT degree. The corresponding figure for female nurses is 5. One nurse is selected at random. Find the probability that the selected one (i) is male or having BPT degree, (ii) is female having BPT degree, (iii) is male under the condition that he has BPT degree.

[**Ans.** (i) $\frac{1}{2}$, (ii) $\frac{1}{10}$, (iii) $\frac{12}{17}$]

16. In a hospital 70% male doctors. Among doctors 60% have foreign degrees. One doctor is selected at random. Find the probability that the selected doctor (i) is female or having foreign degree, (ii) is female under the condition that she has foreign degree.

[**Ans.** (i) 0.72, (ii) 0.3]

17. Forty per cent eligible couples of a village visit the health clinic during pregnancy of mother. Sixty per cent mothers are literate. Once one mother is selected at random. Find the probability that she is illiterate under the condition that she visits the clinic.

[**Ans.** 0.4]

18. In a village 60% mothers are literate. Forty per cent illiterate mothers visit the health clinic during pregnancy. The corresponding figure for literate mothers is 0.8. Once one mother visits the health clinic. Find the probability that she is illiterate. [**Ans.** 0.25]

19. In a block development office 20 male and 30 female agents work to motivate eligible couples for family planning adoption. Sixty per cent male agents and 80% female agents become successful in motivating couples. Once a successful motivated case is identified. Find the probability that it is done by female agent. [**Ans.** 0.67]

20. Fifty per cent people of a locality suffer from diabetes. Ten people are selected at random. Find the probability that, out of 10 people, (i) two suffer from diabetes, (ii) none suffers from diabetes, (iii) at least one suffers from diabetes, (iv) at best one suffers from diabetes.

[**Ans.** (i) $\frac{45}{1024}$, (ii) $\frac{1}{1024}$, (iii) $\frac{1023}{1024}$, (iv) $\frac{11}{1024}$]

21. Twenty per cent people of a locality suffer from hypertension. Six people are selected at random. Find the probability that, out of six people, (i) one suffers from hypertension, (ii) none suffers from hypertension, (iii) at least one suffers from hypertension.

[**Ans.** (i) 0.3932, (ii) 0.2621, (iii) 0.7379]

22. Five per cent operated cases of a hospital are infected. the hospital operates 100 patients per month. Find the probability that in a randomly selected month (i) two patients are infected, (ii) none is infected, (iii) at least one is infected.

[**Ans.** (i) 0.0842, (ii) 0.0067, (iii) 0.9933]

23. On an average 2 medicine packets of an industry are found defective in each hour of production. Find the probability that in a randomly selected hour (i) two packets are defective, (ii) no packet is defective, (iii) at least one packet is defective.

[**Ans.** (i) 0.2707, (ii) 0.1353, (iii) 0.8647]

24. The amount of haemoglobin of human being follows normal distribution with mean 11.0 and variance 4. Find the probability that the haemoglobin of a randomly selected man is (i) more than 12.5, (ii) less than 10.5, (iii) between 10 to 14.

[**Ans.** (i) 0.2266, (ii) 0.4013, (iii) 0.6247]

25. The amount of FBS follows normal distribution with mean 6.5 and variance 4.8. Find the probability that the FBS of a randomly selected person (i) is more than 8.5, (ii) is less than 5.5, (iii) is from 4.8 to 7.8. [**Ans.** (i) 0.1814, (ii) 0.3228, (iii) 0.5047]

26. The average time needed to take a blood sample from a person is 2 minutes. Find the probability that a blood sample will be collected (i) before 1.5 minutes, (ii) after 3 minutes, (iii) between 2 to 4 minutes, (iv) if a man is in queue for 10 minutes, what is the probability that his blood sample will be collected after 13 minutes?

[**Ans.** (i) 0.4724, (ii) 0.7769, (iii) 0.23254, (iv) 0.7769]

27. The average time of gestation period is 265 days. Find the probability that a baby will be born (i) before 260 days, (ii) after 272 days, (iii) between 255 to 265 days.

[**Ans.** (i) 0.3749, (ii) 0.6417, (iii) 0.0183]

28. Two per cent mothers who attend the clinic during pregnancy give still births. Hundred mothers attend the clinic per month. Find the probability that in a randomly selected month (i) 3 mothers give still births, (ii) no mother gives still birth, (iii) at least one mother gives still birth. [**Ans.** (i) 0.1804, (ii) 0.1353, (iii) 0.8647]

Chapter 6

Sampling and Sampling Distribution

6.1 Introduction

The objective of statistical analysis is to infer about the population characteristic (parameter) which is usually unknown. The inference is drawn on the basis of collected data from all population units or from a representative part of population units.

The technique of data collection from all population units is known as census survey. For example, 'Agriculture Census of a Country', 'Population Census of a Country', etc.

The technique of data collection from a representative part of population units is know as sample survey and a representative part of population units is called a sample.

Sampling is a technique to select a representative part of population units.

There are different methods of sampling. Some of the important sampling methods are:

(i) Simple random sampling,
(ii) Systematic sampling
(iii) Circular systematic sampling,
(iv) Stratified sampling
(v) Cluster sampling,
(vi) Two-state sampling
(vii) Double sampling,
(viii) Purposive selection
(ix) Quota sampling, etc.

6.2 Objective of Sampling

The data collection in a census involves lot of resources in terms of money, man power, trained personnel, time, etc. To cover a large population, larger administrative set-up is needed to conduct the survey. The larger is the administrative set-up, the weaker is the efficiency of management. But efficiency in data collection and statistical analysis are needed for accuracy and precision in statistical inference. The accuracy in data collection can be achieved if data are collected by experienced investigator supervised under efficient supervisor and management. Since in a sample survey the data are collected from sample units which are comparatively smaller in size, it is easier to manage efficient staffs to conduct the survey. Therefore, the objective of the sample survey is expressed as follows:

(i) The objective of sampling is to collect data within a short period of time utilizing minimum resources in terms of money and trained manpower.

(ii) The sample survey provides more accurate information of population units and hence most efficient estimate of population parameter is available.

6.3 Terms Related to Sampling

Frame. It has already been mentioned that total units which are under investigation constitute a population. The complete list of population units is known as frame. For example, the opinion of the patients admitted in a hospital is needed to improve the hospital management. In that case, all the patients admitted during the survey period are to be investigated. The list of those patients is a frame.

The frame must be completed and updated. The frame for finite population is usually available. Frame cannot be prepared for infinite population. Even for finite population, it is difficult to prepare an up-to-date frame. For example, a survey will be conducted to study family planning performance. In that case the adopter couples are the ultimate units to be investigated. But the up-to-date data list of adopter couples is not, usually, available.

Sampling unit. The population units which are to be included in a sample for investigation are known as sampling units.

Sampling fraction. Let a population be of size N. From this population a sample of size $n\,(\leq N)$ is to be selected. Then $f = \frac{n}{N}$ is called sampling fraction.

Sampling design. The method of selection of sample units from any specified population is known as sampling design. The sampling designs are mentioned in Section 6.1. A good sampling can be planned if up-to-date frame is available. However, the spread of population units in the survey area must be considered to plan a sampling design. The sampling design must be consistent with the objective of the survey. For example, a survey is needed to study the 'socio-economic background of the outdoor patients' of a hospital. Patients come, usually, from middle class families and from poor class families. But there is no fixed day or time for the particular class of families to visit the outdoor. Thus, sampling design is to be planned in such a way that patients of all types and classes are to be included in the sample. One of the sampling design will be the selection of a visiting patient after every 15 minutes or 10 minutes or 20 minutes, etc. Again, the selection of sample units is related to formulation of research problem.

Formulation of research problem. The identification of population and selection of sampling units are related to the specification of research problem. The researcher first identify the problem of research and formulate a specific plan to conduct the research activity in a scientific way.

Methods of data collection. After selection of sample design, the researcher needs a decision to collect information from sampling units. The information to be collected must be relevant to the objective of the research.

A decision is also needed to apply the mode of data collection. There are different modes of data collection, depending on the nature and objectives of the survey and availability of resources.

The most commonly used mode of data collection is the direct interview of the sampling units. For this a pre-tested questionnaire with closed or open or mixed types of responses are needed. Data can also be collected through mailed questionnaire, where the questionnaire will be filled up by the respondents themselves during investigation of the survey.

Questionnaire. In any sample survey or even in census the information on many aspects related to the objective of the survey are collected from each sampling unit. The information are collected according to some pre-designed questions with closed or open responses. A set of questions for any survey constitute the questionnaire.

The questionnaire is so planned that any investigator is not hurt and can answer all the questions easily. Each question should be self-explanatory. Politically or religiously motivated questions must be avoided.

Pilot survey. To conduct the survey efficiently, it is essential to know the existence of population units, the mode of communication, the suitable time to visit the respondents, etc. These information are to be collected before conducting the final survey. The survey which is conducted to collect preliminary information about population units is called pilot survey.

Random Sampling. In any sampling method, if the population units are selected with associated probability, the sampling is called random sampling.

Random sampling is of two types, viz., (a) each sampling unit is included in the sample with equal probability, (b) different sampling units are included in the sample with different probabilities. The first technique is called simple random sampling.

6.4 Simple Random Sampling

Let there be N units in a population. We need to select a sample of n ($\leq N$) units. The possible number of samples are ${}^nC_n$. If each sample is selected with equal probability $\frac{1}{{}^nC_n}$, then the sampling is called simple random sampling. For example, let us consider a population of size $N = 3$ units, where the units are x : 2, 4, 6. We can select a sample of size $n = 2$. The possible number of samples are ${}^3C_2 = 3$. The samples are (2,4), (2,6) and (4,6). Now, out of these 3 samples if any one is selected with probability $\frac{1}{{}^3C_2} = \frac{1}{3}$, then the sampling is called simple random smpling. Again, out of 3 units, any one unit can be selected with probability 1/3. In that case also the selection method is known as simple random samping. Thus, simple random sampling can also be defined as a method of selection of every unit with probability $1/N$, if there are N units in the population.

Method of selection of simple random sample

There are two methods of selection of simple random sample. These are:

(a) Lottery Method of Selection

(b) Use of Random Number Table

Both the methods can be explained by example.

EXAMPLE 6.1 The following are the s. amylase (x, U/L) of some persons:

x :	25	70	50	52	45	66	78	80	58	49	65	80	78	32	40	48	62	66	70
Sl. No. :	01	02	03	04	05	06	07	08	09	10	11	12	13	14	15	16	17	18	19

Select a simple random sample of size $n = 5$ by (a) Lottery method of selection, (b) Using random number table.

Solution. We have $N = 19$, $n = 5$.

(a) The units are identified by serial number 01 to 19. These numbers can be written in 19 pieces of papers of same size, same colour and same shape. These paper can be folded uniformly and can be put in a box. After well mixing; one by one 5 pieces of papers are taken. A piece of paper contains a serial numhber and against that serial number there is an observation of x. The value of x according to selected papers constitutes a sample. For example, let us consider that the pieces of papers contain serial numbers 08, 12, 13, 14, 19. Then the sample is as follows:

Selected serial number	08	12	13	14	19
Selected x	80	80	78	32	70

(b) The last serial number of the observations is of 2 digits. For the selection of the sample, we need to select 2-digit random numbers from the random number table provided in Appendix 1. In the table there are digits arranged in rows and columns. Using these digits we can form number of any digits as we need. Using first column of the random digits, we have

Random number	16	11	10	19	17
Selected x	48	65	49	70	62

Method of estimation of parameter

The population means, population variance, and other population characteristics are parameters. These parameters are to be estimated using sample observations, where population mean ($\bar{X}$) is calculated by

$$\bar{X} = \frac{1}{N}\Sigma x,$$

where $x_1, x_2, \ldots, x_N$ are population units.

The finite population variance (S^2) is measured by

$$S^2 = \frac{1}{N-1}\left[\Sigma x^2 - \frac{(\Sigma x)^2}{N}\right].$$

For example, if $N = 3$ and $x : 2, 4, 6$, then

$$\bar{X} = \frac{1}{N}\Sigma x = \frac{1}{3}(2+4+6) = 4$$

$$\text{and } S^2 = \frac{1}{N-1}\left[\Sigma x^2 - \frac{(\Sigma x)^2}{N}\right] = \frac{1}{3-1}\left[56 - \frac{(12)^2}{3}\right] = 4.$$

The sample mean ($\bar{x}$) is an unbiased estimate of population mean ($\bar{X}$) and the sample variance (s^2) is an unbiased estimate of population variance (S^2), where

$$\bar{x} = \frac{1}{n}\Sigma x$$

$$s^2 = \frac{1}{n-1}\left[\Sigma x^2 - \frac{(\Sigma x)^2}{n}\right].$$

If $N = 3$, $n = 2$ and $x : 2, 4, 6$, then samples are:

Sl. No.	Samples	Sample means $\bar{x}_i$	Sample variances s_i^2
1	2, 4	$\bar{x}_1 = 3$	$s_1^2 = 2$
2	2, 6	$\bar{x}_2 = 4$	$s_2^2 = 8$
3	4, 6	$\bar{x}_3 = 5$	$s_3^2 = 2$

Now, mean of sample means is $\frac{3+4+5}{3} = 4 = \bar{X}$.

Mean of sample variances is $\frac{2+8+2}{3} = 4 = S^2$.

These two characteristics valid, if sample is selected by simple random sampling method. As mean of sample means is equal to population mean ($\bar{X}$), then sample mean is an unbiased estimate of population mean.

Again, as mean of sample variances is equal to population variance (S^2), the sample variance (s^2) is an unbiased estimate of population variance.

As there are ^nC_n samples, ^nC_n sample means can be calculated. These means has also variance. The variance of the sample means can be estimated by

$$v(\bar{x}) = \frac{N-n}{Nn}s^2, \text{ where } s^2 = \frac{1}{n-1}\left[\Sigma x^2 - \frac{(\Sigma x)^2}{n}\right]$$

$$= \frac{1-\frac{n}{N}}{n}s^2, \text{ where } f = \frac{n}{N} \text{ is sampling fraction}$$

$$= \frac{s^2}{n}, \text{ if } f = \frac{n}{N} \text{ is neglected.}$$

This $f = \frac{n}{N}$ is also called finite population correction.

The population total (X) is estimated by

$$\hat{X} = N\bar{x}$$

and the variance of this estimate ($\hat{X}$) is estimated by

$$v(\hat{X}) = N^2 v(\bar{x}).$$

The estimate of standard error (s.e.) of $\bar{x}$ and $\hat{X}$ are given by

$$\text{s.e. } (\bar{x}) = \sqrt{v(\bar{x})} \text{ and s.e. } (\hat{X}) = \sqrt{v(\hat{X})}, \text{ respectively.}$$

EXAMPLE 6.2 The following are the s. Lipase (x, U/L) of some persons:

x	15	18	35	42	58	17	20	29	33	45	18	55	50	44	36	27
Sl. No.	01	02	03	04	05	06	07	08	09	10	11	12	13	14	15	16
x	22	18	60	50	37	34	41	33	45	42	25	36	39	40	42	28
Sl. No.	17	18	19	20	21	22	23	24	25	26	27	28	29	30	31	32

(i) Select a simple random sample of 6 persons.

(ii) Estimate mean amount of s. Lipase.

(iii) Estimate standard error of estimated mean s. Lipase.

(iv) Estimate total amount of s. Lipase.

(v) Estimate variance of estimated total s. Lipase.

(vi) Find 95% confidence interval of mean s. Lipase.

Solution. We have $N = 32, n = 6$.

(i) As $N = 32$, the last serial number of population unit is of 2-digit. We need to select number of 2-digits from the random number table. Using second column of random numbers, we have

Random numbers	10	19	23	01	20	27
Selected x	45	60	41	15	50	25

(ii) Estimate of mean amount of s. Lipase is:

$$\bar{x} = \frac{1}{n}\Sigma x = \frac{1}{6}[45 + 60 + 41 + 15 + 50 + 25] = 39.33.$$

(iii) Estimate of standard error of estimated mean is:

$$\text{s.e.}\ (\bar{x}) = \sqrt{v(\bar{x})} = \sqrt{37.19} = 6.10.$$

$$v(\bar{x}) = \frac{N-n}{Nn}s^2 = \frac{32-6}{32 \times 6}274.67 = 37.19.$$

$$s^2 = \frac{1}{n-1}\left[\Sigma x^2 - \frac{(\Sigma x)^2}{n}\right] = \frac{1}{6-1}\left[10656 - \frac{(236)^2}{6}\right] = 274.67.$$

(iv) Estimate of total s. Lipase is:

$$\hat{X} = N\bar{x} = 32 \times 39.33 = 1258.56.$$

(v) Estimate of variance of estimated total s. Lipase is:

$$v(\hat{X}) = N^2 v(\bar{x}) = (32)^2 37.19 = 38082.56.$$

(vi) 95% confidence interval of mean s. Lipase is given by

$$\hat{\bar{X}}_L = \bar{x} - t_{0.05,5}\,\text{s.e.}\ (\bar{x}) = 39.33 - 2.571 \times 6.10 = 23.65.$$

$$\text{and } \hat{\bar{X}}_U = \bar{x} - t_{0.05,5}\,\text{s.e.}\ (\bar{x}) = 39.33 - 2.571 \times 6.10 = 55.01.$$

Thus $23.65 < \bar{X} < 55.01$.

Here $t_{0.05,5}$ = tabulated value of t at 5% level of significance with

$$(n-1) = (6-1) = 5\,\text{d.f.}$$

Estimation of proportion

In Example 6.2, the variable under study is quantitative. Variable under study may be qualitative. A particular population unit may possess a qualitative character or may not possess that character. We need to estimate proportion of units possessing a particular character. For example, we may be interested to estimate the proportion of persons having s. Lipase more than 40 [Example 6.2]. In the sample, it is seen that $a = 4$ persons have this character out of $n = 6$ persons. So, estimate of proportion of persons having s. Lipase more than 40 is:

$$p = \frac{a}{n} = \frac{4}{6} = 0.67.$$

Let N = population units, n = sample units

A = number of units in the population possessing a character

$P = \frac{A}{N}$ = proportion of units in the population possessing the character

a = number of sample units possessing the character

$p = \frac{a}{n}$ = proportion of units in the sample possessing the character.

This p is an unbiased estimate of population proportion P.

The estimate of variance of p is

$$v(p) = \frac{N-n}{N(n-1)}pq, \text{ where } q = 1-p.$$

In our given example, [Example 6.2] $p = 0.67, q = 1 - p = 0.33, N = 32, n = 6$.

Therefore $v(p) = \frac{32-6}{32(6-1)}0.67 \times 0.33 = 0.0359$ and s.e. $(p) = 0.1895$.

95% confidence interval for P is given by

$$\hat{P}_L = p - 1.96\,\text{s.e.}\,(p) \text{ and } \hat{P}_U = p + 1.96\,\text{s.e.}\,(p)$$

where 1.96 is the 5% value of z from normal probability table. Therefore, we have [Example 6.2]

$$\hat{P}_L = p - 1.96\,\text{s.e.}\,(p) = 0.67 - 1.96 \times 0.1895 = 0.30.$$

$$\hat{P}_U = p + 1.96\,\text{s.e.}\,(p) = 0.67 + 1.96 \times 0.1895 \approx 1.00.$$

6.5 Systematic Sampling

Let there be $N = nk$ units in a population, where $k = N/n$ is a nearest integer. Now, if first unit is selected from first k units in the population using 'simple random sampling' technique and then every k-th unit is selected, the resultant sample is called systematic sample. Here n is the sample size. For example, let us consider the population observations of S. GGT of some persons, where the amount of S. GGT (x, U/L) are:

x	25	58	150	42	68	92	55	25	50	38	110	162	76	88	90	62	78
Sl. No.	01	02	03	04	05	06	07	08	09	10	11	12	13	14	15	16	17
x	50	38	48	86	97	62	77	104	108	72	53	48	64	72	115	120	70
Sl. No.	18	19	20	21	22	23	24	25	26	27	28	29	30	31	32	33	34

Let us select a sample of $n = 8$ persons, where $k = \frac{N}{n} = \frac{34}{8} \approx 4$. According to systematic sampling method, the first observation is to be selected from first 4 observations in the population. Then every 4-th observation is selected. Using second column of random numbers, we have first one digit number 3. Therefore, the first selected unit will be of serial number 3 and than 7th, 11th, 15th, 19th, 23rd, 27th, 31st units are to be selected. The selected systematic sample is as follows:

Random number	03	07	11	15	19	23	27	31
Selected (x)	150	55	110	90	38	62	72	72

This method of sampling provides k systematic samples. However, all samples may not be of size n, specially, the last sample may be of size $(n - 1)$ depending on the value of N. For example, let $N = 30$ and $n = 8$, where $k = 4$. In this case, if first selected unit is of serial number 03, then 8-th unit in the sample will not be available as 8th unit will be of serial number 31.

The problem of sample size in all k samples can be shown by selecting the $k = 4$ samples in the above case. This can be shown assuming the first selected unit is of serial number 1 or 2 or 3 or 4 as follows:

Random number	**01**	**05**	**09**	**13**	**17**	**21**	**25**	**29**
Selected x	**25**	**68**	**50**	**76**	**78**	**86**	**104**	**48**
Random number	**02**	**06**	**10**	**14**	**18**	**22**	**26**	**30**
Selected x	**58**	**92**	**38**	**88**	**50**	**97**	**108**	**64**
Random number	**03**	**07**	**11**	**15**	**19**	**23**	**27**	**31**
Selected x	**150**	**55**	**110**	**90**	**38**	**62**	**72**	**72**
Random number	**04**	**08**	**12**	**16**	**20**	**24**	**28**	**32**
Selected x	**42**	**25**	**162**	**62**	**48**	**77**	**53**	**115**

In all $k = 4$ cases sample size $n = 8$. But, if we select sample of size $n = 9$, then $k = \frac{34}{9} = 3.78 \approx 4$. The possible k samples are shown below:

Random number	01	05	09	13	17	21	25	29	33
Selected x	25	68	50	76	78	86	104	48	120
Random number	02	06	10	14	18	22	26	30	34
Selected x	58	92	38	88	50	97	108	64	70
Random number	03	07	11	15	19	23	27	31	—
Selected x	150	55	110	90	38	62	72	72	—
Random number	04	08	12	16	20	24	28	32	—
Selected x	42	25	162	62	48	77	53	115	—

Here the last two samples are not of size $n = 9$. This problem of sample size can be overcome. Here the process of avoiding this problem is not discussed.

There are $i(i = 1, 2, \ldots, k)$ samples. In practice, one sample is selected. From one sample the unbiased estimate of population mean is given by

$$\bar{x}_{\text{sy}} = \frac{1}{n}\sum_j x_{ij}, \; i = 1, 2, \ldots, k$$

Here $\bar{x}_{\text{sy}}$ = systematic sample means and x_{ij} is the j-th observation $(j = 1, 2, \ldots, n)$ of i-th sample $(i = 1, 2, \ldots, k)$. The estimated variance of $\bar{x}_{\text{sy}}$ is:

$$v(\bar{x}_{\text{sy}}) = \frac{N-n}{Nn}s^2, \text{ where } s^2 = \frac{1}{n-1}\left[\Sigma x_{ij}^2 - n\bar{x}_i^2\right] = \frac{1}{n-1}\left[\Sigma x_{ij}^2 - \frac{(\Sigma x_{ij})^2}{n}\right]$$

$$\bar{x}_i = \frac{1}{n}\sum_j x_{ij}, \; i = 1.$$

As we consider only one sample,

$$\bar{x}_{\text{sy}} = \bar{x} = \frac{1}{n}\sum_{j=1}^{n} x_j \text{ and } s^2 = \frac{1}{n-1}\left[\Sigma x_j^2 - \frac{(\Sigma x_j)^2}{n}\right].$$

Let us estimate the average S. GGT of the population considered above. The estimate and its estimated variance are shown from sample of size $n = 8$. The sample is:

Random number	03	07	11	15	19	23	27	31
Selected x	150	55	110	90	38	62	72	72

The estimate of average S. GGT is

$$\bar{x}_{\text{sy}} = \bar{x} = \frac{1}{n}\Sigma x_j = \frac{1}{8}(150 + 55 + 110 + 90 + 38 + 62 + 72 + 72)$$
$$= 81.125\,\text{U/L}$$

The estimate of variance of $\bar{x} = \bar{x}_{\text{sy}}$.

$$v(\bar{x}_{\text{sy}}) = \frac{N-n}{Nn}s^2 = \frac{34-8}{34 \times 8}1247.27 = 119.22.$$

$$s^2 = \frac{1}{n-1}\left[\Sigma x_j^2 - \frac{(\Sigma x_j)^2}{n}\right] = \frac{1}{8-1}\left[61381 - \frac{(649)^2}{8}\right] = 1247.27\,(\text{U/L})^2.$$

Circular systematic sampling

Let $N = nk$ and $k = N/n$. In circular systematic sampling the first observation is selected from all $N = nk$ observations using 'simple random sampling' method. Then every k-th unit is selected in a circular way until n units are selected. The method is explained using the population observations of S. GGT discussed above, where $N = 34, n = 8, k \approx 4$. Using second column of 'random number table' we have first two-digit number 10 [as $N = 34$ is of 2-digit]. The other selected serial numbers are $14, 18, 22, 26, 30, 34, 04$. Finally, the sample is as follows:

Random number	10	14	18	22	26	30	34	04
Selected x	38	88	50	97	108	64	70	42

6.6 Stratified Random Sampling

Sample survey is done to estimate the parameter of the population, where the parameter is an unknown constant of a variable, namely mean or variance of the variable. But in large population the values of the variable may not be homogeneous. For example, the parameter to be estimated is the average ever born children per mother. But the number of ever born children varies with the variation in the level of education of mothers. General assumption is that with the increase in the level of education, number of ever born children decreases. The values of number of ever born children are not homogeneous to all group of mothers. In such cases the population units are divided into several classes, where the values of the variables under study in a class are internally more homogeneous. Each class is known as stratum and population observations are divided into several strata. Then from every stratum a simple random sample is selected. The resultant sample is a stratified random sample and the method of selection of sample is known as stratified random sampling.

Let there be N units in a population. The N units are divided into L strata, where h-th $(h = 1, 2, \dots, L)$ stratum has N_h units and $N = N_1 + N_2 + \cdots + N_L$. A sample of n units is to be selected. Let us consider that from h-th stratum $n_h (h = 1, 2, \dots, L)$ units are to be selected such that $n = n_1 + n_2 + \cdots + n_L$. If n_h units are selected from N_h units by simple random sampling method, then the sampling method is known as stratified random sampling.

At this stage a decision is needed to find the value of n_h. One of the simplest method to decide the value of n_h is known as proportional allocation of sample size, where

$$n_h \propto N_h \text{ or, } n_h = \frac{n}{N} N_h \ (h = 1, 2, \dots, L).$$

The estimate of average in stratified sampling is given by

$$\bar{x}_{\text{st}} = \frac{1}{N} \sum_{h=1}^{L} N_h \bar{x}_h, \text{ where } \bar{x}_h = \frac{1}{n_h} \sum_{i=1}^{n_h} x_{hi}$$

is the sample mean of h-th stratum.

The estimated variance of $\bar{x}_{\text{st}}$ in case of proportional allocation is given by

$$v(\bar{x}_{\text{st}}) = \frac{(1-f)}{n} \sum_{h=1}^{L} W_h s_h^2, \text{ where } f = \frac{n}{N}, W_h = \frac{N_h}{N}$$

and $s_h^2 = \dfrac{1}{n_h - 1}\left[\Sigma x_{hi}^2 - \dfrac{(\Sigma x_{hi})^2}{n_h}\right]$. If $f = \dfrac{n}{N}$ is negligible, then $v(\bar{x}_{\text{st}}) = \dfrac{1}{n} \sum_{h=1}^{L} W_h s_h^2$. The estimate of population total is $\hat{X}_{\text{st}} = N\bar{x}_{\text{st}}$ and estimated variance of $\hat{X}_{\text{st}}$ is $v(\hat{X}_{\text{st}}) = N^2 v(\bar{x}_{\text{st}})$.

EXAMPLE 6.3 The following are the number of ever born children ($x_{hi}; h = 1, 2, 3, 4; i = 1, 2, \ldots, N_h$) of $L = 4$ groups of mothers, where mothers are classified by their level of education.

Level of education	*Number of ever born children to mothers* (x_{hi})
Illiterate	5, 3, 7, 4, 2, 2, 8, 4, 5, 5, 4, 4, 6, 3, 2, 1, 4, 5, 7, 6, 4, 3, 5, 2, 2, 6, 4, 6, 5, 4, 3, 3, 5, 6, 6, 5
Primary	3, 2, 4, 4, 3, 5, 4, 3, 2, 4, 4, 3, 5, 6, 4, 4, 5, 6, 3, 7, 4, 2, 2, 8, 4, 3, 5, 3, 3, 2
Secondary	2, 3, 3, 2, 2, 1, 3, 2, 2, 2, 3, 1, 1, 3, 2, 4
Higher	2, 2, 2, 3, 3, 1, 1, 2, 2, 1, 4, 2, 2

(i) Select a stratified random sample of size $n = 20$ mothers.

(ii) Estimate average number of ever born children per mother.

(iii) Estimate variance of the estimated average children.

(iv) Estimate total number of ever born children of mothers.

(v) Estimate variance of estimated total number of ever born children.

(vi) Estimate proportion of mothers having less than 3 children along with its estimated standard error.

Solution. We have $N_1 = 36, N_2 = 30, N_3 = 16, N_4 = 13$.

$$N = N_1 + N_2 + N_3 + N_4 = 36 + 30 + 16 + 13 = 95. \; n = 20.$$

(i) n_h is to be selected from h-th stratum by proportional allocation, where

$$n_h = \frac{n}{N} N_h.$$

$$\therefore n_1 = \frac{20}{95} \times 36 = 7.6 \approx 8, n_2 = \frac{20}{95} \times 30 = 6.3 \approx 6,$$

$$n_3 = \frac{20}{95} 16 = 3.4 \approx 3, n_4 = \frac{20}{95} \times 13 = 2.7 \approx 3,$$

$$n = n_1 + n_2 + n_3 + n_4 = 8 + 6 + 3 + 3 = 20.$$

We have to select $n_1 = 8$ mothers from illiterate group, $n_2 = 6$ mothers from primary educated group, $n_3 = 3$ mothers from secondary educated group and $n_4 = 3$ mothers from higher educated group. Starting from first column of 'random numbers' the selected random numbers and selected ever born children are shown below [assuming the arrangement of mothers in serial order starting from 01 to N_h].

Level of education	*Number of ever born children* (x_{hi}) *to mothers*								
Illiterate	Random numbers	16	11	10	19	17	09	26	01
	x_{1i}	1	4	5	7	4	5	6	5
Primary	Random numbers	24	16	25	10	19	23		
	x_{2i}	8	4	4	4	3	2		
Secondary	Random numbers	01	15	04					
	x_{3i}	2	3	4					
Higher	Random numbers	05	10	09					
	x_{4i}	3	1	2					

(ii) Estimate of average number of ever born children is:

$$\bar{x}_{st} = \frac{1}{N}\Sigma N_h \bar{x}_h, \text{ where}$$

$$\bar{x}_h = \frac{1}{n_h}\Sigma x_{hi}$$

$$\bar{x}_1 = \frac{1}{n_1}\Sigma x_{1i} = \frac{1}{8}(1+4+5+7+4+5+6+5) = \frac{37}{8} = 4.625$$

$$\bar{x}_2 = \frac{1}{n_2}\Sigma x_{2i} = \frac{1}{6}(8+4+4+4+3+2) = \frac{25}{6} = 4.17$$

$$\bar{x}_3 = \frac{1}{n_3}\Sigma x_{3i} = \frac{1}{3}(2+2+2) = 2$$

$$\bar{x}_4 = \frac{1}{n_4}\Sigma x_{4i} = \frac{1}{3}(3+1+2) = 2$$

$$\therefore \bar{x}_{st} = \frac{1}{95}[36 \times 4.625 + 30 \times 4.17 + 16 \times 2 + 13 \times 2] = \frac{349.6}{95} = 3.68.$$

(iii) Estimate of variance of the estimated average is:

$$v(\bar{x}_{st}) = \frac{(1-f)}{n}\Sigma W_h s_h^2, \; s_h^2 = \frac{1}{n_h - 1}\left[\Sigma x_{hi}^2 - \frac{(\Sigma x_{hi})^2}{n_h}\right]$$

$$s_1^2 = \frac{1}{n_1 - 1}\left[\Sigma x_{1i}^2 - \frac{(\Sigma x_{1i})^2}{n_1}\right] = \frac{1}{8-1}\left[193 - \frac{(37)^2}{8}\right] = 3.125.$$

$$s_2^2 = \frac{1}{n_2 - 1}\left[\Sigma x_{2i}^2 - \frac{(\Sigma x_{2i})^2}{n_2}\right] = \frac{1}{6-1}\left[125 - \frac{(25)^2}{6}\right] = 4.167.$$

$$s_3^2 = \frac{1}{n_3 - 1}\left[\Sigma x_{3i}^2 - \frac{(\Sigma x_{3i})^2}{n_3}\right] = \frac{1}{3-1}\left[12 - \frac{(6)^2}{2}\right] = 0.$$

$$s_4^2 = \frac{1}{n_4 - 1}\left[\Sigma x_{4i}^2 - \frac{(\Sigma x_{4i})^2}{n_4}\right] = \frac{1}{3-1}\left[14 - \frac{(6)^2}{3}\right] = 1.$$

$$W_1 = \frac{N_1}{N} = \frac{36}{95} = 0.379, \; W_2 = \frac{N_2}{N} = \frac{30}{95} = 0.316.$$

$$W_3 = \frac{N_3}{N} = \frac{16}{95} = 0.168, \; W_4 = \frac{N_4}{N} = \frac{13}{95} = 0.137.$$

$$v(\bar{x}_{st}) = \frac{(1-0.21)}{20}[0.379 \times 3.125 + 0.316 \times 4.167 + 0.168 \times 0 + 0.137 \times 1]$$
$$= 0.1042.$$

(iv) Estimate of total number of ever born children is

$$\hat{X}_{st} = N\bar{x}_{st} = 95 \times 3.68 = 349.6 \approx 350.$$

(v) Estimate of variance of estimated total number of ever born children is

$$v(\hat{X}_{st}) = N^2 v(\bar{x}_{st}) = (95)^2 \times 0.1042 = 940.405.$$

(vi) Let a_h = number of units of h-th stratum having a characteristic
= number of mothers of h-th stratum having less than 3 children.

Here $a_1 = 1, a_2 = 1, a_3 = 3, a_4 = 2$.

p_h = proportion of mothers of h-th stratum having a characteristic
$= \frac{a_h}{n_h}$.

Here $p_1 = \frac{a_1}{n_1} = \frac{1}{8} = 0.125,\ p_2 = \frac{a_2}{n_2} = \frac{1}{6} = 0.167$

$p_3 = \frac{a_3}{n_3} = \frac{3}{3} = 1.000,\ p_4 = \frac{a_4}{n_4} = \frac{2}{3} = 0.667.$

The estimate of proportion of mothers having less than 3 children:

$$\hat{P}_{st} = \frac{1}{N}\Sigma N_h p_h$$

$$= \frac{1}{95}[36 \times 0.125 + 30 \times 0.167 + 16 \times 1.000 + 13 \times 0.667] = 0.3598.$$

Estimate of variance of $\hat{P}_{st}$ is:

$$v(\hat{P}_{st}) = \frac{1-f}{n}\Sigma W_h \frac{n_h p_h q_h}{n_h - 1}$$

$$= \frac{1-0.21}{20}\left[\frac{0.379 \times 8 \times 0.125 \times 0.875}{8-1} + \frac{0.316 \times 6 \times 0.167 \times 0.833}{6-1} + \frac{0.168 \times 1.000 \times 0.00 \times 3}{3-1} + \frac{0.137 \times 3 \times 0.667 \times 0.333}{3-1}\right]$$

$$= 0.005758.$$

Estimate of standard error of $\hat{P}_{st}$ is s.e. $(\hat{P}_{st}) = \sqrt{v(\hat{P}_{st})} = 0.0759.$

6.7 Cluster Sampling

In selecting sample by simple random sampling method or by stratified sampling method or by systematic sampling method it needs frame [complete and updated list of population units]. In practical situation, specially when the population size is big or population is infinite, updated frame may not be available. For example, if anybody is interested to study the health index of the people aged 60 years and above, or if any research study is needed to evaluate the family planning activities of a country, or if it is needed to study the health hazard of the people living in different slum areas, then the frame of population units will not be available. Even it is difficult to prepare a frame. For example, if a study is needed to study the health hazard of the couples adopting family planning practices, then population units cannot be identified as there are many adopter couples who do the practices without the knowledge of others. Only those adopter couples are known to the family planning workers who are collecting the family planning materials from the family planning workers either in rural area or in urban area. In such a situation, it is difficult to have the frame and also difficult to apply simple random sampling method or stratified random sampling method or systematic sampling method. In such a case cluster sample is a technique to select a sample, where a cluster is defined with a group of neighbouring population units and with more or less homogeneous units.

For cluster sampling the population units under study are clustered [division into groups or segments]. In practice, a cluster is formed with neighbouring population units. For example, a study is needed to study the health index of people aged 60 years and above in a district. The district has some upazillas. In each upazilla there are unions and in each union there are villages. Each village can be considered as a cluster. The list of villages is available. The households in each village may be considered as ultimate sampling units. Some of the villages (clusters) can be selected at random and the households in the vilages are investigated according to the objective of the survey. This method of selection of sampling units is known as cluster sampling.

In cluster sampling, the size of cluster may be equal or unequal. Due to changes in size of cluster the formula to estimate the population characteristics also changes.

In case of cluster sampling all units in a cluster are investigated. However, some units of the cluster can also be selected at random as ultimate sampling units. In that case, the sampling is called two-stage sampling. In a similar way multi-stage sampling method can be applied to select ultimate sampling units. For example, some upazilla can be selected from a district. Here upazilla is known as primary unit. From the selected upazilla, some unions can be selected, where unions are secondary units. From the selected unions some villages can be selected at random, where the villages are third-stage units. From the selected villages some households are selected at random, where households are ultimate sampling units. This latter method of sampling is known as 4-stage sampling methods.

Method of estimation in cluster sampling

Let there be NM units in a population, where the population units are divided into N clusters each of size M. Let us considered that $n(\leq N)$ clusters are randomly selected from N clusters. This selection of n clusters is known as cluster sampling, where the units in a cluster are ultimate sampling units. Then, we have

N = number of clusters in the population

n = number of selected clusters

M = number of units in a cluster

nM = total number of sample units selected from the population

x_{ij} = value of the variable under study of j-th unit in i-th cluster $(i = 1, 2, \ldots, N; j = 1, 2, \ldots, M)$

The population mean is given by

$$\bar{X} = \frac{1}{NM}\sum_i\sum_j x_{ij}$$

$$\bar{X}_i = \frac{1}{M}\sum_j x_{ij} = \text{population mean of } i\text{-th cluster}$$

$$\bar{X}_N = \frac{1}{N}\sum \bar{X}_i = \text{population mean per cluster.}$$

The estimate of population mean $\bar{X}$ is given by $\bar{x}_c = \frac{1}{nM}\Sigma\Sigma x_{ij} = \frac{1}{n}\Sigma\bar{X}_i$.

Here $\bar{x}_c$ = mean of cluster sample and it is unbiased estimate of population mean.

The estimated variance of $\bar{x}_c$ is:

$$v(\bar{x}_c) = \frac{1-f}{n}s_b^2, \text{ where } f = \frac{n}{N} \text{ and}$$

$$s_b^2 = \frac{1}{n-1}\Sigma(\bar{X}_i - \bar{x}_c)^2 = \frac{1}{n-1}\left[\Sigma\bar{X}_i^2 - \frac{(\Sigma\bar{X}_i)^2}{n}\right]$$

If cluster is selected with replacement by simple random sampling method, then $f = \frac{n}{N}$ can be neglected and $v(\bar{x}_c) = \frac{1}{n}s_b^2$.

Method of estimation if clusters are of different sizes

Let M_i = number of units in i-th cluster ($i = 1, 2, \ldots, N$)

$$M_0 = \sum^{N} M_i = \text{total units in all clusters (population units)}$$

$$\bar{M} = \frac{M_0}{N} = \text{average units per cluster in the population}$$

$$\bar{X}_i = \frac{1}{M_i}\sum_j x_{ij} = \text{population mean of } i\text{-th cluster}$$

$$\bar{X} = \frac{1}{M_0}\Sigma\Sigma x_{ij} = \frac{1}{M_0}\sum^{N} M_i\bar{X}_i = \text{population mean}$$

The unbiased estimate of population mean is given by

$$\bar{x}_{c_1} = \frac{N}{nM_0}\sum_{i=1}^{n} M_i\bar{X}_i.$$

The unbiased estimate of variance of $\bar{x}_{c_1}$ is given by

$$v(\bar{x}_{c_1}) = \frac{1-f}{n}s_{bc}^2, \text{ where } f = \frac{n}{N}.$$

$$s_{bc}^2 = \frac{1}{n-1}\sum_{i=1}^{n}\left(\frac{M_i\bar{X}_i}{\bar{M}} - \bar{x}_{c_1}\right)^2.$$

EXAMPLE 6.4 The following are the number of family planning workers in different villages in a union of a upazilla.

Sl. no. of union	*Number of family planning workers in villages* (x_{ij})								*Total* X_i	*Mean* $\bar{X}_i$
	1	2	3	4	5	6	7	8		
01	5	8	7	4	6	7	8	5	50	6.25
02	6	6	6	5	6	6	5	6	46	5.75
03	6	7	5	5	5	7	7	6	48	6.00
04	7	10	4	10	5	8	8	10	62	7.75
05	5	8	5	8	5	8	7	6	52	6.50
06	4	7	6	7	6	9	6	5	50	6.25
07	5	8	6	8	6	10	8	5	56	7.00
08	6	8	8	8	6	10	8	6	60	7.50
09	8	7	8	10	7	9	8	7	64	8.00
10	7	8	9	8	8	8	7	5	60	7.50

(i) Select a cluster sample of 5 unions to estimate the average number of family planning worker per village.

(ii) Estimate standard error of your estimate.

(iii) Estimate total number of family planning workers in the union along with estimated variance.

Solution. Each union is considered a cluster. There are $N = 10$ clusters. In each cluster, there are $M = 8$ villages. Total population units are $NM = 10 \times 8 = 80$.

Using simple random sampling method the selected clusters are shown below:

Sl. no. of union selected	*Number of family planning workers in villages* (x_{ij})								*Total* X_i	*Mean* $\bar{X}_i$	$\bar{X}_i^2$	*Random number*
	1	2	3	4	5	6	7	8				
01	7	8	9	8	8	8	7	5	60	7.50	56.25	10
02	8	7	8	10	7	9	8	7	64	8.00	64.00	09
03	5	8	7	4	6	7	8	5	50	6.25	39.0625	01
04	7	10	4	10	5	8	8	10	62	7.75	60.0625	04
05	5	8	5	8	5	8	7	6	52	6.50	42.25	05
Total										36.00	261.625	

(i) The estimate of average number of family planning workers per village is

$$\bar{x}_c = \frac{1}{n}\Sigma\bar{X}_i = \frac{36.0}{5} = 7.20.$$

(ii) Estimate of standard error of $\bar{x}_c$ is s.e. $(\bar{x}_c) = \sqrt{v(\bar{x}_c)}$.

$$v(\bar{x}_c) = \frac{1-f}{n}s_b^2 = \frac{1-0.5}{5}0.60625 = 0.060625$$

$$s_b^2 = \frac{1}{n-1}\left[\Sigma\bar{X}_i^2 - \frac{(\Sigma\bar{X}_i)^2}{n}\right] = \frac{1}{5-1}\left[261.625 - \frac{(36.0)^2}{5}\right] = 0.60625.$$

$$f = \frac{n}{N} = \frac{5}{10} = 0.5.$$

$$\therefore \text{s.e. } (\bar{x}_c) = \sqrt{0.060625} = 0.2462.$$

(iii) Estimate of total number of family planning workers is given by

$$\hat{X}_c = NM\bar{x}_c = 10 \times 8 \times 7.2 = 576.$$

(iv) Estimate of variance of estimated total is

$$v(\hat{X}_c) = (NM)^2 v(\bar{x}_c) = (10 \times 8)^2 0.060625 = 388.$$

6.8 Quota Sampling and Purposive Sampling

For selection of stratified sample, the strata sizes must be known and units in strata must be identified. But in practice, either the strata sizes or units in the strata or both may not be pre-identified. For example, a survey is needed to study the status of health services provided in an area. Health services are provided by (i) government organisations and (ii) non-government organisations. In both cases, doctors, nurses and other trained personnels are involved in providing health services. Moreover, private doctors also provide health services. Beside these, another group of people who have no medical degree and training, specially, the owner or workers of a pharmaceutical shop [in developing or underdeveloped countries] are involved in providing health services. These third group of health workers cannot be identified properly. The private doctors are also providing health services. Sometimes the doctors of government/non-government health centres are also doing private practices. In such cases, if each group of health workers are considered as stratum units, the units cannot be property identified. So, the stratified random sampling method cannot be applied properly.

To avoid the problem a requisite number of sample units, decided prior to the survey, from each stratum can be investigated. Investigation is continued until the quota of sample units of a stratum are fulfilled. The investigated units of a particular stratum are the ultimate sample units. This method of selection of sample is known as *quota sampling.* It is not a probability sampling as sample units are not selected with associated probability. Rather, it is a purposive method of sample selection.

Purposive sampling

It is also called judgement sampling. This technique of sampling is applied if frame is not available or population units are not identified along with their location. For example, a survey is needed to estimate the number of people suffering from diabetes or from heart problem or from any type of disease which is injurious to health status of the country. In many cases, people suffering from diabetes in initial state do not consult a doctor, or people without any major problem of heart do not consult a doctor either due to economic condition or due to ignorance of health problem. In such a case, the frame of people of a particular disease cannot be prepared or population units cannot be identified. But for the purpose of survey the population units are needed to be investigated. Here population units cannot be selected randomly.

The survey can be conducted in an area/locality by asking the related questions to the available people. The units are selected purposively without any technique of random selection. This type of sampling is known as purposive selection. The purposive selection of sample units may provide satisfactory information, if sample size is sufficiently large

6.9 Capture Re-capture Model

The population size or the location of population units is not known when the population is infinite. For example, the location or the number of patients having HIV positive are usually not known. The amount of bacteria grown in an wound may not be known. The researcher needs to estimate the infinite population size. For example, the health authority needs to estimate the number of patients having HIV positive. The estimate of population size can be obtained by a technique known as Capture re-capture model.

In this model some of the units are first selected by any of the sampling methods and the units are identified by giving an identifiction symbol. Then these identified units are allowed to be mixed up with all the population units. After some times again a sample is selected and observed the required characteristic. Some of the units selected later on (re-capture) may be selected during first survey (capture). That is why this type of samping is known as capture re-capture model. For example, an estimate of baboon population size is needed. Some of the baboons can be captured by any of the sampling techniques and the selected baboons are given an identification symbol. Then those are allowed to mix up with their normal population. After a sufficient time, again some the baboons are selected and investigated the number of baboons selected earlier. Here baboons are captured first and after sufficient time some are re-captured. The selection of sample in this way is called capture re-capture model of sample selection.

The estimate of population size is measured by an index known as *Lincoln Index*, where the index is given by

$$\text{L.I.} = \frac{N \times n}{r}.$$

The estimated standard error of this index is s.e. (L.I.) $= \sqrt{(n^2N(n-r))/r^3}$, where

$n =$ the sample size selected first (capture)
$N =$ the sample size selected during re-capture
$r =$ number of units of first sample selected in the second sample (re-capture)

EXAMPLE 6.5 Let 10 patients are identified from a diagnostic centre whose HIV is positive. After some time by another sample from the same centre 20 are identified having HIV positive. Among these 20 patients 2 were identified in the first sample. Estimate the number of patients along with estimated standard error for whom HIV positive.

Solution. We have $n = 10, N = 20, r = 2$. The estimate of patients having HIV positive is

$$\text{L.I.} = \frac{N \times n}{r} = \frac{20 \times 10}{2} = 100.$$

The estimated standard error of L.I. is

$$\text{s.e. (L.I.)} = \sqrt{[n^2 N(N-r)/r^3]} = \sqrt{[10^2 \times 20(20-2)/2^3]} = 67.08.$$

Estimate of density of population units

This estimate is needed to study in case of infinite population. For example, among the people of a country what is the density of the diabetic patients?

Different diabetic patients visit different diabetic centre. A sample of diabetic patients is selected and investigated their sex. Then the density of the diabetic patients of a particular sex can be studied by an index known as Shanon-Weaver index, where the index is given by

$$H = -\sum_{i=1}^{k} p_i \ln p_i, \text{ where}$$

k = the type of units observed in the sample

n_i = the units of i-th type ($i = 1, 2, \ldots, k$) in the sample

n = sample size

$p_i = \frac{n_i}{n}$

ln = natural logarithm

If H is approximately 1, then all types of units are uniformly distributed. If $H > 1$, all types of units are not uniformly distributed. The density of a particular type is more compared to others.

EXAMPLE 6.6 Let $n = 100$ diabetic patients are selected and found that, out of 100, 60 are females. Investigate the density of diabetic patients among male and female.

Solution.

Type of units :	*Male*	*Female*	*Total (n)*
Amount of units :	$n_1 = 40$	$n_2 = 40$	$n = 100$
p_i :	$p_1 = \frac{40}{100} = 0.4$	$p_2 = \frac{60}{100} = 0.6$	

$k = 2$

The Shanon-Weaver index is

$$H = -\sum_{i=1}^{k=2} p_i \ln p_i = -[p_1 \ln p_1 + p_2 \ln p_2]$$

$$= -[0.4 \times \ln 0.4 + 0.6 \ln 0.6]$$

$$= -[-0.36652 - 0.3065] = 0.67302.$$

As $H < 1$, the densities of diabetic patients among male and female are similar.

6.10 Sampling Distribution

Let there be N units in a population, we can select a sample of size $n(\leq N)$. The possible number of samples are ${}^nC_n$. These samples provide $\bar{x}_1, \bar{x}_2, \ldots, \bar{x}_{{}^nC_n}$ as sample means and $s_1^2, s_2^2, \ldots, s_{{}^nC_n}^2$ as sample variances. Here $\bar{x}_i (i = 1, 2, \ldots, {}^nC_n)$ and s_i^2 are called sample statistics. The distribution of sample means or sample variances or any function of these is known as sampling distribution.

The important sampling distributions are (i) Student's t-distribution, (ii) Chi-square (χ^2) distribution, (iii) F-distribution or distribution of variance ratio.

Let us discuss these sampling distributions in short. For this let us consider that a sample of size n is selected from $N(\mu, \sigma^2)$, where the sample observations are $x_1, x_2, \ldots, x_n$. The sample mean and sample variance of the observations are

$$\bar{x} = \frac{1}{n}\Sigma x_i \text{ and } s_i^2 = \frac{1}{n-1}\Sigma(x_i - \bar{x})^2 \text{ respectively.}$$

Then the distribution of sample statistic:

$$\chi^2 = \frac{(n-1)s^2}{\sigma^2}$$

is distributed as chi-square with $(n-1)$ degrees of freedom (d.f.).

This chi-square distribution is used in the following cases:

(i) To test the significance of a specified population variance.

(ii) To test the goodness of fit of a distribution.

(iii) To test the independence of attributes.

(iv) To test the equality of several corelation coefficients.

(v) To test the homogeneity of several test.

(vi) To test the homogeneity of several population variances.

The statistic:

$$t = \frac{\bar{x} - \mu}{\sqrt{\frac{s^2}{n}}}$$

is distributed as Student's t-distribution with $(n-1)$ d.f.

This distribution is used in the following cases:

(i) To test the significance of population mean.

(ii) To test the significance of the difference between two population means.

(iii) To test the equality of two correlated means.

(iv) To test the significance of population correlation coefficient.

(v) To test the significance of population regression coefficient.

(vi) To test the equality of two independent regression coefficients.

Again, let us consider two samples selected from two populations, where sample observations are $x_{11}, x_{12}, \ldots, x_{1n_1}$, selected from $N(\mu_1, \sigma^2)$. From this sample, we have

$$\bar{x}_1 = \frac{1}{n_1}\sum_j x_{1j} \text{ and } s_1^2 = \frac{1}{n_1 - 1}\left[\Sigma x_{1j}^2 - \frac{(\Sigma x_{1j})^2}{n_1}\right].$$

Another set of sample observations are $x_{21}, x_{22}, \ldots, x_{2n_2}$, selected from $N(\mu_2, \sigma^2)$, where

$$\bar{x}_2 = \frac{1}{n_2}\Sigma x_{2j} \text{ and } s_2^2 = \frac{1}{n_2 - 1}\left[\Sigma x_{2j}^2 - \frac{(\Sigma x_{2j})^2}{n_2}\right].$$

Now, the sample statistic

$$F = \frac{s_1^2}{s_2^2}$$

is distributed as variance ratio (F) with $(n_1 - 1), (n_2 - 1)$ d.f.

This distribution is used in the following cases:

(i) To test the equality of several population means.

(ii) To test the equality of two population variances.

(iii) To test the significance of regression.

(iv) To test the significance of multiple correlation coefficient.

EXERCISE – 6

1. The following are the amount of haemoglobin of some persons:

Sl. no. of persons	01	02	03	04	05	06	07	08	09	10	11	12	13
Amount of haemoglobin (x g/dL)	11.6	14.7	10.5	10.0	11.7	15.8	16.0	14.0	13.2	14.6	9.7	10.7	11.7
Sl. no. of persons	14	15	16	17	18	19	20	21	22	23	24	25	26
Amount of haemoglobin (x g/dL)	12.2	13.3	14.8	15.1	16.2	17.0	9.8	10.0	10.5	11.6	13.4	11.7	12.8
Sl. no. of persons	27	28	29	30	31	32	33	34	35	36	37	38	39
Amount of haemoglobin (x g/dL)	17.0	15.0	14.2	13.8	11.6	11.0	10.7	9.0	9.8	14.4	13.0	13.6	16.4
Sl. no. of persons	40	41	42	43	44	45	46	47	48	49	50	51	52
Amount of haemoglobin (x g/dL)	11.7	12.7	13.0	14.7	14.0	15.0	15.2	16.0	16.5	11.8	10.6	11.2	12.7

(a) Select a sample of size $n = 10$ by (i) simple random sampling method, (ii) circular systematic sampling method, (iii) systematic sampling method.

(b) Estimate mean amount of haemoglobin and total amount of haemoglobin of persons.

(c) Estimate standard error of estimated mean.

(d) Estimate variance of estimated total

(e) Find 95% confidence interval of mean of haemoglobin

(f) Estimate proportion of persons who have haemoglobin less than 12.00.

(g) Estimate variance of the estimated proportion.

2. The following are the sample observations on number of cancer patients in different hospitals. The hospitals are selected by simple random sampling method from a population of $N = 100$ hospitals.

Number of cancer patients (x) : 25, 18, 32, 20, 15, 12, 25, 17, 15, 15.

(a) Estimate average number of cancer patients per hospital along with its estimated variance.

(b) Find 95% confidence interval of average.

(c) Estimate total cancer patients along with its estimated standard error.

3. (a) Distinguish between census and sample survey. Why do we need sample survey?

(b) Explain how simple random sample is selected.

(c) Show, by an example, that sample mean is an unbiased estimate of population mean and sample variance is an unbiased estimate of population variance.

(d) The following are the number of nurses working in different clinics of a city.

Number of nurses (x) : 15, 18, 25, 30, 12, 18, 20, 25, 26, 30, 18, 16, 20, 20, 20, 25, 26, 17, 10, 12, 15, 17, 16, 14, 24, 22, 25, 25, 28, 27, 23, 21, 20, 20.

Select a simple random sample of 5 clinics and estimate average number of nurses per clinic along with its estimated standard error.

4. (a) Define simple random sampling. How does it differ from systematic sampling?

(b) The following are the S. GGT (x, U/L) of some persons:

x : 16, 18, 20, 30, 55, 10, 50, 48, 32, 30, 25, 20, 27, 40, 42, 50, 52, 48, 42, 27, 25, 17, 32, 50, 13, 17, 19, 21, 24, 23, 27, 25, 25, 20, 20, 18, 32, 28.

(i) Select a random sample of 6 persons by (a) systematic sampling method and by (b) circular systematic sampling method.

(ii) Estimate proportion of persons who have S. GGT less than 25 along with its estimated variance.

5. (a) Define sample and sampling. What do you mean by sampling fraction? Explain the necessity of sampling.

(b) Distinguish between random sampling and purposive selection.

(c) The following are the blood pressure (x mmHg) levels of persons of different socio-economic conditions:

Socio-economic conditions	*Blood pressure levels*
Low	70, 75, 80, 65, 78, 80, 75, 70, 65, 65, 70, 74, 85, 76, 70
Medium	80, 82, 75, 85, 85, 80, 88, 90, 75, 74, 78, 80, 85, 85, 80, 80, 84, 85, 88, 85, 85, 80, 82, 84, 85
High	90, 92, 95, 80, 85, 75, 70, 85, 88, 95, 90, 90, 90, 88, 85, 86, 80, 80, 75, 80, 85, 88

Select a stratified sample of size 10 persons and estimate average blood pressure per person along with its estimated variance.

6. (a) Write short notes on (i) random sampling, (ii) quota sampling, (iii) purposive selection, (iv) capture re-capture model.

(b) A survey is conducted to identify the number of patients suffering from T.B. One hundred people are identified and suggested the treatment and are asked to visit the clinic after 6 months. After 6 months another 300 people are identified who are suffering. Among these latter group 15 are identified earlier. Estimate the number of patients of T.B.

(c) Among the latter group of 300 persons, 180 are males. Investigate density of T.B. patients among males and females.

7. The following are the number of births in a week in different clinics of a city.

Clinics	*Births* (x_{ij}) *in days*
1	2, 4, 3, 2, 1, 4, 5
2	4, 3, 3, 3, 0, 2, 3
3	3, 3, 4, 4, 4, 2, 1
4	0, 5, 3, 4, 2, 2, 2
5	1, 6, 4, 7, 3, 3, 4
6	0, 2, 5, 6, 4, 3, 2
7	1, 4, 5, 5, 4, 3, 3
8	2, 2, 2, 2, 5, 4, 4
9	3, 3, 3, 3, 4, 3, 4
10	5, 5, 3, 2, 4, 2, 3
11	4, 3, 2, 2, 2, 4, 3
12	6, 6, 3, 4, 5, 4, 3
13	4, 5, 3, 3, 4, 2, 2
14	2, 3, 4, 2, 3, 4, 4
15	5, 3, 3, 3, 2, 2, 4

Select 5 clinics at random and estimate total births in the clinics in a week. Estimate standard error of estimated total births.

8. One hundred patients of heart diseases are investigated and found that, out of 100 patients, 25 are of ages below 40 years, 55 are of age group 40–50 years and 20 are of ages 50+. Investigate density of patients of heart diseases among different age groups.

9. A group of 100 T.B. patients are under treatment at the initial stage in a centre. They are asked to visit the centre after 3 months. After 3 months another group of 200 patients are under investigation. In this latter group 10 are found who visited the centre earlier also. Estimate number of T.B. patients along with its estimated standard error.

Chapter 7

Test of Hypothesis

7.1 Introduction

The objective of the statistical analysis is to infer about the population parameter, where parameter is an unknown constant in a population. For example, one may be intersted to know the average age of people of Bangladesh in 2014. This average is constant but it is unknown and not easy to calculate due to resource constraint. However, this unknown constant can be estimated, where the estimate is a sample statistic ($\bar{x}$). On the basis of this estimate an inference can be drawn about population mean age using some statistical procedure. The statistical procedure involves some assumption about population mean. This assumption is usually known as *hypothesis.* The conclusion can be drawn about this assumption using some statistical technique. This technique of verification of the assumption is known as *test of hypothesis.*

7.2 Terms Related to Test of Hypothesis

Parameter. The unknown characteristic of a population is known as *parameter.* For example, let $x_1, x_2, \ldots, x_n$ be the sample observations selected at random from a normal population having mean μ and variance σ^2. Here μ and σ^2 are not usually known. So, these two are parameters.

Statistic. Any function of sample observations is called statistic. For example, if $x_1, x_2, \ldots,$ x_n are n selected sample observations, then

$$\bar{x} = \frac{1}{n}\sum x, \; s^2 = \frac{1}{n-1}\sum(x-\bar{x})^2$$

are sample statistics as these are functions of sample observations.

Hypothesis. Any statement, description or specification about population parameter or population is known as hypothesis. The hypothesis is of two types, namely (i) parametric hypothesis and, (ii) non-parametric hypothesis. Each of these hypotheses is again classified into two. These are: (a) null hypothesis and (b) alternative hypothesis.

Parametric hypothesis. Any statement or description regarding parameter of population is known as parametric hypothesis.

Non-parametric hypothesis. The hypothesis which does not specify anything about the parameter of a population, rather it specifies some other characteristics of the population is known as non-parametric hypothesis.

Null hypothesis. The hypothesis which is usually verified for possible rejection by test, even it is true, is known as null hypothesis. It is denoted by H_0. For example, let $x_1, x_2, \ldots, x_n$ be n sample observations selected at random from normal population having mean μ and variance σ^2. Assume that μ has a specified value μ_0 (say) and σ^2 has a specified value σ_0^2 (say). One needs to verify whether these assumptions about μ and σ^2 are true or false. Then the null hypothesis regarding μ is

$$H_0 : \mu = \mu_0.$$

The null hypothesis regarding σ^2 is

$$H_0 : \sigma^2 = \sigma_0^2 \text{ or, } H_0 : \sigma^2 = 0.$$

Again, let us consider that $x_{11}, x_{12}, \ldots, x_{1n_1}$ be the n_1 sample observations from a population, where population parameters are not specified. Independently another sample of n_2 observations $x_{21}, x_{22}, \ldots, x_{2n_2}$ can be selected from another population having unspecified parameters. However, one may assume that both samples are selected from same population. To verify this assumption null hypothesis can be specified as

H_0 : Samples are selected from a population having same median.

This hypothesis is a non-parametric null hypothesis.

Alternative hypothesis. Any statement against null hypothesis is known as alternative hypothesis. Thus, if null hypothesis is $H_0 : \mu = \mu_0$, then alternative hypothesis may be

$$H_A : \mu \neq \mu_0 \text{ or, } H_A : \mu < \mu_0 \text{ or, } H_A : \mu > \mu_0.$$

Again, if H_0 is $\sigma^2 = 0$, then

$$H_A : \sigma^2 > 0.$$

Test of hypothesis. The statistical procedure to verify the null hypothesis is known as test of hypothesis. The procedure also signifies the evidence regarding null hypothesis and hence, it is also called test of significance.

Test statistic. The function of sample observations which is used to verify null hypothesis is called test statistic. For example, if $x_1, x_2, \ldots, x_n$ are randomly selected n observations from normal population with mean μ and variance σ^2 and if it is needed to verify the null hypothesis

$$H_0 : \mu = \mu_0, \text{ against } H_A : \mu \neq \mu_0,$$

then a function of sample observations

$$t = \frac{\bar{x} - \mu_0}{\sqrt{s^2/n}}, \text{ where } \bar{x} = \frac{1}{n}\sum x,\ s^2 = \frac{1}{n-1}\sum(x - \bar{x})^2,$$

can be used to verify null hypothesis. This t is called test statistic and it is distributed as Student's t-distribution with $(n-1)$ degrees of freedom when $n < 30$ and σ^2 is not known.

If $n \geq 30$ or σ^2 is known, the null hypothesis is verified by a statistic

$$z = \frac{\bar{x} - \mu_0}{\sqrt{\sigma^2/n}}.$$

This z is also a test statistic. It is distributed as normal distribution with mean zero and variance 1. If $n \geq 30$, but σ^2 is not known, then

$$z = \frac{\bar{x} - \mu}{\sqrt{s^2/n}},$$

where s^2 is an unbiased estimate of σ^2.

Degrees of freedom. The number of independent values which are used in deriving a test statistic is known as degrees of freedom (d.f.). Let $x_1, x_2, \ldots, x_n$ be n sample observations from normal population with mean μ and variance σ^2 [$x \sim N(\mu, \sigma^2)$]. Then

$$z = \frac{x-\mu}{\sigma} \sim N(0,1) \text{ and } z^2 = \left(\frac{x-\mu}{\sigma}\right)^2 \text{ is } \chi^2 \text{ with 1 d.f.}$$

Also, $\sum z^2 = \sum \left(\frac{x-\mu}{\sigma}\right)^2$ is χ^2 with n d.f. Here χ^2 is derived using n independent values of x, where σ^2 is a specified value.

Again. if μ is not known, then $\chi^2 = \sum \frac{(x-\bar{x})^2}{\sigma^2}$ has $(n-1)$ d.f. Here χ^2 is derived using one estimated value of μ. Thus d.f. is the number of independent values of x minus one estimated value of parameter.

Assumption. The statements which are specified about population and/or sample in deriving test statistic is known as assumption. For example, in deriving the test statistic

$$t = \frac{\bar{x}-\mu_0}{\sqrt{s^2/n}},$$

it is assumed that $x \sim N(\mu, \sigma^2)$, where σ^2 is not known.

Again, $z = \frac{\bar{x}-\mu_0}{\sqrt{\sigma^2/n}}$ is used on the assumption that σ^2 is known.

Level of significance. The small probability for which the null hypothesis is rejected even when it is true is called level of significance. It is denoted by α. Usually, unless otherwise mentioned, α is taken 0.05.

One-tailed and Two-tailed test. Each null hypothesis is opposed by an alternative hypothesis. For example, if

$$H_0 : \mu = \mu_0,$$

it is opposed by (i) $H_A : \mu \neq \mu_0$, (ii) $H_A : \mu > \mu_0$, (iii) $H_A : \mu < \mu_0$.

Here the test of H_0 against H_A (i) is known as two-tailed test, on the other hand, if H_A is (ii) or (iii), the test is known as one-tailed test. The test against H_A (ii) is called right-tailed test and test against H_A (iii) is called left-tailed test.

For two-tailed test critical values of test statistic is obtained from both sides, where $\alpha = 0.05$ is divided by 2. For example, in testing

$$H_0 : \mu = \mu_0, \text{ against } H_A : \mu \neq \mu_0$$

by z-statistic, the critical value of z is taken for $\alpha/2$. If $\alpha = 0.05$, then critical value of z is 1.96 [From normal probability table]. But, if $H_A : \mu < \mu_0$, then the critical value of z is 1.64 for $\alpha = 0.05$.

Critical region and acceptance region. The probability region in a probability curve in which the value of test statistic falls and leads us to reject null hypothesis is known as critical region. For example, in testing $H_0 : \mu = \mu_0$, against $H_A : \mu \neq \mu_0$ by

$$z = \frac{\bar{x}-\mu_0}{\sqrt{\sigma^2/n}},$$

the null hypothesis can be rejected if z-value is big. Here z will be big, if $\bar{x} - \mu_0$ is bigger and it will happen, if estimated mean $\bar{x}$ is very different than the hypothetical mean μ_0. This bigger value of z will lead us to reject the null hypothesis.

But. rejection will be done at certain level of significance. If level of significance is $\alpha = 0.05$ for alternative $H_A : \mu \neq \mu_0$, then the bigger value of z is allowed up to less than 1.96. This 1.96 is the critical value of z. Any value beyond this 1.96 or -1.96 [shown in the graph of z] indicates that there is a big difference between $\bar{x}$ and μ_0. In such a case, we reject null hypothesis. The value 1.96 is the critical value and the probability area after 1.96 or below -1.96 is known as critical or rejection region. The area within -1.96 to 1.96 is acceptance region as z will fall in this area if $\bar{x} - \mu_0$ is smaller and smaller.

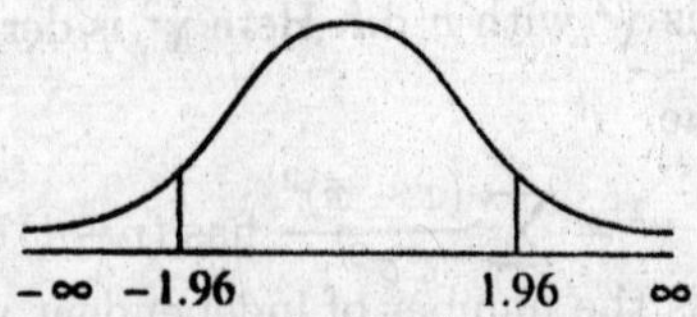

***P*-value.** It is the probability calculated for test statistic under H_0. For example, let us consider that the following are the blood sugar levels of some randomly selected persons:

Blood sugar level (x) : 8.5, 6.3, 6.2, 6.0, 4.5, 5.5, 4.8, 5.7, 7.0

We need to test whether the mean blood sugar level is $\mu = 6.3(\mu_0)$. Assume that blood sugar level follows normal distribution with mean μ and variance $\sigma^2 = 3$. Here as σ^2 is asumed known, we can use the test statistic

$$z = \frac{\bar{x} - \mu_0}{\sqrt{\sigma^2/n}}$$

to test the null hypothesis $H_0 : \mu = \mu_0 = 6.3$, against $H_A : \mu \neq \mu_0$.

Here $n = 9$, $\bar{x} = \frac{1}{n}\sum x = \frac{54.5}{9} = 6.05$.

Then $z = \frac{\bar{x} - \mu_0}{\sqrt{\sigma^2/n}} = \frac{6.05 - 6.3}{\sqrt{3/9}} = -0.43$.

Using normal probability table $P(z < -0.43) = 0.3336$.

Again, $P(z > -0.43) = 1 - P(z < -0.43) = 1 - 0.3336 = 0.6664$. These probabilities can be shown in graph as follows:

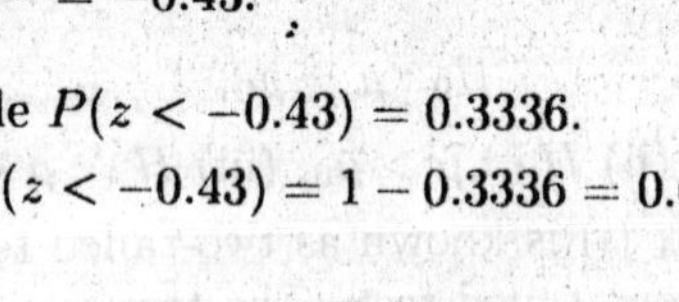

As H_A is two-sided, we need to consider

$$P(z < -0.43) + P(z > 0.43) = 0.3336 + 0.3336 = 0.6672.$$

This probability is the P-value and as P-value $> \alpha = 0.05$, H_0 is accepted at 5% level of significance. For $H_A : \mu > \mu_0$, the P-value is $P(z > 0.43) = 0.3336$. The null hypothesis is rejected at $100 \times \alpha\%$ level, if P-value is less than α.

Type-I error and Type-II error. The null hypothesis is rejected if calculated value of test statistic is greater than or equal to tabulated statistic at $100\alpha\%$ level of significance, usually $\alpha = 0.05$. Also, H_0 is rejected if P-value is less than 0.05. For example, let a z-value is 1.96 in testing $H_0 : \mu = \mu_0$ against $H_A : \mu \neq \mu_0$. This 1.96 is the 5% value of z from probability table (or P-value is 0.05). As calculated, $z = 1.96$ which is equal to tabulated z-value, H_0 is rejected. But, if $z = 1.95$, we accept the null hypothesis (P-value $= 0.0512$).

If $z = 1.955$, P-value will be less than 0.0512 but more than 0.05. In this case also H_0 is accepted. Again, one can approximate $z = 1.955 \approx 1.96$, where P-value is 0.05. In the latter case H_0 is rejected. In reality we reject null hypothesis when it is true. Here we are doing an error known as Type-I error [H_0 is rejected when it is true]. Type-I error is denoted by α and $P(\text{Type-I error}) = \alpha$.

Again, H_0 may be accepted and H_A is rejected even H_0 is not true. This is also an error, known as Type-II error. It is denoted by β.

Power of test. The power of the test is $1 - \beta$, where β is the probability of Type-II error.

7.3 Test Regarding Mean

Let $x_1, x_2, \ldots, x_n$ be n sample observations from $N(\mu, \sigma^2)$, where

$$\bar{x} = \frac{1}{n}\sum x, \; s^2 = \frac{1}{n-1}\sum (x - \bar{x})^2.$$

We need to test $H_0 : \mu = \mu_0$ against $H_A : \mu \neq \mu_0$.

The test statistic is:

$$t = \frac{\bar{x} - \mu_0}{\sqrt{s^2/n}}, \text{ if } \sigma^2 \text{ is not known and } n < 30.$$

The test statistic is:

$$z = \frac{\bar{x} - \mu_0}{\sqrt{\sigma^2/n}}, \text{ if } \sigma^2 \text{ is known or } n \geq 30.$$

If σ^2 is not known, $z = \dfrac{\bar{x} - \mu_0}{\sqrt{s^2/n}}$, if $n \geq 30$.

EXAMPLE 7.1 The following are the S. GGT of some male persons:

S. GGT (x U/L) : 15, 18, 22, 30, 38, 45, 17, 48, 32, 26.

Do you think that the average S. GGT is 25 U/L?

Solution. Assume that $x \sim N(\mu, \sigma^2)$, where σ^2 is not known.

We need to test $H_0 : \mu = \mu_0 = 25$, against $H_A : \mu \neq \mu_0$.

Here $n = 10$,

$$\bar{x} = \frac{1}{n}\sum x = \frac{291}{10} = 29.1$$

$$s^2 = \frac{1}{n-1}\left[\sum x^2 - \frac{(\sum x)^2}{n}\right] = \frac{1}{10-1}\left[9695 - \frac{(291)^2}{10}\right] = 136.32.$$

As σ^2 is not known and $n = 10 < 30$, the test statistic is

$$t = \frac{\bar{x} - \mu_0}{\sqrt{s^2/n}} = \frac{29.1 - 25}{\sqrt{\frac{136.32}{10}}} = 1.11.$$

Comment. As $|t| < t_{0.05,9} = 2.262$ (P-value > 0.05), H_0 is accepted. The average S. GGT can be considered as 25.

EXAMPLE 7.2 The following are the s. magnesium of some selected persons:

s. magnesium (x mmol/l) : 0.8, 1.1, 1.2, 0.7, 0.9, 1.0, 1.0, 0.8

It is assumed that the variance of s. magnesium is $\sigma^2 = 0.3\,(\text{mmol/l})^2$. Is the mean of s. magnesium 0.7?

Solution. Let $x \sim N(\mu, \sigma^2)$, where $\sigma^2 = 0.3$. We need to test

$$H_0 : \mu = \mu_0 = 0.7, \text{ against } H_A : \mu \neq \mu_0.$$

We have $n = 8$,

$$\bar{x} = \frac{1}{n}\sum x = \frac{7.5}{8} = 0.9375.$$

$$s^2 = \frac{1}{n-1}\left[\sum x^2 - \frac{(\sum x)^2}{n}\right] = \frac{1}{8-1}\left[7.23 - \frac{(7.5)^2}{8}\right] = 0.0284.$$

As σ^2 is known, the test statistic is

$$z = \frac{\bar{x} - \mu_0}{\sqrt{\frac{\sigma^2}{n}}} = \frac{0.9375 - 0.7}{\sqrt{\frac{0.3}{8}}} = 1.23.$$

Comment. As $|z| < 1.96$ [P-value $= 0.2186 > 0.05$], where 5% value of $z = 1.96$ for two-sided test, H_0 is accepted.

EXAMPLE 7.3 The following are the birth-weight of some newborn babies:

Birth-weight (x kg) : 3.2, 3.0, 2.8, 2.5, 1.9, 2.1, 2.2, 2.5, 3.5, 3.0, 3.6, 2.9, 2.7, 2.5, 1.8, 1.7, 1.8, 2.0, 2.3, 2.5, 2.4, 2.6, 2.5, 3.3, 3.4, 3.0, 3.6, 3.0, 2.8, 1.8, 3.2, 1.9.

Is the average birth-weight 2.5 kg per baby?

Solution. We have $n = 32$.

$$\therefore \bar{x} = \frac{1}{n}\sum x = \frac{84}{32} = 2.625$$

$$s^2 = \frac{1}{n-1}\left[\sum x^2 - \frac{(\sum x)^2}{n}\right] = \frac{1}{32-1}\left[230.62 - \frac{(84)^2}{32}\right] = 0.3265.$$

As $n = 32 > 30$ and σ^2 is not known, the test statistic is

$$z = \frac{\bar{x} - \mu_0}{\sqrt{\frac{s^2}{n}}} = \frac{2.625 - 2.5}{\sqrt{\frac{0.3265}{32}}} = 1.24.$$

Comment. As $|z| < 1.96$ (P-value $= 0.267 > 0.05$), H_0 is accepted. The average birth-weight can be considered 2.5 kg.

Let $x_{11}, x_{12}, \dots, x_{1n_1}$ be n_1 randomly selected sample observations from $N(\mu_1, \sigma^2)$. Also, consider another set of sample observations $x_{21}, x_{22}, \dots, x_{2n_2}$ from $N(\mu_2, \sigma^2)$.

We have

$$\bar{x}_1 = \frac{1}{n_1}\sum x_{1j}, \; s_1^2 = \frac{1}{n_1 - 1}\sum (x_{1j} - \bar{x}_1)^2$$

$$\bar{x}_2 = \frac{1}{n_2}\sum x_{2j}, \; s_2^2 = \frac{1}{n_2 - 1}\sum (x_{2j} - \bar{x}_2)^2$$

We need to test $H_0 : \mu_1 = \mu_2$, against $H_A : \mu_1 \neq \mu_2$ under the assumption that two samples are independently selected. Then the test statistic is

$$t = \frac{\bar{x}_1 - \bar{x}_2}{\sqrt{s^2\left(\frac{1}{n_1} + \frac{1}{n_2}\right)}}, \text{ where } s^2 = \frac{1}{n_1 + n_2 - 2}[(n_1 - 1)s_1^2 + (n_2 - 1)s_2^2]$$

is distributed as Student's t-distribution with $(n_1 + n_2 - 2)$ d.f. under the assumption that σ^2 is not known.

If σ^2 is known and/or both n_1 and n_2 are greater than or equal to 30, then the test statistic is

$$z = \frac{\bar{x}_1 - \bar{x}_2}{\sqrt{\frac{s_1^2}{n_1} + \frac{s_2^2}{n_2}}} \sim N(0, 1).$$

However, if σ_1^2 and σ_2^2 are known, then

$$z = \frac{\bar{x}_1 - \bar{x}_2}{\sqrt{\frac{\sigma_1^2}{n_1} + \frac{\sigma_2^2}{n_2}}} \sim N(0, 1),$$

where σ_1^2 and σ_2^2 are the variances of first and second population, respectively.

EXAMPLE 7.4 The following are the number of selected persons from villages of two unions, who are suffering from chronic dysentry. The persons are observed independently from two unions.

Union	*Number of persons having dysentry in villages* (x_{ij})
1	17, 12, 18, 20, 15, 25, 35, 24
2	38, 27, 40, 42, 36, 50, 52, 48, 35

$i = 1, 2$

Are the unions similar in respect of people suffering from dysentry?

Solution. We have $n_1 = 8, n_2 = 9$; assume that $x_{1j} \sim N(\mu_1, \sigma^2)$ and $x_{2j} \sim N(\mu_2, \sigma^2)$.

We need to test $H_0 : \mu_1 = \mu_2$, against $H_A : \mu_1 \neq \mu_2$.

$$\bar{x}_1 = \frac{1}{n_1}\sum x_{1j} = \frac{166}{8} = 20.75.$$

$$s_1^2 = \frac{1}{n_1 - 1}\left[\sum x_{1j}^2 - \frac{(\sum x_{1j})^2}{n_1}\right] = \frac{1}{8-1}\left[3808 - \frac{(166)^2}{8}\right] = 51.9.$$

$$\bar{x}_2 = \frac{1}{n_2}\sum x_{2j} = \frac{368}{9} = 40.89.$$

$$s_2^2 = \frac{1}{n_2 - 1}\left[\sum x_{2j}^2 - \frac{(\sum x_{2j})^2}{n_2}\right] = \frac{1}{9-1}\left[15566 - \frac{(368)^2}{9}\right] = 64.8.$$

$$s^2 = \frac{1}{n_1 + n_2 - 2}[(n_1 - 1)s_1^2 + (n_2 - 1)s_2^2]$$

$$= \frac{1}{8 + 9 - 2}[(8 - 1)51.93 + (9 - 1)64.86] = 58.826.$$

As σ^2 is not known and both n_1 and n_2 are less than 30, the test statistic is

$$t = \frac{\bar{x}_1 - \bar{x}_2}{\sqrt{s^2\left(\frac{1}{n_1} + \frac{1}{n_2}\right)}} = \frac{20.75 - 40.89}{\sqrt{58.826\left(\frac{1}{8} + \frac{1}{9}\right)}} = -5.40.$$

Comment. As $|t| > t_{0.05,15} = 2.131$ [P-value < 0.0005], H_0 is rejected. Number of people suffering from dysentry in two unions are not similar.

EXAMPLE 7.5 The following are the number of outdoor patients of different days of a month in two rural community health clinics.

Clinic	*Patients (x_{ij}) in different days*
1	20, 25, 18, 27, 15, 22, 22, 20, 18, 19, 22, 20, 25, 27, 26, 25, 24, 20, 20, 20, 18, 15, 15, 20, 25, 25, 20, 18, 25, 24
2	38, 42, 35, 27, 25, 40, 35, 37, 30, 30, 30, 28, 28, 25, 40, 40, 40, 40, 36, 35, 35, 45, 50, 40, 40, 36, 32, 34, 28, 27.

Are the clinics similar in serving outdoor patients?

Solution. We have $n_1 = 30,\ n_2 = 30;\ i = 1, 2$.

Assume that $x_{1j} \sim N(\mu_1, \sigma^2), x_{2j} \sim N(\mu_2, \sigma^2)$. We need to test

$$H_0 : \mu_1 = \mu_2, \text{ against } H_A : \mu_1 \neq \mu_2.$$

Here

$$\bar{x}_1 = \frac{1}{n_1}\sum x_{1j} = \frac{640}{30} = 21.33.$$

$$s_1^2 = \frac{1}{n_1 - 1}\left[\sum x_{1j}^2 - \frac{(\sum x_{1j})^2}{n_1}\right] = \frac{1}{30-1}\left[14434 - \frac{(640)^2}{30}\right] = 26.92.$$

$$\bar{x}_2 = \frac{1}{n_2}\sum x_{2j} = \frac{1048}{30} = 34.93.$$

$$s_2^2 = \frac{1}{n_2 - 1}\left[\sum x_{2j}^2 - \frac{(\sum x_{2j})^2}{n_2}\right] = \frac{1}{30-1}\left[37734 - \frac{(1048)^2}{30}\right] = 38.75.$$

As both n are large, the test statistic is

$$z = \frac{\bar{x}_1 - \bar{x}_2}{\sqrt{\frac{s_1^2}{n_1} + \frac{s_2^2}{n_2}}} = \frac{21.33 - 34.93}{\sqrt{\frac{26.92}{30} + \frac{38.75}{30}}} = -9.19.$$

Comment. As $|z| > 1.96$ [P-value < 0.0000], H_0 is rejected. The clinics are not similar in serving outdoor patients.

7.4 Test of Equality of Several Means (One-way Analysis of Variance)

Let us consider that we have k samples selected independently from k normal populations, where x_{ij} $(i = 1, 2, \ldots, k;\ j = 1, 2, \ldots, n_i)$ be the j-th observation of i-th sample. It is assumed that

$$x_{ij} \sim N(\mu_i, \sigma^2);\ i = 1, 2, \ldots, k.$$

The problem is to test the hypothesis

$$H_0 : \mu_1 = \mu_2 = \cdots = \mu_k$$

against H_A : At least one equality sign does not hold good.

The sample observations of different sample can be shown as follows:

Sample	*Observations* (x_{ij})						*Total* x_i	*Mean* $\bar{x}_i$
1	x_{11}	x_{12}	$\cdots$	x_{1j}	$\cdots$	x_{1n_1}	x_1	$\bar{x}_1$
2	x_{21}	x_{22}	$\cdots$	x_{2j}	$\cdots$	x_{2n_1}	x_2	$\bar{x}_2$
$\vdots$	$\vdots$	$\vdots$	$\cdots$	$\vdots$	$\cdots$	$\vdots$	$\vdots$	$\vdots$
i	x_{i1}	x_{i2}	$\cdots$	x_{ij}	$\cdots$	x_{in_i}	x_i	$\bar{x}_i$
$\vdots$	$\vdots$	$\vdots$	$\cdots$	$\vdots$	$\cdots$	$\vdots$	$\vdots$	$\vdots$
k	x_{k1}	x_{k2}	$\cdots$	x_{kj}	$\cdots$	x_{kn_k}	x_k	$\bar{x}_k$
Total							G	

Here $x_i = \sum_j x_{ij}$, $\bar{x}_i = \frac{1}{n_i}\sum_j x_{ij}$, $G = \sum\sum x_{ij} = \sum x_i$.

If null hypothesis is true, all samples follow the same distribution and sample means ($\bar{x}_i$'s, $i = 1, 2, \ldots, k$) should be homogeneous. Heterogeneity in sample means will be more, if null hypothesis is not true. So, variance of sample means may provide some idea about the acceptance or rejection of null hypothesis. Again, a set of x_{ij} observations is selected from a population and these are expected to be internally homogeneous. But, in practice, the x_{ij} observations of i-th sample vary from observation to observation. The variance of any sample observations can be measured by

$$\frac{1}{n_i - 1}\sum(x_{ij} - \bar{x}_i)^2,\ i = 1, 2, \ldots, k.$$

We have in total $n = (n_1 + n_2 + \cdots + n_k)$ observations. The variance of all these observations can be measured by

$$\frac{1}{n-1}\sum\sum(x_{ij} - \bar{x})^2, \text{ where } \bar{x} = \frac{1}{n}\sum\sum x_{ij} = \frac{1}{n}\sum n_i\bar{x}_i.$$

Here $\sum\sum(x_{ij} - \bar{x})^2$ is called total sum of squares. It can be partitioned as follows:

$$\begin{aligned}\sum\sum(x_{ij} - \bar{x})^2 &= \sum\sum[(x_{ij} - \bar{x}_i) + (\bar{x}_i - \bar{x})]^2\\ &= \sum\sum(x_{ij} - \bar{x}_i)^2 + \sum n_i(\bar{x}_i - \bar{x})^2.\end{aligned}$$

Total sum of squares = sum of square within sample observations + sum of squares between sample observations

S.S. (total) = S.S. (within) + S.S. (between)

[S.S. = Sum of Squares]

Here total variation of data set is decomposed into two components of variation, viz., S.S. (between) and S.S. (within). The main variation is due to one criteria, i.e., variation due to sample. This decomposition of total variation is mainly for one factor (sample) except the S.S. (within). So, it is called one-way classification. Here S.S. (within) is also called S.S. (error) or S.S. (uncontrolled source).

Analysis of variance. According to R.A. Fisher, "The analysis of variance is a technique to partition the total variation of a data set caused by a set of causes into components of variation according to pre-identified sources of variation".

Under the assumption $x_{ij} \sim N(\mu_i, \sigma^2)$ the sum of squares of each component is distributed as chi-square. We also know that ratio of two independent chi-squares is distributed as F. Therefore, to test null hypothesis we can apply F-test.

The different sum of squares are calculated as follows:

$$\text{S.S. (total)} = \sum\sum(x_{ij} - \bar{x})^2 = \sum\sum x_{ij}^2 - \text{C.T.}$$

where

$$\text{C.T.} = \frac{G^2}{n},\ G = \sum\sum x_{ij} = \text{grand total}, n = \sum n_i.$$

If $n_1 = n_2 = \cdots = n_k$, the $\sum n_i = nk$ = total observations in the sample.

$$S_1 = \text{S.S. (between)} = \sum n_i(\bar{x}_i - \bar{x})^2 = \sum \frac{x_i^2}{n_i} - \text{C.T.}$$

$$= \sum \frac{x_i^2}{n} - \text{C.T., if } n_1 = n_2 = \cdots = n_k.$$

These sum of squares and their corresponding d.f. are shown in the analysis of variance (ANOVA) table as below:

ANOVA Table

Sources of variation	*d.f.*	*S.S.*	*M.S.* $= \frac{S.S.}{d.f.}$	*Calculated F*	*Tabulated F* $[F_{0.05}]$
Between	$k-1$	S_1	s_1	s_1/s_2	
Error	$n-k$	S_2	s_2		
Total	$n-1$				

EXAMPLE 7.6 A doctor has prescribed medicines of hypertension produced by different companies to 10 patients every working day. Some of the patients observed positive results due to the use of medicine. The number of improved patients, out of 10, of different days are shown below:

Company	*Number of improved patients* (x_{ij}) *of a day*						*Total* x_i	*Mean* $\bar{x}_i$
C_1	5	6	5	5	6	6	33	5.50
C_2	3	3	4	4	3	3	20	3.33
C_3	8	8	8	8	7	8	47	7.83
C_4	4	4	4	2	2	3	19	3.17
C_5	9	8	8	10	9	9	53	8.83
Total							172	

(i) Are the medicines of companies similar?

(ii) Is there any difference between the medicines of companies C_2 and C_4?

(iii) It is known that medicine of company C_3 is standard one. Are medicines of other companies better than the medicine of company C_3?

Solution. We have $k = 5, n_1 = n_2 = n_3 = n_4 = n_5 = n_6 = 6.$.

Total observations are $nk = \sum n_i = 6 \times 5 = 30, G = 172$

$$\text{C.T.} = \frac{G^2}{nk} = \frac{(172)^2}{6 \times 5} = 986.13$$

$$\text{S.S. (total)} = \sum\sum x_{ij}^2 - \text{C.T.}$$
$$= 1156 - 986.13 = 169.87$$

$$\text{S.S. (between)} = \sum \frac{x_i^2}{n} - \text{C.T.}$$
$$= \frac{6868}{6} - 986.13 = 158.54$$

$$\text{S.S. (within)} = \text{S.S. (total)} - \text{S.S. (between)}$$
$$= 169.87 - 158.54 = 11.33.$$

ANOVA Table

Sources of variation	*d.f.*	*S.S.*	*M.S.* $= \frac{S.S.}{d.f.}$	*Calculated F*	*Tabulated F* $[F_{0.05}]$
Between	4	158.54	39.635	87.46	2.76
Within	25	11.33	0.4532		
Total	29				

(i) Assume that $x_{ij} \sim N(\mu_i, \sigma^2), i = 1, 2, \ldots, k = 5$

We need to test $H_0 : \mu_1 = \mu_2 = \cdots = \mu_5$

H_A : At least one of the equality does not hold good.

Comment. As calculated $F >$ tabulated F, H_0 is rejected. The medicines of different companies are not similar.

(ii) We need to test $H_0 : \mu_2 = \mu_4$, against $H_A : \mu_2 \neq \mu_4$.

The test statistic is:

$$t = \frac{\bar{x}_2 - \bar{x}_4}{\sqrt{\frac{2 \text{ M.S. (within)}}{n}}} = \frac{3.33 - 3.17}{\sqrt{\frac{2 \times 0.4532}{6}}} = 0.41.$$

Comment. As $|t| < t_{0.05,25} = 2.06$, H_0 is accepted. There is no difference between the medicines of companies C_2 and C_4.

(iii) As the medicine of company C_3 is standard one, it is to be compared with other medicines. The medicine of company C_3 is considered as control treatment. The comparison of control treatment with other treatments is done by Dunnett's test, where the test statistic is:

$$D = d_{0.05,\ \text{error d.f.},\ k}\sqrt{\frac{2 \text{ M.S. (within)}}{n}}$$

where $d_{0.05,\ \text{error d.f.},\ k}$ = 5% value from Dunnett's table for k means (including control treatment) with error d.f.

$$D = 2.605\sqrt{\frac{2 \times 0.4532}{6}} = 1.01.$$

If $|\bar{x}_i - \bar{x}_j| < D$, there is no difference between i-th and j-th means, where $i \neq j = 1, 2, \ldots, k$. In our case, we need the comparison of $|x_3 - \bar{x}_j|$, where $j = 1, 2, 4, 5$ $(j \neq 3)$

$$|\bar{x}_3 - \bar{x}_1| = |7.83 - 5.50| = 2.33 > D, \ \bar{x}_1 \text{ is different from } \bar{x}_3$$
$$|\bar{x}_3 - \bar{x}_2| = |7.83 - 3.33| = 4.5 > D, \ \bar{x}_2 \text{ is different from } \bar{x}_3$$
$$|\bar{x}_3 - \bar{x}_4| = |7.83 - 3.17| = 4.66 > D, \ \bar{x}_4 \text{ is different from } \bar{x}_3$$
$$|\bar{x}_3 - \bar{x}_5| = |7.83 - 8.83| = 1.0 < D, \ \bar{x}_5 \text{ is not different from } \bar{x}_3.$$

The medicine of company C_3 is better than that of companies C_1, C_2 and C_4.

EXAMPLE 7.7 The following data represent the number of ever born children of different mothers having different levels of education.

Levels of education	*Number of ever born children* (x_{ij}) *to mothers*	*Total* x_i	*Mean* $\bar{x}_i$
Illiterate	5, 4, 7, 6, 5, 8, 5, 3, 4, 6	53	5.30
Primary	4, 3, 3, 4, 5, 4	23	3.83
Secondary	3, 3, 2, 2, 3, 2	15	2.50
Higher	2, 2, 2, 1, 3	10	2.00
Total		101	

(i) Are the ever born children to mothers of all levels of education similar?

(ii) Is there any difference in ever born children of secondary and higher educated mothers?

Solution. We have $n_1 = 10, n_2 = 6, n_3 = 6, n_4 = 5; k = 4$.

Total observations are $n = n_1 + n_2 + n_3 + n_4 = 10 + 6 + 6 + 5 = 27$.

$$G = 101, \ \text{C.T.} = \frac{G^2}{n} = \frac{(101)^2}{27} = 377.81.$$

$$\text{S.S. (total)} = \sum\sum x_{ij}^2 - \text{C.T.} = 453 - 377.81 = 75.19.$$

$$\text{S.S. (between)} = \sum \frac{x_i^2}{n_i} - \text{C.T.} = \frac{53^2}{10} + \frac{23^2}{6} + \frac{15^2}{6} + \frac{10^2}{5} - 377.81$$
$$= 426.57 - 377.81 = 48.76.$$

$$\text{S.S. (within)} = \text{S.S. (total)} - \text{S.S. (between)}$$
$$= 75.19 - 48.76 = 26.43.$$

ANOVA Table

Sources of variation	*d.f.*	*S.S.*	*M.S.* $= \frac{S.S.}{d.f.}$	*Calculated F*	*Tabulated F* $[F_{0.05}]$
Between	3	48.76	16.253	14.15	3.03
Within	23	26.43	1.149		
Total	26				

(i) Assume that $x_{ij} \sim N(\mu_i, \sigma^2), i = 1, 2, \ldots, k = 4$.

We need to test $H_0 : \mu_1 = \mu_2 = \mu_3 = \mu_4$, against

H_A : At least one equality does not hold good.

Comment. As calculated $F >$ tabulated F, H_0 is rejected. Number of ever born children varies with the variation of levels of education of mothers.

(ii) We need to test $H_0 : \mu_3 = \mu_4$, against $H_A : \mu_3 \neq \mu_4$.

The test statistic is:

$$t = \frac{\bar{x}_3 - \bar{x}_4}{\sqrt{\text{M.S. (within)}\left[\frac{1}{n_3} + \frac{1}{n_4}\right]}} = \frac{2.50 - 2.00}{\sqrt{1.149\left(\frac{1}{6} + \frac{1}{5}\right)}} = 0.77.$$

Comment. As $|t| < t_{0.05,23} = 2.069$, H_0 is accepted. There is no difference in number of ever born children of mothers having secondary and higher level of education.

7.5 Test of Equality of Two Correlated Means

Let $x_1, x_2, \ldots, x_n$ be n randomly selected observations from $N(\mu_1, \sigma^2)$. Consider that $y_1, y_2, \ldots, y_n$ are the n observations recorded from the previously selected n sample points and $y \sim N(\mu_2, \sigma^2)$. As (x_i, y_i) is a pair of observation $(i = 1, 2, \ldots, n)$ recorded from the same i-th sample point, x_i and y_i are assumed to be correlated. For example, let x_i be the blood sugar level of i-th person before using a medicine and y_i be the blood sugar level of the same person after using a medicine against blood sugar level. Here x_i and y_i are expected to be correlated.

The objective of this analysis is to verify whether medicine is effective in reducing blood sugar level. The medicine will be effective if $\mu_1 > \mu_2$. Thus, we need to test the hypothesis:

$$H_0 : \mu_1 = \mu_2, \text{ against } H_A : \mu_1 > \mu_2 \ [\text{or, } H_A : \mu_1 \neq \mu_2]$$

The test statistic to test this hypothesis is:

$$t = \frac{\bar{d}}{\sqrt{\frac{s_d^2}{n}}}, \text{ where } \bar{d} = \frac{1}{n}\sum d_i,\ d_i = x_i - y_i$$

$$s_d^2 = \frac{1}{n-1}\sum (d_i - \bar{d})^2.$$

This t has $(n - 1)$ d.f.

EXAMPLE 7.8 The following are the s. uric acid of some persons before using a medicine (x_i mg/dl) and after using a medicine (y_i mg/dl).

x_i	9.5	7.8	10.6	9.0	8.5	10.0	9.5	11.7	12.5	10.5
y_i	4.5	8.0	6.2	7.0	9.0	8.5	9.0	10.2	13.6	11.8

Is medicine effective in reducing s. uric acid?

Solution. Let $x_i \sim N(\mu_1, \sigma^2)$ and $y_i \sim N(\mu_2, \sigma^2)$.

We need to test $H_0 : \mu_1 = \mu_2$, against $H_A : \mu_1 > \mu_2$.

Here $n = 10$,

$$d_i = x_i - y_i : \ 5.0 \quad -0.2 \quad 4.2 \quad 2.0 \quad -0.5 \quad 1.5 \quad 0.5 \quad 1.5 \quad -1.1 \quad -1.3$$

$$\bar{d} = \frac{1}{n}\sum d_i = \frac{11.6}{10} = 1.16.$$

$$s_d^2 = \frac{1}{n-1}\left[\sum d_i^2 - \frac{(\sum d_i)^2}{n}\right] = \frac{1}{10-1}\left[54.58 - \frac{(11.6)^2}{10}\right] = 4.569.$$

The test statistic is: $t = \dfrac{\bar{d}}{\sqrt{\frac{s_d^2}{n}}} = \dfrac{1.16}{\sqrt{\frac{4.569}{10}}} = 1.72.$

Comment. As $|t| < t_{0.05,9} = 1.833$ (for one-sided test with $\alpha = 0.05$), H_0 is accepted. The medicine is not effective. If $H_A : \mu_1 \neq \mu_2$, $|t| < 2.262$ (for two-sided test with $\alpha = 0.05$). H_0 is accepted.

7.6 Test of One Proportion

The variable under study may be qualitative. For example, if health condition of people is investigated, some will be observed with ill health and some will be of good health. Here health condition of people is qualitative variable. Couples may be investigated to measure the family planning adoption rate. Some couples may adopt the family planning method and some may not. The adoption behaviour is a qualitative character. In such cases, the population characteristic (parameter) to be estimated is population proportion.

Let $x_1, x_2, \ldots, x_n$ be the n values observed from n randomly selected sample points, where x_i values are the values of a qualitative variable. For example, one may be interested in estimating the proportions of treated patients admitted in a hospital. If treatment is the characteristic under study, one can take $x_i = 1$, if a patient is cured and $x_i = 0$, if he is not cured. Let us consider that out of n patients 'a' number of patients are cured and $(n - a)$ number of patients are not cured. Here

$$\sum_{i=1}^{n} x_i = a.$$

The proportion of cured patient is $p = a/n$. This p is an unbiased estimate of P, where

$$P = \frac{A}{N}, \; A = \sum_{i=1}^{N} x_i.$$

This P is the population proportion.

We need to test $H_0 : P = P_0$ (a specific value) against $H_A : P \neq P_0$. The test statistic is:

$$z = \frac{p - P_0}{\sqrt{\frac{P_0 Q_0}{n}}} \sim n(0, 1).$$

EXAMPLE 7.9 In a locality the family planning implementation authority made a target to increase adoption rate among eligible couple to 80%. Some family planning service providers are instructed to visit the locality and motivate the couples. Hundred couples are purposively selected and are motivated. But, finally 65 are agreed to adopt family planning. Is the target fulfilled?

Solution. Let P = proportion of targetted adopter. We need to test

$$H_0 : P = P_0 = 0.8, \text{ against } H_A : P \neq P_0.$$

Given

$$n = 100, \; a = 65, \; p = \frac{a}{n} = \frac{65}{100} = 0.65.$$

The test statistic is:

$$z = \frac{p - P_0}{\sqrt{\frac{P_0 Q_0}{n}}} = \frac{0.65 - 0.8}{\sqrt{\frac{0.8 \times 0.2}{100}}} = -3.75, \; Q_0 = 1 - P_0 = 0.2.$$

Comment. As $|z| > 1.96$, H_0 is rejected. The target is not fulfilled.

EXAMPLE 7.10 It is assumed that 45% people suffer from diabetes. In a diabetic centre one day 400 people gave their blood sample for test. It is found that, out of 400 people, 225 people have their blood sugar level higher than the normal level. Do the information support the assumption?

Solution. Let P = proportion of diabetic patient. We need to test

$$H_0 : P = P_0 = 0.45, \text{ against } H_A : P \neq P_0.$$

Given $n = 400$, $a = 225$. So, $p = \frac{a}{n} = \frac{225}{400} = 0.5625$.

The test statistic is:

$$z = \frac{p - P_0}{\sqrt{\frac{P_0 Q_0}{n}}} = \frac{0.5625 - 0.45}{\sqrt{\frac{0.5625 \times 0.4375}{400}}} = 4.54.$$

Comment. $|z| > 1.96$, H_0 is rejected. The assumption is not correct.

7.7 Test of Equality of Several Proportions

The qualitative character of population units can be investigated from two or more populations. Let us consider that the i-th population has N_i units ($i = 1, 2, \ldots, k$). Out of N_i units A_i units possess a character. Then

$$P_i = \frac{A_i}{N_i} = \text{proportion of units possessing the character in } i\text{-th population.}$$

Let n_i be the sample units selected from i-th population. All samples are independently selected. Out of n_i units a_i units in the sample possess the character. Then

$$p_i = \frac{a_i}{n_i} = \text{proportion of units in the sample possessing the character.}$$

The problem is to test the hypothesis

$$H_0 : P_1 = P_2 = \cdots = P_k = P$$

against H_A : At least one equality does not hold good.

When $k = 2$, we have $H_0 : P_1 = P_2$, against $H_A : P_1 \neq P_2$. The test of this latter hypothesis is done by z-statistic, where

$$z = \frac{p_1 - p_2}{\sqrt{pq\left(\frac{1}{n_1} + \frac{1}{n_2}\right)}} \sim N(0, 1). \text{ Here } p = \frac{a_1 + a_2}{n_1 + n_2},\ q = 1 - p.$$

The sample information of two samples can be presented in row and column as follows:

Characteristic	*Sample*		*Total* R_i
	1	2	
Present	$a_1 = O_{11}$	$a_2 = O_{12}$	R_1
Absent	$n_1 - a_1 = O_{21}$	$n_2 - a_2 = O_{22}$	R_2
Total, C_j	$n_1 = C_1$	$n_2 = C_2$	$n_1 + n_2 = n$

Here O_{ij} – observation of i-th row and j th column ($i - 1, 2,;\ j - 1, 2$)

Under H_0, we can write:

$$\frac{O_{11}}{C_1} = \frac{O_{12}}{C_2} = \frac{R_1}{n} \Rightarrow O_{11} = \frac{C_1 R_1}{n}.$$

This O_{11} under H_0 can be written as E_{11}, where

$$E_{ij} = \frac{R_i C_j}{n} = \text{observation of } i\text{-th row and } j\text{-th column under } H_0.$$

Then the test statistic is $\chi^2 = \sum\sum \frac{O_{ij}^2}{E_{ij}} - n.$

This χ^2 has 1 d.f. If χ^2 calculated $\geq$ tabulated value of χ^2 at 5% level of significance with 1 d.f., H_0 is rejected. In testing this hypothesis the assumptions are:

(i) Samples are independently selected,

(ii) Row totals and column totals are fixed.

If $a_1 = a$, $a_2 = b$, $n_1 - a_1 = c$ and $n_2 - a_2 = d$ are considered, then the test statistic takes the form:

$$\chi^2 = \frac{n(ad - bc)^2}{(a+b)(a+c)(b+d)(c+d)}.$$

When the sample observations of two samples are arranged in 2×2 table shown above, then any one value of E_{ij} can be calculated independently by the formula:

$$E_{ij} = \frac{R_i C_j}{n} \ (i = 1, 2, ; \ j = 1, 2).$$

For example, if $E_{11} = \frac{R_1 C_1}{n}$, then

$$E_{12} = O_1 - E_{11}, \ E_{21} = C_1 - E_{11}, E_{22} = C_2 - E_{12}.$$

Out of two values in a row, we calculate $(2 - 1)$ value independently. Out of two values in a column, we calculate $(2 - 1)$ value independently. Thus, we calculate $(2 - 1) \times (2 - 1)$ value independently in rows and columns. If there are r rows and c columns, we can calculate $(r - 1)(c - 1)$ values independently and hence the d.f. of χ^2 statistic is $(r - 1)(c - 1)$.

This χ^2 statistic is applied to test the former hypothesis, where there are k samples. Sample observations are arranged in two rows and k columns. In such a case, the d.f. of χ^2 becomes $(2-1)(k-1) = k-1$. Again, if k samples are selected independently, where sample observations of any sample is not classified into different classes, these sample observations have been utilized to verify that P_i is the specific proportion of i-th sample observation among all samples. We have $n = \sum n_i$ = total observations of k samples. Now, under H_0 the sample observation of i-th sample becomes

$$E_i = \frac{n}{k}.$$

Let O_i be the observed sample observations of i-th sample. Then the test statistic is

$$\chi^2 = \sum_{i=1}^{k} \frac{O_i^2}{E_i} - n, \text{ where } \sum^{k} O_i = \sum^{k} E_i = n.$$

This χ^2 has $(k - 1)$ d.f. The χ^2-statistic provides satisfactory information, if $n \geq 10$, $k \geq 3$, $\frac{n^2}{k} \geq 10$ and $E_i > 0.25$. If $E_i < 5$ for any sample, it is to be added to the nearest E_i value. The d.f. of χ^2 will then be $(k_1 - 1)$, where k_1 is the number of samples after merging any sample to the nearest sample or class.

EXAMPLE 7.11 In a village there are 125 people of ages 70 years and above. Among them 35 are with good health. In another village the corresponding figures are 90 and 28. Are the proportions of people with good health in two villages similar?

Solution. Let P_1 = proportion of people with good health in village-1.

P_2 = proportion of people with good health in village-2.

We need to test $H_0 : P_1 = P_2$, against $H_A : P_1 \neq P_2$.

We have $n_1 = 125, a_1 = 35, n_2 = 90$ and $a_2 = 28$.

$$p_1 = \frac{a_1}{n_1} = \frac{35}{125} = 0.28,\ p_2 = \frac{a_2}{n_2} = \frac{28}{90} = 0.31.$$

$$p = \frac{a_1 + a_2}{n_1 + n_2} = \frac{35 + 28}{125 + 90} = 0.29,\ q = 1 - p = 0.71.$$

The test statistic is:

$$z = \frac{p_1 - p_2}{\sqrt{pq\left(\frac{1}{n_1} + \frac{1}{n_2}\right)}} = \frac{0.28 - 0.31}{\sqrt{0.29 \times 0.71\left(\frac{1}{125} + \frac{1}{90}\right)}} = -0.48.$$

Comment. As $|z| < 1.96$, H_0 is accepted [P-value = 0.6312]. The proportions of people in both villages are similar.

Alternative. The sample data can be presented in a 2 × 2 table as follows:

Good health	*Village*		*Total*
	1	2	
Yes	$35 = a$	$28 = b$	63
No	$90 = c$	$62 = d$	152
Total	125	90	$n = 215$

The test statistic is:

$$\chi^2 = \frac{n(ad - bc)^2}{(a+b)(a+c)(b+d)(c+d)} = \frac{215(35 \times 62 - 28 \times 90)^2}{63 \times 125 \times 90 \times 152} = 0.24.$$

Comment. As calculated $\chi^2 <$ 5% value of χ^2 with 1 d.f. = 3.84, H_0 is accepted.

Alternative. The sample data can be presented in a 2 × 2 table as follows:

Observed frequency, O_{ij}

Good health	*Village*		*Total R_i*
	1	2	
Yes	35	28	63
No	90	62	152
Total C_j	125	90	215 n

Under H_0

$$E_{11} = \frac{R_1 C_1}{n} = \frac{63 \times 125}{215} = 36.6.$$

Frequencies under H_0, E_{ij}

Good health	*Village*		*Total R_i*
	1	2	
Yes	36.6	26.4	63
No	88.4	63.6	152
Total C_j	125	90	215 n

The test statistic is:

$$\chi^2 = \sum\sum \frac{O_{ij}^2}{E_{ij}} - n = \frac{35^2}{36.6} + \frac{28^2}{26.4} + \frac{90^2}{88.4} + \frac{62^2}{63.6} - 215 = 0.24.$$

In the above example the null hypothesis is accepted. It indicates that the proportions of people with good health in both villages are similar. In other words, we can say that the health condition (A) does not depend on village (B). Here the test of equality of proportions is equivalent to the test of independence of two characters A and B, where A has r levels and B has c levels. The hypothesis to be mentioned is:

H_0 : A does not depend on B.

H_A : A depends on B.

EXAMPLE 7.12 In a government hospital there are 100 doctors. Among them 52 have foreign degrees. In a private hospital there are 45 doctors. Among them 18 have foreign degrees. Are the proportions of doctors having foreign degrees in two hospitals similar?

Solution. Let P_1 = proportion of doctors having foreign degrees in government hospital.

P_2 = proportion of doctors having foreign degrees in private hospital.

We need to test $H_0 : P_1 = P_2$, against $H_A : P_1 \neq P_2$.

We have $n_1 = 100, a_1 = 52, n_2 = 45, a_2 = 18$.

$$p_1 = \frac{a_1}{n_1} = \frac{52}{100} = 0.52,\ p_2 = \frac{a_2}{n_2} = \frac{18}{45} = 0.40$$

$$p = \frac{a_1 + a_2}{n_1 + n_2} = \frac{52 + 18}{100 + 45} = 0.48,\ q = 1 - p = 0.52.$$

The test statistic is:

$$z = \frac{p_1 - p_2}{\sqrt{pq\left(\frac{1}{n_1} + \frac{1}{n_2}\right)}} = \frac{0.52 - 0.40}{\sqrt{0.48 \times 0.52\left(\frac{1}{100} + \frac{1}{45}\right)}} = 1.34.$$

Comment. As $|z| < 1.96$, H_0 is accepted. The proportions of doctors having foreign degrees in two hospitals are similar.

EXAMPLE 7.13 The following are the number of infected patients after operation in differnt hospitals in a year:

Hospital	1	2	3	4	5	Total n
Number of infected patients, O_i	136	62	96	70	152	516

Are the proportions of infected patients in all hospitals similar?

Solution. Let P_i = proportion of infected patients in i-th hospital.

$i = 1, 2, \ldots, 5$.

We need to test $H_0 : P_1 = P_2 = P_3 = P_4 = P_5$,

against H_A : At least one of the equality sign does not hold good.

We have $n = 516$. Under H_0, $E_i = \frac{n}{k}$, $k = 5$.

$$\therefore E_i = \frac{516}{5} = 103.2\ (i = 1, 2, 3, 4, 5).$$

The test statistic is:

$$\chi^2 = \sum \frac{O_i^2}{E_i} - n = \frac{1}{103.2}[136^2 + 62^2 + 96^2 + 70^2 + 152^2] - 516 = 61.13.$$

Comment. $\chi^2 = 61.13 >$ tabulated value of χ^2 at 5% level of significance with $(5-1) = 4$ d.f. is 9.488. H_0 is rejected. The proportions of infected patients in different hospitals are not similar.

EXAMPLE 7.14 The following are the number of adopter females having different levels of education.

Level of education	Illiterate	Primary	Secondary	Higher	Total
Adopter females	208	190	45	557	1000

It is assumed that the proportions of adopter females of different levels of education are in the ratio

Illiterate : Primary : Secondary : Higher :: 1 : 1 : 2 : 3.

Is the assumption correct?

Solution. Let P_i = proportion of adopter females of i-th level of education; $i = 1, 2, 3, 4$.

We need to test $H_0 : P_1 : P_2 : P_3 : P_4 :: 1 : 1 : 2 : 3$.

H_A : These ratios are not true.

The null hypothesis indicates that out of $(1 + 1 + 2 + 3) = 7$ adopters 1 is illiterate, 1 is primary educated, 2 are secondary educated and 3 are higher educated. Therefore, from 1000 adopters the adopters of i-th level of education is calculated as follows:

$$E_1 = \frac{1000}{7} = 142.9, \ E_2 = \frac{1000}{7} = 142.9, \ E_3 = \frac{2 \times 1000}{7} = 285.7.$$

But, as 1000 is fixed, we have $E_4 = 1000 - 142.9 - 142.9 - 285.7 = 428.5$.

The test statistic is:

$$\chi^2 = \sum \frac{O_i^2}{E_i} - n = \frac{208^2}{142.9} + \frac{190^2}{142.9} + \frac{45^2}{285.7} + \frac{557^2}{428.5} - 1000 = 286.5.$$

Comment. As calculated $\chi^2 >$ 5% value of χ^2 with $(k-1) = 3$ d.f. is 7.81, H_0 is rejected. The assumed proportions of adopter couples of different levels of education are not correct.

7.8 Test of Independence

In Section 7.7, we have discussed the test of equality of several proportions, where k samples are independently selected from k populations. It is assumed that n_i is the number of observations of i-th sample $(i = 1, 2, \ldots, k)$ and a_i of n_i observations possess a characteristic and $p_i = \frac{a_i}{n_i}$ $(i = 1, 2, \ldots, k)$ is the proportion of observations of i-th sample possessing the character. If P_i $(i = 1, 2, \ldots, k)$ is the population proportion of i-th population possessing the characteristic, then H_0 to be tested is

$$H_0 : P_1 = P_2 = \cdots = P_k$$

against H_A : At least one equality does not hold good. This hypothesis indicates that the presence of a characteristic in the sample observations does not depend on sample. If the presence of a characteristic is attributed by A and the sample is attributed by B, the above hypothesis is equivalent to the hypothesis of independence of the attributes A and B, where A may have 2 or more levels and B has 2 or more levels. The statistic is

$$\chi^2 = \sum\sum \frac{O_{ij}^2}{E_{ij}} - n, \; n = \sum\sum O_{ij} = \sum\sum E_{ij}.$$

If A has r levels and B has c levels, this χ^2 has $(r-1)(c-1)$ d.f. The hypothesis is

H_0 : The attributes A and B are independent.

H_A : The attributes A and B are not independent.

If $r = 2$ and $c = 2$, then χ^2 takes the form:

$$\chi^2 = \frac{n(ad-bc)^2}{(a+b)(a+c)(b+d)(c+d)},$$

where the total n observations can be presented in a 2 × 2 table as follows:

A \ B	B_1	B_2	Total
A_1	$O_{11} = a$	$O_{12} = b$	$a+b = R_1$
A_2	$O_{21} = c$	$O_{22} = d$	$c+d = R_2$
Total	$a+c = C_1$	$b+d = C_2$	$n = a+b+c+d$

If $n < 40$, Yates suggested a correction in test statistic, where the corrected test statistic is:

$$\chi^2 = \frac{n\left[|ad-bc| - \frac{n}{2}\right]^2}{(a+b)(a+c)(b+d)(c+d)}.$$

Upton (1982) suggested another corrected form of the test statistic, where the corrected one is:

$$\chi^2 = \frac{(n-1)(ad-bc)^2}{(a+b)(a+c)(b+d)(c+d)}.$$

The assumptions for this test are:

(i) The total observations of any row or any column are fixed.

(ii) Samples are independently selected.

(iii) Total observations n are fixed.

When $r = 2$ and $c = 2$, there are two conditions for test. These are (i) $n > 40$, (ii) if $20 < n < 40$, then no $O_{ij} < 5$. If second condition is not satisfied, Fisher's exact probability test is applied, where the probability is calculated by

$$P(T = a) = \frac{(a+b)!\,(a+c)!\,(b+d)!\,(c+d)!}{n!\,a!\,b!\,c!\,d!},$$

where $a = 0, 1, 2, \ldots, \min(a+b, a+c)$. Here T is the test statistic. For two-sided alternative hypothesis (H_A : The attributes A and B are not independent), the probability $P(T \le a)$ or $P(T \ge a)$ are calculated. Out of these probabilities the smallest one is multiplied by 2 to find the P-value of the test statistic. If P-value $\le \alpha$ (where using $\alpha = 0.05$), H_0 is rejected.

Overall (1990) suggested that Yate's correction can be applied in the adjusted 2 × 2 contin-gency table, where adjustment is done by adding 1 to a value O_{ij} if any $O_{ij} > E_{ij}$.

Here $E_{ij} = \dfrac{R_i \times C_j}{n}$.

Connover (1990) suggested z-test in such a situation, where

$$z = \frac{a - \frac{(a+b)(a+c)}{n}}{\sqrt{\left[\frac{(a+b)(a+c)(n-a-b)(n-a-c)}{n^2(n-1)}\right]}} \sim N(0,1).$$

This z-test provides satisfactory results, if row total or column total or both are random. For 2×2 contingency table z-test is also suggested, where

$$z = \frac{\sqrt{n}(ad - bc)}{\sqrt{[(a+b)(a+c)(b+d)(c+d)]}} \sim N(0,1).$$

If $|z| \geq 1.96$, H_0 is rejected at $\alpha = 0.05$ level of significance.

EXAMPLE 7.15 Anti-hepatitis B vaccine has been given in a locality. After some time a sample of 50 people are investigated and they are classified as follows:

Vaccine given	*Affected by hepatitis B*		*Total R_i*
	Yes	***No***	
Yes	7	23	30
No	15	5	20
Total C_j	22	28	50

Is vaccine effective against hepatitis B?

Solution. We need to test

H_0 : Affected by hepatitis B is independent of vaccination.

H_A : Vaccination is effective against hepatitis B.

The test statistic is:

$$\chi^2 = \frac{n(ad - bc)^2}{(a+b)(a+c)(b+d)(c+d)} = \frac{50(7 \times 5 - 15 \times 23)^2}{30 \times 22 \times 28 \times 20} = 13.00.$$

Comment. As calculated $\chi^2 > \chi^2_{0.05,1} = 3.84$, H_0 is rejected. Vaccination is effective against hepatitis B.

EXAMPLE 7.16 A group of outdoor patients of a hospital are classified according to their residential origin and type of disease. The classified results are shown below:

Type of disease	*Residential origin*		*Total R_i*
	Slum area	*Non-slum area*	
Stomach problem	10	7	17
Others	12	6	18
Total C_j	22	13	35

Do you think that more people of slum area suffer from stomach problem?

Solution. We need to test

H_0 : Stomach problem does not depend on residential origin.
H_A : Stomach problem depends on residential origin.

Here $n < 35$. But no $E_{ij} < 5$ [$E_{11} = 10.7$, $E_{12} = 6.3$, $E_{21} = 11.3$, $E_{22} = 6.7$].
χ^2-static with Yate's correction can be applied, where

$$\chi^2 = \frac{n\left[|ad - bc| - \frac{n}{2}\right]^2}{(a+b)(a+c)(b+d)(c+d)} = \frac{35\left[|10 \times 6 - 7 \times 12| - \frac{35}{2}\right]^2}{17 \times 22 \times 13 \times 18} = 0.0175.$$

Comment. As calculated $\chi^2 < \chi^2_{0.05,1} = 3.84$, H_0 is accepted. Stomach problem does not depend on residential origin.

EXAMPLE 7.17 A group of 20 patients who visited two Upazilla health complexes are investigated and asked whether they are satisfied with the service of the health complexes. The patients are classified as follows:

Satisfaction	*Upazilla health complex*		*Total R_i*
	1	2	
Yes	$7 = a$	$5 = b$	$a+b$ 12
No	$3 = c$	$5 = d$	$c+d$ 8
Total C_j	10 $a+c$	10 $b+d$	20 n

Is satisfaction independent of service of Upazilla health complexes?

Solution. We need to test

H_0 : Satisfaction does not depend on the service of Upazilla health complexes.
H_A : Satisfaction depends on the service of Upazilla health complexes.

Here $n = 20$, so we can use Fisher's exact test,

where $P(T = a) = \dfrac{(a+b)!\ (a+c)!\ (b+d)!\ (c+d)!}{n!\ a!\ b!\ c!\ d!}$.

Here $a = 0, 1, 2, \ldots, \min(a+b)(a+c) = 3$. So, we need $P(T \leq 3)$.

$$P(T \leq 3) = P(T = 0) + P(T = 1) + P(T = 2) + P(T = 3)$$
$$= \frac{12!\ 10!\ 10!\ 8!}{20!\ 10!\ 5!\ 0!\ 8!} + \frac{12!\ 10!\ 10!\ 8!}{20!\ 9!\ 6!\ 1!\ 7!} + \frac{12!\ 10!\ 10!\ 8!}{20!\ 8!\ 4!\ 2!\ 6!} + \frac{12!\ 10!\ 10!\ 8!}{20!\ 7!\ 5!\ 3!\ 5!}$$
$$= 0.3152$$

Comment. As $P(T \leq 3) = 0.3152 > 0.05$, there is no evidence against null hypothesis at 5% level of significance.

Alternative. Here $E_{11} = \frac{10 \times 12}{20} = 6$, $E_{12} = 6$, $E_{21} = 4$, $E_{22} = 4$. As $O_{11} > E_{11}$, we can add 1 to O_{11} [according to the suggestion of Overall (1990)] and then χ^2-statistic with Yate's correction can be applied,

where $$\chi^2 = \frac{n\left[|ad - bc| - \frac{n}{2}\right]^2}{(a+b)(a+c)(b+d)(c+d)} = \frac{20\left[8 \times 6 - 4 \times 2| - \frac{20}{2}\right]^2}{12 \times 10 \times 10 \times 8} = 1.875.$$

The new 2 × 2 contingency table becomes:

Yes	8	4	12
No	2	6	8
	10	10	20

Comment. As calculated $\chi^2 < \chi^2_{0.05,1} = 3.84$, H_0 is accepted. Satisfaction does not depend on service of Upazilla health complexes.

Alternative. As 20 patients are randomly selected, the row total and column total can be considered as random. So, we can apply z-test also,

$$\text{where } z = \frac{a - \frac{(a+b)(a+c)}{n}}{\sqrt{\left[\frac{(a+b)(a+c)(n-a-b)(n-a-c)}{n^2(n-1)}\right]}} = \frac{7 - \frac{10\times 12}{20}}{\sqrt{\left[\frac{12\times 10\times 8\times 10}{(20)^2(20-1)}\right]}} = \frac{1}{1.1239} = 0.89.$$

Comment. As $|z| < 1.96$, H_0 is accepted.

EXAMPLE 7.18 Forty patients visiting an Endoscopic unit of a diagnostic centre are investigated and classified according to their social status and Helicobacter Pylori infection [Positive = 1 or Negative = 0]. The classified results are shown below:

H. Pylori infection	*Social status*		*Total R_i*
	Low	*High*	
Yes	18	7	25
No	6	9	15
Total C_j	24	16	40

Does H. Pylori infection depend on social status?

Solution. We need to test

H_0 : H. Pylori infection does not depend on social status.

H_A : H. Pylori infection depends on social status.

Here z-test can be appied,

$$\text{where } z = \frac{\sqrt{n}\,(ad - bc)}{\sqrt{[(a+b)(a+c)(b+d)(c+d)}} \sim N(0,1) = \frac{\sqrt{40}(18\times 9 - 7\times 6)}{\sqrt{[25\times 24\times 16\times 15]}} = 2.00.$$

Comment. As $|z| > 1.96$, H_0 is rejected. H. Pylori infection depends on social status.

EXAMPLE 7.19 A group of outdoor patients of a hospital are investigated to observe whether they are suffering from stomach problem. These patients are classified according to use of drinking water they use. The classified results are shown below:

Stomach problem	*Sources of drinking water*				*Total R_i*
	Boiled water	*Tubewell water*	*Supply water*	*Bottled water*	
Yes	10	35	75	3	123
No	68	47	30	12	157
Total C_j	78	82	105	15	280 (n)

Do you think that the stomach problem is associated with source of drinking water?

Solution. We need to test

H_0 : **Stomach problem is independent of the source of drinking water.**

H_A : **Stomach problem depends on the source of drinking water.**

The frequency (number of patients) of different cells under $H_0[E_{ij}]$ are shown below, where

$$E_{ij} = \frac{R_i C_j}{n}.$$

Stomach problem	*Sources of drinking water*				*Total* R_i
	Boiled water	*Tubewell water*	*Supply water*	*Bottled water*	
Yes	34.3	36.0	46.1	6.6	123
No	43.7	46.0	58.9	8.4	157
Total C_j	78	82	105	15	$n = 280$

The test statistic is:

$$\chi^2 = \sum\sum \frac{O_{ij}^2}{E_{ij}} - n$$

$$= \frac{10^2}{34.3} + \frac{35^2}{36.0} + \frac{75^2}{46.1} + \frac{3^2}{6.6} + \frac{68^2}{43.7} + \frac{47^2}{46.0} + \frac{30^2}{58.9} + \frac{12^2}{8.4} - 280$$

$= 66.58.$

Comment. As calculated $\chi^2 > \chi^2_{0.05,3} = 7.815$, H_0 is rejected. Stomach problem depends on the source of drinking water.

EXAMPLE 7.20 The following data represent the number of persons who are classified according to their socio-economic condition and body condition score [Body condition score is measured by the values of variables: age, height, weight, calorie consumption, treatment received during disease, etc.] socio-economic condition is measured by income, occupation, residential status, electrical/electronic items used, furniture available in the house, etc. The classified number of persons are shown below:

Socio-economic condition	*Body condition score*			*Total* R_i
	Low	*Medium*	*High*	
Low	42	28	7	77
Medium	18	30	10	58
High	8	20	15	43
Total C_j	68	78	32	178 (n)

Do you think that body condition score is associated with socio-economic condition?

Solution. We need to test

H_0 : Body condition score does not depend on socio-economic status.

H_A : Body condition score depends on socio-economic status.

Here $E_{ij} = \dfrac{R_i \times C_j}{n}$.

The E_{ij} values of different cells are shown below:

E_{ij}-values

Socio-economic condition	*Body condition score*			*Total*
	Low	*Medium*	*High*	R_i
Low	29.4	33.7	13.9	77
Medium	21.8	25.4	10.8	58
High	16.8	18.9	7.3	43
Total C_j	68	78	32	$n = 178$

The test statistic is:

$$\chi^2 = \sum\sum \frac{O_{ij}^2}{E_{ij}} - n$$

$$= \frac{42^2}{29.4} + \frac{28^2}{33.7} + \frac{7^2}{13.9} + \frac{18^2}{21.8} + \frac{30^2}{25.4} + \frac{10^2}{10.8} + \frac{8^2}{16.8} + \frac{20^2}{18.9} + \frac{15^2}{7.3} - 178 = 22.72.$$

Comment. As calculated $\chi^2 > \chi^2_{0.05,4} = 9.488$, H_0 is rejected. Body condition score depends on socio-economic condition.

Likelihood ratio test. This test is also performed in testing the independence of two attributes in case of contingency table. Wilks (1935) suggested this test, where the test statistic is

$$G^2 = 2\sum\sum O_{ij} \ln \frac{O_{ij}}{E_{ij}}, \text{ where } E_{ij} = \frac{R_i \times C_j}{n}, \; n = \sum\sum O_{ij}.$$

Here O_{ij} = observed frequency of j-th column corresponding to i-th row

R_i = total frequency of i-th row

C_j = total frequency of j-th column.

This G^2 asymptotically follows chi-square distribution with $(r-1)(c-1)$ d.f. However, according to Fowler and Cohen (1997) G^2 can be adjusted as shown below:

Adjusted $G^2 = \dfrac{G^2}{F}$, where

$$F = 1 + \frac{1}{6n}\left[\frac{n}{a+b} + \frac{n}{c+d} - 1\right]\left[\frac{n}{a+c} + \frac{n}{b+d} - 1\right].$$

This adjustment is done for 2×2 contingency table, where

$$O_{11} = a,\; O_{12} = b,\; O_{21} = c \text{ and } O_{22} = d.$$

EXAMPLE 7.21 Fifty male and 50 female patients are prescribed ACE tablet for headache. After some time they are investigated to know the response of the medicine against headache. Their responses are recorded and they are classified as follows:

Sex	*Response*		*Total* R_i
	Positive	*Negative*	
Male	32	18	50
Female	20	30	50
Total C_j	52	48	$n = 100$

Is ACE tablet responsive to headache of male and female similarly?

Solution. H_0 : Response of ACE tablet is independent of sex.

H_A : Response of ACE tablet is not independent of sex.

Under H_0 the numbers of male and female patients according to response of medicine are as follows:

Sex	*Response*		*Total*
	Positive	*Negative*	
Male	26	24	50
Female	26	24	50
Total	52	48	$n = 100$

The test statistic is:

$$G^2 = 2\sum\sum O_{ij} \ln \frac{O_{ij}}{E_{ij}} = 2\left[32\ln\frac{32}{26} + 18\ln\frac{18}{24} + 20\ln\frac{20}{26} + 30\ln\frac{30}{24}\right] = 11.07$$

$$F = 1 + \frac{1}{6\times 100}\left[\frac{100}{50} + \frac{100}{50} - 1\right]\left[\frac{100}{52} + \frac{100}{48} - 1\right] = 1.015.$$

$$\text{Adjusted } G^2 = \frac{G^2}{F} = \frac{11.07}{1.015} = 10.91.$$

Comment. As calculated $\chi^2 > \chi^2_{0.05,1} = 3.84$, H_0 is rejected. Response of the ACE tablet is not independent of sex.

7.9 Coefficient of Association

The test of independence of two attributes (characteristics) is done by χ^2-test, where the null hypothesis of independence of attribute is rejected, if calculated $\chi^2 \geq$ tabulated value of χ^2 or if P-value of calculated $\chi^2 \leq 0.05$. The rejection of null hypothesis implies that there is association between two attributes. So, we need to measure the degree of association of the attributes. The degree of association of two attributes is measured by coefficient of association.

The different coefficients of association are:

(i) Contingency coefficient, $C = \sqrt{\frac{\chi^2}{n+\chi^2}}$.

(ii) Cramer's contingency coefficient, $C = \sqrt{\frac{\chi^2}{n(q-1)}}$,
where n = total frequency, $q = \min(r, c)$.

(iii) Pearson's mean square contingency coefficient, $C = \frac{\chi^2}{n}$.

(iv) The phi-coefficient for 2×2 contingency table, $\varphi = \frac{ad-bc}{\sqrt{[(a+b)(a+c)(b+d)(c+d)]}}$.

For Ex. 7.20, the contingency coefficient of body condition score and socio-economic status is $C = \sqrt{\frac{\chi^2}{n+\chi^2}} = \sqrt{\frac{22.72}{178+22.72}} = 0.34$.

For Ex. 7.18, the phi-coefficient is:

$$\varphi = \frac{ad-bc}{\sqrt{[(a+b)(a+c)(b+d)(c+d)]}} = \frac{18\times 9 - 6\times 7}{\sqrt{[25\times 24\times 16\times 15]}} = \frac{120}{379.47} = 0.32.$$

7.10 More Applications of t-test, z-test and χ^2-test

7.10.1 Test of variance

Let $x_1, x_2, \ldots, x_n$ be n sample observations selected from $N(\mu, \sigma^2)$. We need to test

$$H_0 : \sigma^2 = \sigma_0^2, \text{ against } H_0 : \sigma^2 \neq \sigma_0^2.$$

The test statistic is:

$$\chi^2 = \frac{(n-1)s^2}{\sigma_0^2}, \text{ where } s^2 = \frac{1}{n-1}\left[\sum x^2 - \frac{(\sum x)^2}{n}\right].$$

This χ^2 has $(n-1)$ d.f.

EXAMPLE 7.22 The following are the data on fasting blood sugar (x mmol/l) level of some persons:

x : 8.6, 6.2, 5.5, 7.8, 10.2, 5.8, 5.0, 7.5, 6.3, 6.0, 7.5

Do you think that the variance of fasting blood sugar levels is 3.5?

Solution. Let $x \sim N(\mu, \sigma^2)$. We need to test

$$H_0 : \sigma^2 = \sigma_0^2 = 3.5, \text{ against } H_A : \sigma^2 \neq \sigma_0^2.$$

Here $n = 11$,

$$s^2 = \frac{1}{n-1}\left[\sum x^2 - \frac{(\sum x)^2}{n}\right] = \frac{1}{11-1}\left[554.36 - \frac{(76.4)^2}{11}\right] = 2.37.$$

The test statistic is:

$$\chi^2 = \frac{(n-1)s^2}{\sigma_0^2} = \frac{(11-1)2.37}{3.5} = 6.77.$$

Comment. As calculated $\chi^2 < \chi^2_{0.05,10} = 18.307$, H_0 is accepted. The variance of fasting blood sugar levels can be considered as 3.5.

7.10.2 Test of equality of two variances

Let $x_{11}, x_{12}, \ldots, x_{1n_1}$, be n_1 sample observations selected from $N(\mu_1, \sigma_1^2)$. Consider another sample observations $x_{21}, x_{22}, \ldots, x_{2n_2}$ which are selected from $N(\mu_2, \sigma_2^2)$. We need to test

$$H_0 : \sigma_1^2 = \sigma_2^2, \text{ against } H_A : \sigma_1^2 \neq \sigma_2^2.$$

The test statistic is: $F = \dfrac{s_1^2}{s_2^2}$, if $s_1^2 > s_2^2$.

This F has $(n_1 - 1)$ and $(n_2 - 1)$ d.f. Here

$$s_1^2 = \frac{1}{n_1 - 1}\left[\sum x_{ij}^2 - \frac{(\sum x_{ij})^2}{n_1}\right]$$

$$s_2^2 = \frac{1}{n_2 - 1}\left[\sum x_{2j}^2 - \frac{(\sum x_{ij})^2}{n_2}\right].$$

EXAMPLE 7.23 The birth-weight of two groups of newborn babies are as follows:

Group-1 (x_{1j}) : 3.2, 3.0, 2.8, 2.9, 3.0, 3.5

Group-2 (x_{2j}) : 4.0, 2.5, 4.1, 3.6, 3.0, 3.0, 2.8, 2.7

Are the variances of birth-weights of two groups of babies similar?

Solution. Let $x_{1j} \sim N(\mu_1, \sigma_1^2)$ and $x_{2j} \sim N(\mu_2, \sigma_2^2)$. We need to test

$$H_0 : \sigma_1^2 = \sigma_2^2, \text{ against } H_A : \sigma_1^2 \neq \sigma_2^2.$$

We have $n_1 = 6, n_2 = 8$.

$$s_1^2 = \frac{1}{n_1 - 1}\left[\sum x_{1j}^2 - \frac{(\sum x_{1j})^2}{n_1}\right] = \frac{1}{6-1}\left[56.74 - \frac{(18.4)^2}{6}\right] = 0.063.$$

$$s_2^2 = \frac{1}{n_2 - 1}\left[\sum x_{2j}^2 - \frac{(\sum x_{2j})^2}{n_2}\right] = \frac{1}{8-1}\left[85.15 - \frac{(25.7)^2}{8}\right] = 0.3698.$$

The test statistic is:

$$F = \frac{s_2^2}{s_1^2}\ (\text{as } s_2^2 > s_1^2) = \frac{0.3698}{0.063} = 5.87.$$

This F has 7 and 5 d.f.

Comment. As calculated F is greater than tabulated value of F with 7 and 5 d.f. (4.88), H_0 is rejected. The variances of the birth-weights are not similar.

7.10.3 Test of equality of several variances

Let $s_1^2, s_2^2, \ldots, s_k^2$ be the sample variances of k independent samples selected from k populations, where the population variance of i-th population $(i = 1, 2, \ldots, k)$ is σ_i^2. The hypothesis to be tested is

$$H_0 : \sigma_1^2 = \sigma_2^2 = \cdots = \sigma_k^2, \text{ against}$$

H_A : At least one equality sign does not hold good.

The test statistic is:

$$\chi^2 = \frac{(n-k)\ln s^2 - \sum(n_i - 1)\ln s_i^2}{1 + \frac{1}{3(k-1)}\left[\sum \frac{1}{n_i - 1} - \frac{1}{n-k}\right]}.$$

This χ^2 has $(k-1)$ d.f. Here

$$s^2 = \frac{1}{n-k}\sum(n_i - 1)s_i^2.$$

EXAMPLE 7.24 The following are the s. triglyceride (mg/dl) of 4 groups of patients selected from 4 hospitals:

Group	*s. triglyceride, x_{ij}*										*Total*
1	120	135	70	75	100	122	125	128			875
2	150	155	170	60	92	120	120	145	150		1162
3	200	195	85	70	95	100	100	125	75		1045
4	50	75	88	92	110	115	88	117	150	161	1046

Are the variances of s. triglyceride of groups of patients similar?

Solution. Let $x_{ij} \sim N(\mu_i, \sigma_i^2)$, $i = 1, 2, 3, 4, \ldots, k = 4$. We need to test

$$H_0 : \sigma_1^2 = \sigma_2^2 = \sigma_3^2 = \sigma_4^2, \text{ against}$$

H_A : At least one equality sign does not hold good.

We have $n_1 = 8, n_2 = 9, n_3 = 9, n_4 = 10;\ n = \sum n_i = 36.$

$$s_1^2 = \frac{1}{n_1 - 1}\left[\sum x_{1j}^2 - \frac{(\sum x_{1j})^2}{n_1}\right] = \frac{1}{8-1}\left[10043 - \frac{(875)^2}{8}\right] = 619.982.$$

$$s_2^2 = \frac{1}{n_2 - 1}\left[\sum x_{2j}^2 - \frac{(\sum x_{2j})^2}{n_2}\right] = \frac{1}{9-1}\left[159814 - \frac{(1162)^2}{9}\right] = 1223.361.$$

$$s_3^2 = \frac{1}{n_3 - 1}\left[\sum x_{3j}^2 - \frac{(\sum x_{3j})^2}{n_3}\right] = \frac{1}{9-1}\left[140425 - \frac{(1045)^2}{9}\right] = 2386.111.$$

$$s_4^2 = \frac{1}{n_4 - 1}\left[\sum x_{4j}^2 - \frac{(\sum x_{4j})^2}{n_4}\right] = \frac{1}{10-1}\left[119512 - \frac{(1046)^2}{10}\right] = 1122.267.$$

$$\begin{aligned} s^2 &= \frac{1}{n-k}\left[\sum (n_i - 1)s_i^2\right] \\ &= \frac{1}{36-4}[(8-1)619.982 + (9-1)1223.361 + (9-1)2386.111 + (10-1)1122.267] \\ &= 1353.627. \end{aligned}$$

$$\begin{aligned} &(n-k)\ln s^2 - \sum (n_i - 1)\ln s_i^2 \\ &\quad = (36-4)\ln 1353.627 - [(8-1)\ln 619.982 + (9-1)\ln 1223.361 \\ &\qquad\qquad + (9-1)\ln 2386.111 + (10-1)\ln 1122.267] \\ &\quad = 230.737 - 227.310 = 3.427. \end{aligned}$$

$$1 + \frac{1}{3(k-1)}\left[\sum \frac{1}{n_i - 1} - \frac{1}{n-k}\right] = 1 + \frac{1}{3(4-1)}\left[\frac{1}{7} + \frac{1}{8} + \frac{1}{8} + \frac{1}{9} - \frac{1}{36-4}\right] = 1.0525.$$

$$\chi^2 = \frac{(n-k)\ln s^2 - \sum (n_i - 1)\ln s_i^2}{1 + \frac{1}{3(k-1)}\left[\sum \frac{1}{n_i - 1} - \frac{1}{n-k}\right]} = \frac{3.427}{1.0525} = 3.26.$$

Comment. As calculated $\chi^2 < \chi^2_{0.05,3} = 7.81$, H_0 is accepted. The variances of s. triglyceride of 4 groups of patients are similar.

7.10.4 Test of correlation coefficient and regression coefficient

Let $(x_1, y_1), (x_2, y_2), \ldots, (x_n, y_n)$ be n pairs of values observed from n sample points, where y is the dependent variable and x is the independent variable. The variable y and x are linearly related by a model

$$y = A + Bx + e,$$

where $B = B_{yx}$ = regression coefficient of y on x,
A = the value of y in absence of x,
e = random component.

The estimate of A and B are respectively $a = \bar{y} - b\bar{x}$, $b = \dfrac{\text{S.P.}(xy)}{\text{S.S.}(x)}$.

Here $\bar{y} = \frac{1}{n}\sum y$, $\bar{x} = \frac{1}{n}\sum x$

$$\text{S.P.}(xy) = \sum xy - \frac{\sum x \sum y}{n}, \quad \text{S.S.}(x) = \sum x^2 - \frac{(\sum x)^2}{n}.$$

The fitted regression line is $\hat{y} = a + bx$.

The correlation coefficient of x and y is:

$$r = \frac{\text{S.P.}(xy)}{\sqrt{\text{S.S.}(x)\ \text{S.S.}(y)}}, \text{ where S.S.}(y) = \sum y^2 - \frac{(\sum y)^2}{n}.$$

Here we need to test the hypotheses

$$H_0 : B = 0, \text{ against } H_A : B \neq 0.$$
$$H_0 : P = 0, \text{ against } H_A : P \neq 0,$$

where P = population correlation coefficient.

The test statistic for the first hypothesis is:

$$t = \frac{b}{\sqrt{\frac{s^2}{\text{S.S.}(x)}}}, \text{ where } s^2 = \frac{1}{n-2}\left[\text{S.S.}(y) - b\,\text{S.P.}(xy)\right].$$

This t has $(n-2)$ d.f.

The test statistic for the second hypothesis is: $t = \frac{r\sqrt{n-2}}{\sqrt{1-r^2}}$.

This t has $(n-2)$ d.f.

EXAMPLE 7.25 The following are the level of education of currently married mothers (x in completed years of schooling) and their number of ever born children (y):

x	5	10	0	0	12	10	14	0	5	5
y	7	4	5	4	3	2	2	4	3	4

(i) Do you think that the impact of level of education on number of ever born children significant?

(ii) Are x and y significantly correlated?

Solution.

(i) We need to test the hypothesis $H_0 : B = 0$, against $H_A : B \neq 0$.

The test statistic is $t = \frac{b}{\sqrt{\frac{s^2}{\text{S.S.}(x)}}}$, $b = \frac{\text{S.P.}(xy)}{\text{S.S.}(x)} = \frac{-37.8}{242.9} = -0.156$

$$\text{S.P.}(xy) = \sum xy - \frac{\sum x \sum y}{n}, \quad n = 10$$
$$= 194 - \frac{61 \times 38}{10} = -37.9.$$
$$\text{S.S.}(x) = \sum x^2 - \frac{(\sum x)^2}{n} = 615 - \frac{(61)^2}{10} = 242.9.$$
$$\text{S.S.}(y) = \sum y^2 - \frac{(\sum y)^2}{n} = 164 - \frac{(38)^2}{10} = 19.6.$$
$$s^2 = \frac{1}{n-2}[\text{S.S.}(y) - b\,\text{S.P.}(n-1)] = \frac{1}{10-2}[19.6 - 0.156 \times 37.8] = 1.7129.$$
$$t = \frac{-0.156}{\sqrt{\frac{1.7129}{242.9}}} = -1.86.$$

Comment. As $|t| < t_{0.05,8} = 2.306$, H_0 is accepted. The impact of level of educaton of mother in number of ever born children is not significant.

(ii) $H_0 : P = 0$, $H_A : P \neq 0$.

The test statistic is $t = \dfrac{r\sqrt{n-2}}{\sqrt{1-r^2}}$.

$$r = \frac{\text{S.P.}(xy)}{\sqrt{\text{S.S.}(x)\ \text{S.S.}(y)}} = \frac{-37.8}{\sqrt{242.9 \times 19.6}} = -0.55.$$

$$t = \frac{-0.55\sqrt{10-2}}{\sqrt{1-(-0.55)^2}} = \frac{-1.5556}{0.8352} = -1.86.$$

Comment. As $|t| < t_{0.05,8} = 2.306$, H_0 is accepted. Level of education of mother and number of ever born children are not significantly correlated.

7.10.5 Test of equality of two regression coefficients and two correlation coefficients

Let $(x_{11}, y_{11}), (x_{12}, y_{12}), \ldots, (x_{1n_1}, y_{1n_1})$ be n_1 pairs of values of two variables X and Y observed from n_1 sample points. It is assumed that the dependent variable Y is linearly related to X by the model

$$y_1 = B_{10} + B_1 x_1 + e_1.$$

Also, assume that P_1 is the population correlation coefficient of x_1 and y_1.

Here the estimate of B_1 is $b_1 = \dfrac{\text{S.P.}(x_1 y_1)}{\text{S.S.}(x_1)}$.

$$\text{S.P.}(x_1 y_1) = \sum x_1 y_1 - \frac{\sum x_1 \sum y_1}{n_1}$$

$$\text{S.S.}(x_1) = \sum x_1^2 - \frac{(\sum x_1)^2}{n_1}.$$

Consider another n_2 pairs of observations of the same variables X and Y observed from an independent sample. The variables X and Y follow the linear model

$$y_2 = B_{20} + B_2 x_2 + e,$$

where the pairs of observations are $(x_{21}, y_{21}), (x_{22}, y_{22}), \ldots, (x_{2n_2}, y_{2n_2})$.

The estimate of B_2 is $b_2 = \dfrac{\text{S.P.}(x_2 y_2)}{\text{S.S.}(x_2)}$.

$$\text{S.P.}(x_2 y_2) = \sum x_2 y_2 - \frac{\sum x_2 \sum y_2}{n_2}$$

$$\text{S.S.}(x_2) = \sum x_2^2 - \frac{(\sum x_2)^2}{n_2}.$$

Here we need to test the hypothesis $H_0 : B_1 = B_2$, against $H_A : B_1 \neq B_2$.

The test statistic is: $t = \dfrac{b_1 - b_2}{\sqrt{\left\{s^2\left[\frac{1}{\text{S.S.}(x_1)} + \frac{1}{\text{S.S.}(x_2)}\right]\right\}}}$,

where $s^2 = \dfrac{1}{2}\left[s_1^2 + s_2^2\right]$

$$s_1^2 = \frac{1}{n_1 - 2}\left[\text{S.S.}(y_1) - b_1 \text{S.P.}(x_1 y_1)\right]$$

$$s_2^2 = \frac{1}{n_2 - 2}\left[\text{S.S.}(y_2) - b_2 \text{S.P.}(x_2 y_2)\right]$$

This t has $(n_1 - 2) + (n_2 - 2)$ d.f.

We also need to test the hypothesis

$$H_0 : P_1 = P_2, \text{ against } H_A : P_1 \neq P_2,$$

where P_2 is the population correlation coefficient of the population from which second sample is selected. The test statistic for this hypothesis is d, where

$$d = \frac{z_1 - z_2}{\sqrt{\left[\frac{1}{n_1-3} + \frac{1}{n_2-3}\right]}}.$$

$$z_1 = \frac{1}{2}\ln\frac{1+r_1}{1-r_1}, \quad z_2 = \frac{1}{2}\ln\frac{1+r_2}{1-r_2}.$$

$$r_1 = \frac{\text{S.P.}(x_1y_1)}{\sqrt{\text{S.S.}(x_1)\ \text{S.S.}(y_1)}}, \quad r_2 = \frac{\text{S.P.}(x_2y_2)}{\sqrt{\text{S.S.}(x_2)\ \text{S.S.}(y_2)}}.$$

This d is normally distributed with mean zero and variance $1[N(0,1)]$. If $|d| \leq 1.96$, H_0 is rejected or if P-value of d is less than 0.05, H_0 is rejected.

EXAMPLE 7.26 The following are the ages (x_1 years) and systolic blood pressures (y_1 mmHg) of a group of males:

x_1	25	35	40	42	38	50	55	58	65	60
y_1	75	75	80	80	85	85	90	75	80	80

The observations of these variables for females are as follows:

x_2	40	45	44	48	50	55	60	58	62	70
y_2	70	78	85	90	90	85	80	85	85	85

Are the impacts of age on blood pressure similar for both male and female? Do you think that the correlation of age and blood pressure is same for both male and female?

Solution. We have $n_1 = 10, n_2 = 10$.

$$\text{S.P.}(x_1y_1) = \sum x_1y_1 - \frac{\sum x_1 \sum y_1}{n_1} = 37840 - \frac{468 \times 805}{10} = 166.$$

$$\text{S.S.}(x_1) = \sum x_1^2 - \frac{(\sum x_1)^2}{n_1} = 23372 - \frac{(468)^2}{10} = 1469.6.$$

$$\text{S.S.}(y_1) = \sum y_1^2 - \frac{(\sum y_1)^2}{n_1} = 65025 - \frac{(805)^2}{10} = 222.5.$$

$$b_1 = \frac{\text{S.P.}(x_1y_1)}{\text{S.S.}(x_1)} = \frac{166}{1469.6} = 0.113.$$

$$r_1 = \frac{\text{S.P.}(x_1y_1)}{\sqrt{[\text{S.S.}(x_1)\ \text{S.S.}(y_1)]}} = \frac{166}{\sqrt{1469.6 \times 222.5}} = 0.29.$$

$$\text{S.P.}(x_2y_2) = \sum x_2y_2 - \frac{\sum x_2 \sum y_2}{n_2} = 44495 - \frac{532 \times 833}{10} = 179.4.$$

$$\text{S.S.}(x_2) = \sum x_2^2 - \frac{(\sum x_2)^2}{n_2} = 29098 - \frac{(532)^2}{10} = 795.6.$$

$$\text{S.S.}(y_2) = \sum y_2^2 - \frac{(\sum y_2)^2}{n_2} = 69709 - \frac{(833)^2}{10} = 320.1.$$

$$b_2 = \frac{\text{S.P.}(x_2y_2)}{\text{S.S.}(x_2)} = \frac{179.4}{795.6} = 0.225.$$

$$r_2 = \frac{\text{S.P.}(x_2y_2)}{\sqrt{[\text{S.S.}(x_2)\ \text{S.S.}(y_2)]}} = \frac{179.4}{\sqrt{795.6 \times 320.1}} = 0.36.$$

$$s_1^2 = \frac{1}{n_1 - 2}[\text{S.S.}(y_1) - b_1\text{S.P.}(x_1y_1)] = \frac{1}{10-2}[222.5 - 0.113 \times 166] = 25.47.$$

$$s_2^2 = \frac{1}{n_2 - 2}[\text{S.S.}(y_2) - b_2\text{S.P.}(x_2y_2)] = \frac{1}{10-2}[320.1 - 0.225 \times 179.4] = 34.97.$$

$$s^2 = \frac{1}{2}\left[s_1^2 + s_2^2\right] = \frac{1}{2}[25.47 + 34.97] = 29.72.$$

We need to test $H_0 : B_1 = B_2$, against $H_A : B_1 \neq B_2$.

The test statistic is:

$$t = \frac{b_1 - b_2}{\sqrt{\left[s^2\left\{\frac{1}{\text{S.S.}(x_1)} + \frac{1}{\text{S.S.}(x_2)}\right\}\right]}} = \frac{0.113 - 0.225}{\sqrt{\left[29.72\left(\frac{1}{1469.5} + \frac{1}{795.6}\right)\right]}} = -0.47.$$

Comment. As $|t| < t_{0.05,16} = 2.12$, H_0 is accepted. The impacts of age on blood pressure for both male and female are similar.

We need to test $H_0 : P_1 = P_2$, against $H_A : P_1 \neq P_2$.

The test statistic is: $d = \dfrac{z_1 - z_2}{\sqrt{\left[\frac{1}{n_1-3} + \frac{1}{n_2-3}\right]}} = \dfrac{0.2986 - 0.3769}{\sqrt{\left[\frac{1}{10-3} + \frac{1}{10-3}\right]}} = -0.15;$

where

$$z_1 = \frac{1}{2}\ln\frac{1 + r_1}{1 - r_1} = \frac{1}{2}\ln\frac{1 + 0.29}{1 - 0.29} = 0.2986.$$

$$z_2 = \frac{1}{2}\ln\frac{1 + r_2}{1 - r_2} = \frac{1}{2}\ln\frac{1 + 0.36}{1 - 0.36} = 0.3769.$$

Comment. As $|d| < 1.96$, H_0 is accepted.

Or $P(z < -0.15) + P(z > 0.15) = 0.4404 + 0.4404 = 0.8808 > 0.05$, H_0 is accepted. The correlation of age and blood pressure is similar for both male and female.

7.10.6 Equality of several correlation coefficients

Let us consider that samples of sizes $n_1, n_2, \ldots, n_k$ are independently selected from k bi-variate populations, where the population correlation coefficients are $P_1, P_2, \ldots, P_k$. We need to test the hypothesis:

$$H_0 : P_1 = P_2 \cdots = P_k$$

against H_A : At least one equality sign does not hold good.

Let r_i be the sample correlation coefficient of i-th sample ($i = 1, 2, \ldots, k$).

Then the test statistic is: $\chi^2 = \displaystyle\sum_{i=1}^{k}(n_i - 3)z_i^2 - \frac{[\sum(n_i - 3)z_i]^2}{\sum(n_i - 3)}$,

where $z_i = \dfrac{1}{2}\ln\dfrac{1 + r_i}{1 - r_i}$, $i = 1, 2, \ldots, k$.

This χ^2 has $(k - 1)$ d.f.

EXAMPLE 7.27 Four independent samples of sizes $n_1 = 20$, $n_2 = 25$, $n_3 = 30$ and $n_4 = 35$ are selected from patients of 4 hospitals. The amount of s. cholesterol (mg/dl) and amount of s. triglyceride (mg/dl) of selected patients are recorded. These variables provide correlation coefficients for 4 samples as $r_1 = 0.52$, $r_2 = 0.67$, $r_3 = 0.45$ and $r_4 = 0.75$. Are the correlations similar for all patients?

Solution. We need to test the hypothesis:

$$H_0 : P_1 = P_2 = P_3 = P_4 \text{ against } H_A : P_1 \neq P_2 \neq P_3 \neq P_4,$$

where $P_i (i = 1, 2, 3, 4)$ is the population correlation coefficient of s. triglyceride and s. cholesterol of i-th group of patients.

$$z_1 = \frac{1}{2}\ln\frac{1+r_1}{1-r_1} = \frac{1}{2}\ln\frac{1+0.52}{1-0.52} = 0.576.$$

$$z_2 = \frac{1}{2}\ln\frac{1+r_2}{1-r_2} = \frac{1}{2}\ln\frac{1+0.67}{1-0.67} = 0.8107.$$

$$z_3 = \frac{1}{2}\ln\frac{1+r_3}{1-r_3} = \frac{1}{2}\ln\frac{1+0.45}{1-0.45} = 0.4847.$$

$$z_4 = \frac{1}{2}\ln\frac{1+r_4}{1-r_4} = \frac{1}{2}\ln\frac{1+0.75}{1-0.75} = 0.9729.$$

$$\sum(n_i-3)z_i^2 = (20-3)(0.576)^2 + (25-3)(0.8107)^2 + (30-3)(0.4847)^2 + (35-3)(0.9729)^2$$
$$= 56.73.$$

$$\sum(n_i-3)z_i = (20-3)(0.576) + (25-3)(0.8107) + (30-3)(0.4847) + (35-3)(0.9729)$$
$$= 71.8471.$$

$$\sum(n_i-3) = (20-3) + (25-3) + (30-3) + (35-3)$$
$$= 98.$$

The test statistic is: $\chi^2 = \sum(n_i-3)z_i^2 - \dfrac{[\sum(n_i-3)z_i]^2}{\sum(n_i-3)} = 56.73 - \dfrac{(71.8471)^2}{98} = 4.06.$

Comment. As calculated $\chi^2 < \chi^2_{0.05,3} = 7.81$, H_0 is accepted. The correlation between s. cholesterol and s. triglyceride for all groups of patients is similar.

7.11 Non-parametric Test

Let $x_1, x_2, \ldots, x_n$ be n sample observations selected from normal population with mean μ and variance σ^2 $[x \sim N(\mu, \sigma^2]$.

Here $\bar{x} = \dfrac{1}{n}\sum x$, $s^2 = \dfrac{1}{n-1}\sum(x-\bar{x})^2$.

We need to test $H_0 : \mu = \mu_0$ (a specific value) or $H_0 : \sigma^2 = \sigma_0^2$ (a specific value). These two hypothesis are related to the parameters of the distribution and hence these are called parametric hypothesis. The tests related to these hypothesis are called parametric tests.

Again, if it is assumed that $x_1, x_2, \ldots, x_n$ are sample observations from a population, where population or population parameter(s) is/are not mentioned. One may be interested to test the hypothesis

H_0 : The samples are randomly selected.

H_A : The samples are not randomly selected.

or H_0 : The sample observations follow normal distribution.

H_A : The sample observations do not follow normal distribution.

These two hypothesis are not related to the parameter of the population. These are called non-parametric hypotheses and the tests related to these are known as non-parametric test.

Different non-parametric tests are:

(a) Test of randomness
(b) Binomial test
(c) One-sample and Two-sample sign test
(d) Friedman test
(e) Page test
(f) Chochran test
(g) χ^2-test
(h) Mantel-Haenszel test
(i) Likelihood ratio test
(j) Kruskal-Wallis test
(k) Mann-Whitney U-test.

7.11.1 Test of randomness

It is done by using runs of the observations, where a run is a sequence of similar observations. For example, patient is coming to the doctor's chamber in the following sequence, when M = Male and F = Female

M M F M F M M M F F

Here run $r = 6$, where first run is a run of two males, second run is a run of one female. If males and females are coming at random, the number of runs will be more. The number of runs will be less, if the patients are not coming at random, rather they are coming following a system. Run can also be measured with respect to median, where an observation below or equal to median will form a run and an observation above median will form another run. For example,

Observations (x) : 12, 18, 13, 10, 16, 18, 20, 15, 17 $(n = 9)$

Array (x) : 10, 12, 13, 15, 16, 17, 18, 18, 20

$$\text{Median} = \text{value of } \frac{n+1}{2}\text{th observation in the array}$$
$$= \text{value of } \frac{9+1}{2}\text{th} \cdots\cdots\cdots\cdots$$
$$= \text{value of 5th} \cdots\cdots\cdots\cdots$$
$$= 16$$

$x - M: - + - - + + - + \; (r = 6)$

When $n \geq 30$, the test is done by z-test, where

$$z = \frac{r - \text{Mean}\ (r)}{\sqrt{v(r)}} \sim N(0,1), \text{Mean}\ (r) = 1 + \frac{2n_1n_2}{n_1 + n_2}$$
$$v(r) = \frac{2n_1n_2(2n_1n_2 - n_1 - n_2)}{(n_1 + n_2)^2(n_1 + n_2 - 1)}.$$

n_1 = number of observations above median.

n_2 = number of observations below median.

In the above example, $n_1 = 4$, $n_2 = 4$ [No. of observations equal to median is not counted].

EXAMPLE 7.28 The following are the number of family planning workers in 30 consecutive unions of a district.

Number of family planning workers (x) : 5, 8, 10, 7, 5, 6, 5, 8, 9, 10, 12, 10, 8, 7, 12, 10, 9, 8, 7, 10, 11, 9, 5, 7, 11, 10, 9, 6, 8, 9

Are the family planning workers in unions randomly appointed?

Solution. We need to test

H_0 = The workers are appointed in the unions randomly.

H_A = The workers are not appointed randomly in the unions.

Array (x) : 5, 5, 5, 5, 6, 6, 7, 7, 7, 7, 8, 8, 8, 8, 8, 9, 9, 9, 9, 9,
10, 10, 10, 10, 10, 10, 11, 11, 12, 12 ($n = 30$)

$$\text{Median Me} = \text{value of } \frac{1}{2}\left[\frac{n}{2}\text{th} + \left(\frac{n}{2}+1\right)\text{th}\right] \text{ observations in the array, as } n \text{ is even}$$

$$= \frac{8+9}{2} = 8.5.$$

The observations below median and above median are arranged by − and + signs, respectively, as follows:

$$- - + - - - - - - + + + + - - + + + - - + + + - - - + + + - - +.$$

We have $r = 12$, $n_1 = 15$, $n_2 = 15$.

$$\text{Mean }(r) = 1 + \frac{2n_1n_2}{n_1+n_2} = 1 + \frac{2\times 15\times 15}{15+15} = 16.$$

$$V(r) = \frac{2n_1n_2(2n_1n_2 - n_1 - n_2)}{(n_1+n_2)^2(n_1+n_2-1)} = \frac{2\times 15\times 15(2\times 15\times 15 - 15 - 15)}{(15+15)^2(15+15-1)}$$

$$= 7.33.$$

The test statistic is:

$$z = \frac{r - \text{Mean }(r)}{\sqrt{V(r)}} = \frac{12-16}{\sqrt{7.33}} = -1.48.$$

Comment. As $|z| < 1.96$, H_0 is accepted. The workers are appointed randomly. Here,

$$P\text{-value} = P(z < -1.48) + P(z > -1.48) = 0.0694 + 0.0694 = 0.1388.$$

7.11.2 Test of randomness by χ^2-test

In drawing simple random sample, we use random numbers from the random number table. The random numbers are arrangement of digits 0 to 9. If selected numbers are random all digits will be equal in the selected numbers. For example, if $n = 100$ observations are selected from $N = 1000$ units, then every selected number will be of 4 digits and we have $4 \times 1000 = 4000$ digits. There are 10 digits 0 to 9. If selected numbers are random, the probability of any digits to be selected is 0.1. Therefore, under the null hypothesis the expected number of any kind of digits will be $E_i = 0.1 \times n$, where $i = 1, 2, \ldots, 9$. Let O_i be the observed number of i-th digit. Therefore, under H_0 the deviation between O_i and E_i can be observed by χ^2-test, where

$$\chi^2 = \sum_{i=1}^{k} \frac{O_i^2}{E_i} - n,$$

where $k = 10$. So, the d.f. of χ^2 will be $(10 - 1) = 9$, where

H_0 = the sample observations are randomly selected.

H_A = the sample observations are not randomly selected.

If calculated $\chi^2 \geq \chi^2_{0.05,9}$, H_0 is rejected.

EXAMPLE 7.29 A sample of 25 observations is selected from a population of size $N = 200$. For the selected sample, the random numbers are the following:

001	072	115	142	168
008	174	146	140	170
012	185	162	176	070
045	107	179	134	018
186	193	184	056	046

Is the selected sample random?

Solution. Here $n = 3 \times 25 = 75$ digits. We need to test

H_0 : the sample is random.

H_A : the sample is not random.

Under H_0 the expected number of i-th digits is $E_i = 0.10 \times 75$ or $E_i = 7.5$ $(i = 0, 1, 2, 3, 4, \ldots, 9)$. The frequency distribution of the digits in selected numbers is as follow:

Digit	0	1	2	3	4	5	6	7	8	9	Total
Observed frequency, O_i	15	20	4	2	8	4	7	7	6	2	75

The test statistic is:

$$\chi^2 = \sum^k \frac{O_i^2}{E_i} - n$$

$$= \frac{1}{7.5}\left[15^2 + 20^2 + 4^2 + 2^2 + 8^2 + 4^2 + 7^2 + 7^2 + 6^2 + 2^2\right] - 75$$

$$= 115.06 - 75 = 40.07.$$

Comment. As calculated $\chi^2 > \chi^2_{0.05,9} = 16.919$, H_0 is rejected. The selected sample is not random.

7.11.3 Binomial test

The experimental results of many experiments can be classified into two classes. For example, among n currently married couples x couples adopted family planning and $(n - x)$ couples did not adopt family planning; a diagnostic centre tested n samples of blood to investigate the presence of blood sugar, out of these n samples, x samples contain higher level of sugar and $(n - x)$ samples have normal sugar. In the above mentioned cases, the experimental results are classified into two; and among the sample results some results possess a character and some other do not presses the character. If p be the probability that x results possess the character, then the probability of $(n - x)$ results not possessing the character is $q = 1 - p$, where $p + q = 1$. In such a case, out of n results the probability of x results possessing a character is given by

$$P(x) = {}^nc_x p^x q^{n-x}; \ x = 0, 1, 2, \ldots, n.$$

This probability law is called binomial probability law and the test using this law is known as binomial test of classification. Let P be the proportion that a group of results possess a character. Then, the null hypothesis to be tested is:

$H_0 : P = P_0$ (a specific value)

against $H_0 : P \neq P_0$ (two-sided test)

or, $H_0 : P > P_0$ (one-sided test)

or, $H_0 : P < P_0$ (one-sided test).

If $P_0 = \frac{1}{2}$, then

$$P(x) = {}^nc_x \left(\frac{1}{2}\right)^x \left(\frac{1}{2}\right)^{n-x} = {}^nc_x \left(\frac{1}{2}\right)^n, \quad x = 0, 1, 2, \ldots, n.$$

To test the hypothesis the probability calculated is:

$$P(x) = \sum_{i=0}^{x} {}^nc_i \left(\frac{1}{2}\right)^n \quad \text{or, } P(x) = \sum_{i=0}^{x} {}^nc_i p_0^i q_0^{n-i}, \; i = 0, 1, \ldots, x.$$

For two-sided test, H_0 is rejected if $2P(x) \leq 0.05$ and for one-sided test H_0 is rejected if $P(x) \leq 0.05$.

EXAMPLE 7.30 The birth of a boy and a girl is equiprobable. One day 5 new babies born in a clinic and found that, out of 5 babies, 2 are girls. Is the assumption correct?

Solution. Let P be the proportion of birth of a girl. We need to test

$$H_0 : P = P_0 = \frac{1}{2}, \text{ against } H_A : P \neq P_0.$$

Here $x = 2$ (is the minimum number of observations of any class)

$$\therefore P(x) = \sum_{i=0}^{x=2} {}^nc_i p_0^i q_0^{n-i} = \sum_{i=0}^{2} {}^5c_i \left(\frac{1}{2}\right)^5, \quad p_0 = q_0 = \frac{1}{2}$$

$$= \left(\frac{1}{2}\right)^5 [{}^5c_0 + {}^5c_1 + {}^5c_2] = 0.5.$$

Comment. As $2p(x) = 2 \times 0.5 = 1 > 0.05$, H_0 is true. The assumption of equiprobable birth of boy and girl is true.

EXAMPLE 7.31 It is assumed that one per cent operated patients in an O.T. of a hospital are infected. Once 6 patients are operated and found that one is infected. Is the assumption correct?

Solution. Let P be the proportion of infected patients after operation. We need to test

$$H_0 : P = P_0 = 0.1, \text{ against } H_A : P > P_0.$$

Under H_0 the probability to be calculated is:

$$P(x) = \sum_{i=0}^{x} {}^nc_i p^i q^{n-i}, \text{ where under } H_0 : p = 0.1$$

$$= \sum_{i=0}^{1} {}^6c_i (0.1)^i (0.9)^{6-i}, \text{ where } x = \text{minimum of patients (infected)}$$

$$= (0.9)^6 + {}^6c_1 (0.1)(0.9)^{6-1} = 0.8857.$$

Comment. As H_A is one-sided and $P(x) = 0.8857 > 0.05$, H_0 is accepted.

Binomial test for classification of observations into two classes can also be done by z-test, if sample size is large (≥ 30). It can be shown by example, where $z = \dfrac{x - np}{\sqrt{npq}} \sim N(0, 1)$. However, a continuity correction is done by $x + 0.5$ or $x - 0.5$ depending on $x < np$ or, $x > np$, respectively.

EXAMPLE 7.32 Eighty per cent currently married couples are assumed to use contraceptive. Once, out of 100 couples, 60 are found to use contraceptive. Is the assumption correct?

Solution. Let p be the proportion of couples who use contraceptive. We need to test

$$H_0 : p = p_0 = 0.8, \text{ against } H_A : p \neq p_0.$$

Under H_0, mean of x (number of couples using contraceptive) is $nP = 100 \times 0.80 = 80$ and variance of x is $npq = 100 \times 0.8 \times 0.2 = 40$.

The test statistic is:

$$z = \frac{x - np}{\sqrt{npq}} \sim N(0,1)$$

$$= \frac{60 + 0.5 - 100 \times 0.80}{\sqrt{100 \times 0.8 \times 0.2}} = \frac{-19.5}{4} = -4.875 \text{ [as } x < np]$$

Comment. As $|z| > 1.96$, H_0 is rejected. The assumption is not correct.

7.11.4 One sample sign test

This test is also used to classify the experimental results into two classes, where one class of observations are less than or equal to median (M) and another class of observations are greater than median. As median divides the array of the observations into two equal parts, proportion of observations less than or equal to median is 0.5. We need to test

$$H_0 : p = p_0 = \frac{1}{2}, \text{ against } H_A : p \neq p_0.$$

As the experimental results are classified into two classes, viz., observations less than or equal to median (M) or observations above median, one can use binomial test, where the probability function is:

$$P(i) = {}^nc_i p^i q^{n-i} = {}^nc_i \left(\frac{1}{2}\right)^i \left(\frac{1}{2}\right)^{n-i} = {}^nc_i \left(\frac{1}{2}\right)^n$$

To perform the test the sign of $D = X - M$ is observed. If $X \leq M$, the sign of D is negative, otherwise it is positive. Let x be the minimum of positive or negative sign. Then the test statistic is

$$P(x) = \sum_{i=0}^{x} {}^nc_i p^i q^{n-i} = \sum_{i=0}^{x} {}^nc_i \left(\frac{1}{2}\right)^i.$$

If $2P(x) \leq 0.05$, H_0 is rejected against alternative hypothesis $p \neq p_0$.

EXAMPLE 7.33 The following are the systolic blood pressure level (x mmHg) of some persons:

$$x : 70, 75, 85, 82, 68, 70, 74, 82.$$

Do you think that median blood pressure level is 75?

Solution. Let M be the median of blood pressure. We need to test

$$H_0 : M = M_0 = 75$$

$$H_0 : P(x > M) = P(x \leq M) = \frac{1}{2}$$

$$H_A : M \neq 75$$

$$\therefore P(i) = {}^nc_i \left(\frac{1}{2}\right)^n, \quad i = 1, 2, \ldots, x, \text{ where } n = 8.$$

We need to calculate

$$P(x) = \sum_{i=1}^{x} P(i) = \sum_{i=0}^{x} {}^{n}c_i \left(\frac{1}{2}\right)^n,$$

where x is the minimum of sign of $x - M$. Here signs of $x - M$ are:

$$- \; - \; + \; + \; - \; - \; - \; +$$

Minimum sign of $x - M$ is 3

$$P(x) = \sum_{i=0}^{3} {}^{8}c_i \left(\frac{1}{2}\right)^8 = \left(\frac{1}{2}\right)^8 [{}^{8}c_0 + {}^{8}c_1 + {}^{8}c_2 + {}^{8}c_3] = 0.3633.$$

Comment. As H_A is two-sided $2P(x) = 2 \times 0.3633 = 0.7266 > 0.05$. H_0 is accepted. Median blood pressure can be assumed 75. Here also z-test can be applied, if $n \geq 30$. The z-statistic is

$$z = \frac{x \pm 0.5 - np}{\sqrt{npq}} \sim N(0, 1).$$

If $x < np$, 0.5 is added and, if $x > np$, then 0.5 is to be subtracted.

EXAMPLE 7.34 The following are the birth-weight (x kg) of some newborn babies:

x : 2.8, 3.4, 3.2, 3.0, 2.5, 2.6, 3.8, 4.0, 4.2, 2.7, 2.8, 2.9, 3.8, 2.8, 3.1, 3.2, 3.4, 4.0, 4.2, 3.5, 3.6, 3.0, 3.5, 2.5, 2.2, 2.0, 2.8, 3.5, 3.2, 3.1, 4.2, 4.0, 3.0, 2.5

Is the median birth-weight 3.0?

Solution. Let M be the median. We need to test

$$H_0 : M = M_0 = 3 \Rightarrow P(x \leq M) = P(x > M) = \frac{1}{2}. \text{ Here } n = 34$$

$$H_A : M \neq M_0$$

Signs of $x - M$ are

$$- + + - - - + + + - - - + - + + + + + + +$$
$$- + - - - - + + + + + + -$$

Here $x = 14$ minimum sign negative. As $np = 34 \times \frac{1}{2} = 17$, $x < np$, the test statistic is

$$z = \frac{x + 0.5 - np}{\sqrt{npq}} = \frac{14 + 0.5 - 34 \times \frac{1}{2}}{\sqrt{34 \times \frac{1}{2} \times \frac{1}{2}}} = -0.86.$$

Comment. As $|z| < 1.96$, H_0 is accepted. Median birth-weight is 3.0 kg. Here P-value of test is:

$$P\text{-value} = P(z < -0.86) + P(z > 0.86)$$
$$= 0.1949 + 0.1949 = 0.3898 > 0.05.$$

7.11.5 Two samples sign test

Let $x_1, x_2, \ldots, x_n$ be n sample observations observed from n sample points from a population having cumulative density function $F(x)$. Let us consider that some treatment is applied on the sample units and the results after the application of the treatment are recorded, where recorded observations are $y_1, y_2, \ldots, y_n$. Assume that these observations are selected from a population having cumulative density function $F(y)$. We need to test

$$H_0 : F(x) = F(y)$$
$$H_A : F(x) \neq F(y).$$

If the treatment applied has no effect on sample units, $P(x < y) = P(x > y) = \frac{1}{2}$.

In other words, if H_0 is true $D_i = x_i - y_i$ $(i = 1, 2, \ldots, n)$ will show half negative signs and half positive signs. However, if any $D_i = x_i - y_i = 0$, the observation is dropped from the analysis and sample size will be $n_1 < n$. In such a case, if x is the minimum of positive or negative sign, then the test statistic is:

$$P(x) = \sum_{i=0}^{x} {}^nc_i \left(\frac{1}{2}\right)^n.$$

If $2P(x) \leq 0.05$, H_0 is rejected at 5% level of significance.

EXAMPLE 7.35 The following are the systolic blood pressure level of a group of people before using a medicine (x mmHg) and after using a medicine (y mmHg):

x	75	85	90	92	95	80	88	100
y	78	80	86	90	85	82	80	88

Is medicine effective?

Solution. Let $F(x)$ be the cumulative probability density function of x and $F(y)$ be the cumulative probability density function of y. We need to test

$$H_0 : F(x) = F(y), \text{ against } H_A : F(x) \neq F(y).$$

Here $D_i = x_i - y_i = - + + + + - + +$; $n = 8$.

$x = 2$ (minimum number of sign $(-)$)

The test statistic is:

$$P(x) = \sum_{i=0}^{x} {}^nc_i \left(\frac{1}{2}\right)^n = \sum_{i=0}^{2} \left(\frac{1}{2}\right)^8 {}^8c_i$$
$$= \left(\frac{1}{2}\right)^8 \left[{}^8c_0 + {}^8c_1 + {}^8c_2\right]$$
$$= 0.1445.$$

Comment. As $2P(x) = 2 \times 0.1445 = 0.289 > 0.05$, H_0 is accepted. Treatment is not effective.

7.11.6 Friedman test

This test is used to test the homogeneity of several treatments, when treatments are applied on the same sample point repeatedly. After each application of treatment the observation of j-th treatment $(j = 1, 2, \ldots, k)$ of i-th sample unit $(i = 1, 2, \ldots, n)$ is recorded. Let x_{ij} be the observation of j-th treatment of i-th sample unit.

The observation can be arranged as follows:

Sample unit	*Observation of j-th treatment* (x_{ij})					
1	x_{11}	x_{12}	$\cdots$	x_{1j}	$\cdots$	x_{1k}
2	x_{21}	x_{22}	$\cdots$	x_{2j}	$\cdots$	x_{2k}
$\vdots$	$\vdots$	$\vdots$	$\cdots$	$\vdots$	$\cdots$	$\vdots$
n	x_{n1}	x_{n2}	$\cdots$	x_{nj}	$\cdots$	x_{nk}

As all n observations of treatment are coming from same sample points, these are related. Assume that x_{ij} observation is selected at random from a population having cumulative probability density function F_j ($j = 1, 2, \ldots, k$). Here we need to test the hypothesis:

$H_0 : F_1 = F_2 = \cdots = F_k.$

H_A : At least one equality sign does not hold good.

This hypothesis is the hypothesis of equality of k population, where population observations are related as observations are observed from the same sample point. For example, a group of diabetic patients are taking medicine in different doses in different times. Every time the blood sugar level is measured after taking medicine. Here different doses of medicines are treatments. We need to test the equality of treatments. As blood sugar levels of a patient are observed after each dose of medicine, the blood sugar levels are related.

The test is done by Friedman Rank test, where the test statistic is:

$$\chi^2 = \frac{12}{nk(k+1)} \sum R_j^2 - 3n(k+1).$$

This χ^2 has $(k-1)$ d.f.

Here $R_j = \sum_{i=1}^{n} R_{ij}$, where R_{ij} is the rank of j-th observation of i-th sample point. The rank $R_{ij} = 1$, if observation of j-th treatment of i-th sample point is the lowest, and $R_{ij} = k$ if j-th observation of i-th sample point is the highest. If some x_{ij} are equal, their average rank is calculated.

EXAMPLE 7.36 The following are the F.B.S. (mmol/l) of some diabetic patients after using different doses of medicine.

Patients	*F.B.S. after using doses of medicine*			
	D_1	D_2	D_3	D_4
1	8.6	10.2	7.5	8.0
2	10.6	11.4	9.3	9.0
3	9.3	10.5	10.5	10.0
4	7.5	8.5	8.0	7.5
5	6.0	8.0	7.1	6.8

Are the doses of medicine similar?

Solution. Let F_j ($j = 1, 2, 3, 4$) be the cumulative probability density function of j-th population. We need to test

$H_0 : F_1 = F_2 = F_3 = F_4.$

H_A : At least one of the equality sign does not hold good.

Here $n = 5, k = 4$. The rank of the sample observations of $k = 4$ treatments are shown below:

Rank of x_{ij}

Patients	*Doses of medicine*				*Total*
	D_1	D_2	D_3	D_4	
1	3	4	1	2	10
2	3	4	2	1	10
3	1	3.5	3.5	2	10
4	1.5	4	3	1.5	10
5	1	4	3	2	10
Total, R_j	9.5	19.5	12.5	8.5	50

$$\chi^2 = \frac{12}{nk(k+1)}\sum R_j^2 - 3n(k+1)$$

$$= \frac{12}{5 \times 4(4+1)}\left[(9.5)^2 + (19.5)^2 + (12.5)^2 + (8.5)^2\right] - 3 \times 5(4+1) = 8.88.$$

Comment. As calculated $\chi^2 > \chi^2_{0.05,3} = 7.815$, H_0 is rejected. The doses of medicines are not similar.

7.11.7 Cochran test

This test is also done to test the homogeneity of several treatments. The observations of i-th treatment are the sample observations from a population having cumulative probability density function F_i $(i = 1, 2, \ldots, k)$.

The hypothesis to be tested is

$$H_0 : F_1 = F_2 = \cdots = F_k$$

H_A : At least one equality sign does not hold good.

The test is different from Friedman test in respect of collection of sample observations. In Friedman test different treatments are applied on the sample observations and after every application of treatment the observations are recorded. So, the sample observations are related.

In Cochran test, observations of different treatments are assumed independent. Each treatment is applied independently on n_j $(j = 1, 2, \ldots, k)$ sample units. The observations recorded after the application of treatment as 0 or 1. 0 is recorded, if the response of the treatment is not positive and 1 is recorded, if the response of the treatment is positive. For example, a doctor recommended folic acid to a patient to increase the level of haemoglobin. If amount of haemoglobin is increased after use of folic acid, the observation is recorded as 1, otherwise it is recorded as zero.

The test statistic for the hypothesis is $Q = \dfrac{(k-1)\left[k\sum C_j^2 - (\sum C_j)^2\right]}{k\sum R_i - \sum R_i^2}$.

This Q is distributed as chi-square with $(k-1)$ d.f.

Here C_j = total of observations of j-th treatment,

R_i = total of observations of i-th group of units on which k treatments are applied.

A group of sample units on which all treatments are applied is called a block. So, R_i is the i-th block total.

EXAMPLE 7.37 A group of patients of hypertension visited a doctor for treatment. The doctor suggested different medicines for different patients. The amount of systolic blood pressure levels (in mmHg) of patients before and after use of medicine are recorded and given below:

Patients group	*Blood pressure before use of medicine*	*Blood pressure after use of medicines*			
		M_1	M_2	M_3	M_4
1	90	85	88	95	90
2	95	90	85	90	90
3	88	75	80	85	85
4	100	90	95	105	90
5	87	80	80	90	85

Are the medicines similar?

Solution. We need to test H_0 : All medicines are similar.
H_A : Medicines are not similar.

The impacts of medicines are observed 1, if a medicine is effective to reduce blood pressure level. It is observed zero, if due to the use of medicine the blood pressure level is not reduced.

The results are shown below:

Patients group	*Impact of medicine on blood pressure*				*Total R_i*
	M_1	M_2	M_3	M_4	
1	1	1	0	0	2
2	1	1	1	1	4
3	1	1	1	1	4
4	1	1	0	1	3
5	1	1	0	1	3
Total C_j	5	5	2	4	16

The test statistic is:

$$Q = \frac{(k-1)\left[k\sum C_j^2 - (\sum C_j)^2\right]}{k\sum R_i - \sum R_i^2}$$

$$= \frac{(4-1)[4(5^2+5^2+2^2+4^2)-(16)^2]}{4\times 16-(2^2+4^2+4^2+3^2+3^2)}$$

$$= \frac{72}{6} = 12.$$

Comment. As calculated $Q > \chi^2_{0.05,3} = 7.815$, H_0 is rejected. Medicines are not similar.

7.11.8 Kruskal-Wallis test

Let there be k samples selected from k populations. The samples are selected independently. The sample observations are shown as follows:

Sample

1	2	3	j	$\cdots$	k
x_{11}	x_{21}	x_{31}	x_{j1}	$\cdots$	x_{k1}
x_{12}	x_{22}	x_{32}	x_{j2}	$\cdots$	x_{k2}
$\vdots$	$\vdots$	$\vdots$	$\vdots$	$\cdots$	$\vdots$
x_{1n_1}	x_{2n_2}	x_{3n_3}	x_{jn_j}	$\cdots$	x_{kn_k}

It is assumed that j-th sample is selected from a population having cumulative density function F_j $(j = 1, 2, \ldots, k)$. We need to test the hypothesis

$$H_0 : F_1 = F_2 = \cdots = F_k.$$

H_A : At least one equality sign does not hold good.

Kruskal-Wallis suggested the test statistic for this hypothesis, where the test statistic is

$$H = \frac{12}{n(n+1)} \sum \frac{R_j^2}{n_j} - 3(n+1).$$

This H is distributed as chi-square with $(k-1)$ d.f.

Here $n = \sum n_j$ = total observations in the samples, R_j = total of rank of j-th sample.

The rank of sample observation is assigned according to the value of observation. The smallest value of the sample observation is assigned rank 1 and the largest one is assigned rank n.

The test statistic follows χ^2-distribution on the assumption that there is no tie. If there is tie, the test statistic is adjusted, where the adjusted test statistic is:

$$H = \frac{1}{C}\left[\sum \frac{R_i^2}{n_j} - \frac{n(n+1)^2}{4}\right],$$

where

$$C = \frac{1}{n-1}\left[\sum\sum R_{ij}^2 - \frac{n(n+1)^2}{4}\right].$$

EXAMPLE 7.38 Some patients of hypertension consulted a doctor for treatment. The doctor observed that the s. cholesterol of every one is higher. Accordingly, the doctor has prescribed different medicines for different groups of patients, so that the level of s. cholesterol is reduced. After the use of medicines for 3 months, the patients are asked again to test the amount of s. cholesterol. The amount of s. cholesterol after the use of medicines are as follows:

Patients group	*Medicine*				
	M_1	M_2	M_3	M_4	M_5
1	168	200	250	188	140
2	147	125	175	178	130
3	215	180	184	160	150
4	138	145	210	190	135
5	170	144	186	185	120

Are the medicines similar in reducing the amount of cholesterol?

Solution. We need to test

$$H_0 : F_1 = F_2 = \cdots = F_k.$$
$$H_A : \text{At least one equality does not hold good.}$$

To perform the test, we need to assign rank to the sample observations. Here the lowest value is 120 and its rank is 1; the next lowest value is 125 and its rank is 2, and so on. The ranks are shown below:

Rank of the sample observations

Patients group	*Medicine*				
	M_1	M_2	M_3	M_4	M_5
1	12	22	25	20	6
2	9	2	14	15	3
3	24	16	17	11	10
4	5	8	23	21	4
5	13	7	19	18	1
Total, R_j	63	55	98	85	24

$n_1 = n_2 = n_3 = n_4 = n_5 = 5.$
$\sum n_j = 25 = n.$

The test statistic is:

$$H = \frac{12}{n(n+1)} \sum \frac{R_i^2}{n_j} - 3(n+1)$$
$$= \frac{12}{25(25+1)} \left[\frac{63^2}{5} + \frac{55^2}{5} + \frac{98^2}{5} + \frac{85^2}{5} + \frac{24^2}{5} \right] - 3(25+1)$$
$$= \frac{12}{25 \times 26} 4879.8 - 78 = 12.08.$$

Comment. As calculated $H > \chi^2_{0.05,4} = 9.488$, H_0 is rejected. The medicines are not similar.

7.11.9 Mantel-Haenszel test

Let there be n observations which are selected at random. Let us consider that these n observations can be classified according to two attributes, where each attribute has two levels. Then n observations can be presented in a 2×2 contingency table, where the attribute A has two levels A and A_0 and the attribute B has two levels B and B_0. The n observations can be presented as follows:

B \ A	B	B_0	*Total*
A	a_l	b_l	$a_l + b_l$
A_0	c_l	d_l	$c_l + d_l$
Total	$a_l + c_l$	$b_l + d_l$	$n_l = a_l + b_l + c_l + d_l$

Let us consider that any observation of l-th experiment will be included in first row and first column with probability p_{l1}, and p_{l2} is the probability of the observations to be included in second row or second column.

The hypothesis to be tested is that A and B are independent or all the experimental results are homogeneous. Mathematically, the hypothesis is:

$$H_0 : p_{l1} = p_{l2} \text{ and } H_A : p_{l1} \neq p_{l2}.$$

Mantel-Haenszel suggested z-test for this hypothesis, where

$$z = \frac{\sum_{l=1}^{k} a_l - \sum_{l=1}^{k} \frac{(a_l+b_l)(a_l+c_l)}{n_l}}{\sqrt{\left[\sum_{l=1}^{k} \frac{(a_l+b_l)(a_l+c_l)(n_l-a_l-b_l)(n_l-a_l-c_l)}{n_l^2(n_l-1)}\right]}}.$$

This z is normally distributed with mean zero and variance 1 $[N(0,1)]$.

According to Mantel-Haenszel, the test is also possible if row total and column total are random. In that case,

$$z = \frac{\sum_{l=1}^{k} a_l - \sum_{l=1}^{k} \frac{(a_l+b_l)(a_l+c_l)}{n_l}}{\sqrt{\left[\sum \frac{(a_l+b_l)(a_l+c_l)(n_l-a_l-b_l)(n_l-a_l-c_l)}{n_l^3}\right]}}.$$

EXAMPLE 7.39 Some urban and rural newborn babies are vaccinated against whooping cough. After some time those babies and the babies who are not vaccinated are observed and they are classified according to suffering from whooping cough. The classified results are shown below:

Urban babies

Suffered	*Vaccinated*		Total
	Yes	*No*	
Yes	6	10	16
No	32	7	39
Total	38	17	55

Rural babies

Suffered	*Vaccinated*		*Total*
	Yes	*No*	
Yes	10	8	18
No	12	10	22
Total	22	18	40

Is vaccination effective against whooping cough in both urban and rural areas?

Solution. H_0 : Vaccination is independent of suffering in both urban and rural areas.

H_A : Suffering depends on vaccination.

The test statistic is:

$$z = \frac{\sum a_l - \sum \frac{(a_l+b_l)(a_l+c_l)}{n_l}}{\sqrt{\left[\sum \frac{(a_l+b_l)(a_l+c_l)(n_l-a_l-b_l)(n_l-a_l-c_l)}{n_l^2(n_l-1)}\right]}}, \text{ where } n_1 = 56,\ n_2 = 40;\ l = 1, 2.$$

Here $a_1 = 6,\ b_1 = 10,\ c_1 = 32,\ d_1 = 7;\ \ a_1 + b_1 = 16,\ a_1 + c_1 = 38$

$a_2 = 10,\ b_2 = 8,\ \ c_2 = 12,\ d_2 = 18;\ a_2 + b_2 = 18,\ a_2 + c_2 = 22.$

$$z = \frac{(6+10) - \frac{1}{55}(16)(38) - \frac{1}{40}(18)(22)}{\sqrt{\left[\frac{16\times 38\times 39\times 17}{55^2(55-1)} + \frac{18\times 22\times 22\times 18}{40^2(40-1)}\right]}}$$

$$= \frac{-4.9545}{\sqrt{2.4677 + 2.5131}} = -2.22.$$

Comment. As $|z| > 1.96$, H_0 is rejected. Suffering is not independent of vaccination in both rural and urban areas.

Or, As $P(z < -2.22) + P(z > 2.22) = 0.0132 + 0.0132 = 0.0264 < 0.05$ [P-value $= 0.0264$], H_0 is rejected.

7.11.10 Median test

Let x_{ij} be the j-th observation ($j = 1, 2, \ldots, n_i$) of i-th sample selected from ($i = 1, 2, \ldots, k$) i-th population. Let M_i be the median of i-th population. Assume that the samples are selected independently and observations are recorded in ordinal scale.

The hypothesis to be tested is

$$H_0 : M_1 = M_2 = \cdots = M_k.$$

$$H_A : \text{At least one equality sign does not hold good.}$$

The test statistic is:

$$\chi^2 = \sum_{l=1}^{2}\sum_{i=1}^{k}\frac{O_{li}^2}{E_{li}} - n, \; n = \sum_{i=1}^{k} n_i, l = 1, 2.$$

This χ^2 has $(k - 1)$ d.f. Here

O_{1i} = observation of i-th sample which is greater than median of all the sample observations

O_{2i} = observation of i-th sample which is less than or equal to median of all the observations.

Therefore, as a first step of the test, the median of the combined sample observations is to be calculated and the observations are classified into two classes, namely (a) less than or equal to median, (b) greater than median. Accordingly, the observations can be presented in a $2 \times k$ contingency table.

If $k = 2$ and $n \leq 20$, then Fisher's exact test is to be performed instead of χ^2-test. Again, if $20 < n < 40$ and any $E_{1j} \nless 5$, then χ^2 test with Yate's correction can be applied. Here

$$E_{lj} = \frac{R_l \times C_j}{n}. \quad R_l = \text{total of } l\text{-th row}$$

$$C_j = \text{total of } j\text{-th column.}$$

EXAMPLE 7.40 Two independent samples of 20 males and 15 females are selected from the admitted patients in a hospital and their haemoglobins are recorded. The recorded data are as follows:

Haemoglobin of male (x_{1j}) : 12.2, 10.6, 15.0, 16.5, 13.8, 14.7, 14.9, 11.9, 10.7, 13.6, 14.2, 15.5, 12.7, 9.8, 10.7, 12.3, 13.5, 12.0, 10.0, 11.7

Haemoglobin of female (x_{2j}) : 11.7, 10.0, 10.4, 12.8, 9.8, 9.0, 11.3, 12.7, 13.0, 9.8, 10.5, 11.7, 14.2, 15.0, 9.5

Do you think that the distributions of haemoglobin of both male and female are similar?

Solution. Let M_1 = median of male population.

M_2 = median of female population.

The hypothesis to be tested is $H_0 : M_1 = M_2$, $H_A : M_1 \neq M_2$.

We have $n_1 = 20$, $n_2 = 15$, $n = n_1 + n_2 = 35$.

The median of the combined sample observations is Me = 12.7.

The observations above median and less than or equal to median are shown as follows:

Median Me	*Male*	*Female*	*Total*
> Me	9	4	13
≤ Me	11	11	22
Total	20	15	35

Here $20 < n < 40$ and $E_{11} = 7.4$, $E_{12} = 5.6$, $E_{21} = 12.\bar{6}$, $E_{22} = 9.4$. So, χ^2-statistic with Yate's correction can be used,

where $\chi^2 = \dfrac{n\left\{|ad - bc| - \frac{n}{2}\right\}^2}{(a+b)(a+c)(c+d)(b+d)} = \dfrac{35\left\{|9 \times 11 - 4 \times 11| - \frac{35}{2}\right\}^2}{13 \times 20 \times 15 \times 22} = 0.57.$

Comment. As calculated $\chi^2 < \chi^2_{0.05,1} = 3.84$, H_0 is accepted. The distributions of haemoglobin of both male and female are similar.

EXAMPLE 7.41 Four groups of patients suffering from heart disease are independently selected from 4 hospitals and their blood pressure levels (in mmHg) are recorded and given below:

Blood pressure level of patients

Groups of patients	*Blood pressure level*
1	77, 87, 80, 85, 95, 78, 105, 90, 95, 90, 90, 92, 95, 80, 86, 95
2	85, 90, 90, 78, 75, 80, 80, 80, 85, 90, 92, 94, 95, 80, 84, 86, 87, 90, 78, 88
3	90, 95, 92, 88, 85, 80, 75, 78, 78, 80, 88, 98, 100, 87, 85, 87, 90, 95
4	86, 78, 88, 95, 96, 100, 90, 90, 80. 85, 88, 75.

Is the median blood pressure of the patients 85?

Solution. We have $n_1 = 16$, $n_2 = 20$, $n_3 = 18$, $n_4 = 12$. Let M be the median blood pressure of the patients.

We need to test $H_0 : M = M_0 = 85, H_A : M \neq M_0$.

The observations are classified with respect to median, where some are greater than median and some are less than or equal to median.

Median	*Samples*				*Total* R_i
	1	2	3	4	
> Me	11	10	11	4	36
≤ Me	5	10	7	8	30
Total C_j	16	20	18	12	66

The expected frequencies E_{ij} are shown below:

Median	***Samples***				***Total*** R_i
	1	2	3	4	
> Me	8.7	10.9	9.8	6.6	36
≤ Me	7.3	9.1	8.2	5.4	30
Total C_j	16	20	18	12	66

As no $E_{ij} < 5$, usual χ^2-statistic can be used,

$$\text{where } \chi^2 = \sum\sum \frac{O_{ij}^2}{E_{ij}} - n$$

$$= \frac{11^2}{8.7} + \frac{10^2}{10.9} + \frac{11^2}{9.8} + \frac{4^2}{6.6} + \frac{5^2}{7.3} + \frac{10^2}{9.1} + \frac{7^2}{8.2} + \frac{8^2}{5.4} - 66$$

$$= 4.09.$$

This χ^2 has $(k-1) = 3$ d.f.

Comment. As calculated $\chi^2 < \chi^2_{0.05,3} = 7.815$, H_0 is accepted. The median blood pressure level of all patients can be considered 85.

7.12 Additive Property of Chi-square

Let $\chi_1^2, \chi_2^2, \ldots, \chi_k^2$ be the independent chi-square statistics having d.f. $n_1, n_2, \ldots, n_k$. The sum of these χ^2 is also distributed as chi-square with $(n_1 + n_2 + \cdots + n_k) = n$ d.f., where

$$\chi^2 = \chi_1^2 + \chi_2^2 + \cdots + \chi_k^2.$$

EXAMPLE 7.42 To study the impact of polio vaccination on children 3 groups of research workers independently collected data from 3 groups of children. The children are classified as follows:

Groups of children

	1			2			3		
Affected	*Applied polio vaccination*								
by polio	*Yes*	*No*	*Total*	*Yes*	*No*	*Total*	*Yes*	*No*	*Total*
Yes	5	10	15	7	13	20	8	16	24
No	25	10	35	32	18	50	38	13	51
Total	30	20	50	39	31	70	46	29	75

Do you think that polio vaccination is effective against polio?

Solution. The hypothesis to be tested is

H_0 : Attack by polio does not depend on polio vaccination.

H_A : Polio vaccination is effective against polio.

The test statistic with Yate's correction is:

$$\chi^2 = \frac{n\left[|ad - bc| - \frac{n}{2}\right]^2}{(a+b)(a+c)(b+d)(c+d)}.$$

For first group,

$$\chi_1^2 = \frac{50\left[|10 \times 5 - 25 \times 10| - \frac{50}{2}\right]^2}{15 \times 30 \times 20 \times 35} = 2.48.$$

For second group,

$$\chi_2^2 = \frac{70\left[|18 \times 7 - 32 \times 13| - \frac{70}{2}\right]^2}{20 \times 39 \times 31 \times 50} = 3.76.$$

For third group,

$$\chi_3^2 = \frac{75\left[|8 \times 13 - 16 \times 38| - \frac{75}{2}\right]^2}{24 \times 46 \times 29 \times 51} = 10.00.$$

$$\chi^2 = \chi_1^2 + \chi_2^2 + \chi_3^2 = 2.48 + 3.76 + 10.00 = 16.24.$$

Comment. A calculated $\chi^2 > \chi^2_{0.05,3} = 7.81$, H_0 is rejected. Polio vaccination is effective against polio.

EXERCISE – 7

1. (a) The following are the body condition scores of some persons:
Body condition scores (x) : 5.8, 6.7, 7.8, 8.5, 6.7, 8.7, 5.0, 5.5, 8.0.
Is the average body condition score 6.0? [**Ans.** $t = 2.16$]

(b) The following are the body mass index (x kg/cm^2) of some persons:
x : 17.5, 18.6, 20.2, 24.5, 25.6, 18.0, 30.0, 30.5, 35.6, 39.0, 32.2, 18.5, 27.0, 28.5, 25.0, 25.0, 22.5, 20.6, 19.7, 18.5, 26.6, 27.5, 28.0, 18.0, 20.5, 24.4, 20.0, 22.6, 23.5, 29.7, 29.0, 30.0.
Do you think that the average body mass index is 25.0? [**Ans.** $z = -0.48$]

(c) Do you think that the proportion of persons having body mass index [given in (b)] less than 25.0 is 0.5? [**Ans.** $z = -0.34$]

2. (a) The following are the s. acid phosphate (x U/L) of some males and of some females (y U/L):

x : 4.5, 4.0, 6.8, 7.0, 4.5, 5.0, 3.5, 4.0, 7.2, 7.0
y : 4.0, 4.0, 4.5, 5.0, 5.5, 6.0, 6.5, 6.0, 5.8, 6.0

(i) Are the amounts of s. acid phosphate of males and females similar? [**Ans.** $t = 0.04$]

(ii) Are the proportions of males and females having normal s. acid phosphate similar? [For male normal is < 6.5 and for female it is < 5.5]. [**Ans.** $z = 0.89$]

(b) The following are the S. GGT of some males (x U/L) and of some females (y U/L):

x : 15, 18, 50, 32, 38, 40, 55, 60, 45, 16, 42, 18, 60, 55, 40
y : 10, 15, 12, 14, 35, 32, 38, 40, 18, 20, 25, 35, 37, 25, 28, 17, 15

(i) Are the amounts of S. GGT of males and females similar? [**Ans.** $t = 3.09$]

(ii) Are the proportions of males and females having normal S. GGT similar? [Normal S. GGT for male is 11–50, for females, it is 7–32] [**Ans.** $z = -0.126$]

3. The following are the information provided from a sample of 40 persons on their s. amylase (x U/L):

$$\sum x = 3500, \sum x^2 = 306850.$$

Is the average s. amylase per person 75? [**Ans.** $z = 20.15$]

4. (a) In a study it is observed that out of 50 males the S. CPK of 40 is in normal range. The S. CPK of 30 females out of 50 is in normal range. Are the proportions of males and females having normal S. CPK similar? [**Ans.** $z = 2.18$ or, $\chi^2 = 4.76$]

(b) The following are the s. triglyceride of some males (x) and of some females (y):

x : 75, 125, 150, 100, 70, 65, 65, 60, 55, 165, 150, 155, 70, 50, 75, 80, 85

y : 50, 65, 72, 110, 115, 70, 60, 50, 52, 160, 140, 145, 110, 118, 100, 100, 110, 88, 85, 105, 72, 75, 65, 68, 60, 55, 50, 75, 70.

Is the median of s. triglyceride of males and females 100? [**Ans.** $\chi^2 = 0.03$]

5. (a) The following are the number of people classified according to socio-economic status and their visit to different clinics for medical assistance:

Visited clinics	*Socio-economic status*			*Total*
	Low	*Medium*	*High*	
Private	5	20	40	65
Government	20	40	15	75
Charitable	35	15	5	55
Total	60	75	60	195

Are socio-economic status and visit of clinics associated? [**Ans.** $\chi^2 = 69.9$]

(b) A group of people are classified by their age and suffering from hypertension. The classified results are shown below:

Hypertension	*Age (in years)*			*Total*
	< 40	40–50	50+	
Yes	15	50	25	90
No	10	35	15	60
Total	25	85	40	150

Is hypertension associated with age? [**Ans.** $\chi^2 = 0.15$]

6. (a) Distinguish between parametric and non-parametric test. Why do we need non-parametric test?

(b) The following are the information recorded in a hospital about the admission of child patients (C) and adult patients (A) in a day:

C C A A A A C A C A C C C A A A A A C A A C C A A C A C A A C C A

Is the admission of child and adult patients occurs at random?

7. (a) A sample of 30 observations is selected from a population of $N = 500$ observations using the following random numbers:

161 264 192 459 489
111 172 196 461 175
110 270 018 358 173
095 350 009 364 412
017 351 001 287 444
019 211 481 200 308

Is the sample selected at random?

(b) It is assumed that 70% outdoor patients of a hospital come from slum area. Once 8 patients are randomly observed and found that 2 are from slum area. Is the assumption correct?

8. (a) The following are the S. TCO_2 of some persons:

S. TCO_2 (x): 27, 24, 28, 24, 25, 29, 30, 26

Is the median of S. TCO_2 26?

(b) The following are the F.B.S (mmol/l) of some persons before using a medicine (x) and after using a medicine (y):

x	7.5	8.0	9.5	7.8	6.5	6.8	8.0	9.2
y	7.0	7.0	10.0	7.0	6.7	7.0	7.0	8.7

Is medicine effective?

9. (a) The following are the S. TCO_2 of males (x) and females (y):

x : 25, 26, 25, 27, 28, 24, 22, 27, 26, 28, 29, 27, 24, 25

y : 22, 23, 26, 27, 28, 24, 25, 25, 25, 26, 27, 28, 22, 23, 22, 25

Do you think that the S. TCO_2 of males and females are similar?

(b) A group of diabetic patients visited a doctor for treatment. The doctor suggested medicines of different doses to them. Their F.B.S. (mmol/l) before treatment and after treatment are shown below:

Patients	*F.B.S. before treatment*	*F.B.S. after use of different doses of medicine*			
		D_1	D_2	D_3	D_4
1	10.5	7.5	8.5	6.3	5.8
2	9.6	6.3	8.0	6.2	5.5
3	9.0	6.3	9.5	6.0	5.7
4	7.5	7.0	8.0	6.0	5.5
5	8.0	8.0	8.5	5.8	5.4
6	8.5	8.0	9.0	6.0	6.0

Are the doses of medicines similar? Use (i) Cochran's test, (ii) Kruskal-Wallis test.

10. A group of couples are classified by adoption behaviour and education. The classified results are as follows:

Education	*Adoption* Yes	No	*Total*
Illiterate	6	18	24
Literate	17	9	26
Total	23	27	50

Education	*Adoption* Yes	No	*Total*
Illiterate	25	55	80
Literate	32	18	50
Total	57	73	130

Is adoption associated with education? Use Mantel and Haenszel test.

Chapter 8

Design of Experiments and Analysis of Variance

8.1 Introduction

Statistical analysis deals with inference regarding population characteristics (parameter). The inference is drawn from the analysis of data, where data are collected at random through experiment. The experiment is of two types, viz., (a) controlled experiment (collection of data by design of experiment, (b) uncontrolled experiment (collection of data by sampling methods). In this chapter, we shall discuss some design of experiments and the analysis of variance of data collected through design of experiments.

8.2 Terms Related to Design of Experiment

Experiment. It is the controlled act through which data are collected according to some predetermined objective. For example, a researcher needs to know the health care system prevailed in different rural health care centre. According to the objective, information (data) on mode of services provided in the centres are to be collected from different centres. The method of collection of data of the services provided in the centres is known as experiment.

Treatment. The characteristics which are to be investigated through experiment is known as treatment. A treatment must have at least two levels. For example, let there be some health service systems prevailed in different centres, viz., investigating the visited patients and prescribing the medicine, investigating the visited patients and providing medicines with or without money, visiting the patients' residence and providing the medicine or prescribing the medicine, investigated patients may be asked for follow-up study, etc. Here the mode of service is a treatment and it has different levels. These levels may vary from centre to centre. Any one mode may be superior to another. The experiment is done to identify the superior mode or modes. Again, a pharmaceutical company may be producing a medicine against a disease. Medicine may be prescribed in different doses. One may need to identify the best dose against a disease. Here medicine is a treatment and its different doses are the levels of treatment.

Experimental plot. The experimental materials on which the treatment is allocated are called experimental plots. The experimental plots may be homogeneous or may be heterogeneous. For example, there are different modes of implementation of family planning programme. It is needed to identify the best method of family planning practices. Here mode of family planning practices is a treatment and the treatment has different levels. Each mode is to be applied on a couple (either on male or on female). Here a couple is an experimental plot. The couples may vary in different demographic characteristics. So, the couples are heterogeneous. But, the couples may be classified according to homogeneous demographic characteristics. This is done by design of experiment. The design of experiment has three basic principles. These are (a) randomization, (b) replication and (c) local control.

Randomization. It is a method of allocation of treatment to experimental plot(s) using any random process so that personal bias of the researcher may not distort the results of the experiment. For example, let there be three medicines produced by three pharmaceutical companies. But, one of the medicine is best against a disease. If a researcher does not like this medicine, he can recommend this medicine to patients of ill-health index and finally may conclude that the medicine is not good. This type of allocation of treatment to the plot is not randomization. It will be randomnization if all 3 medicines are applied to patients of similar health index without preferring any particular patient.

Replication. Repeated allocation of a level of treatment in different plots is known as replication. It is needed as result of one plot may be influenced by external source of variation. For example, paracetamol 500 mg may be effective for a particular patient to reduce pain and temperature. But, it may not be suitable for all patients. So, using paracetamol 500 mg by one patient, it will not be wise to conclude that it is good for everybody. For the recommending conclusion the medicine must be used by different but similar type of patients. Here application of paracetamol 500 mg on different patients is known as replication.

Local control. It is a technique to identify the homogeneous group of experimental plots for the allocation of a level of treatment in one replication. For example, say, one needs to identify the best medicine against high blood pressure. There are different medicines, viz., (i) Atenolol-50, (ii) Nifedipine B.P. 20 mg, (iii) Metaprolol, etc. Now, to select the best medicine, all these medicines are to be applied on homogeneous type of patients. If patients are not homogeneous in terms of health index, a particular medicine may be good for a particular patient but not for all. Here local control can be used to select homogeneous type of patients.

Experimental results. The reorded observation after the allocation of treatment to experimental plots is known as experimental results.

Experimental error. The experimental result is not influenced only by treatment. It may be affected by some other uncontrolled source of factor. For example, a dose of medicine is to administer a patient against blood pressure. But during the use of medicine the patient experiences a tense situation in family affairs. In that case the effect of medicine may be affected by tesion. This impact of tension on medicine is known as experimental error and due to this, some error will creep in the experimental result.

There are different designs to collect experimental results. These are: (i) completely randomized design, (ii) randomized block design, (iii) Latin square design, (iv) split-plot design, (v) incomplete block design, etc. Besides these designs, there are techniques to identify treatment. One of those techniques is known as factorial experiment. In this chapter, the analysis of the data collected through designs or through factorial experiment will be done.

8.3 Completely Randomized Design

Let there be nk experimental units or plots which are homogeneus. We need to allocate k treatments to these plots, where each treatment is to be replicated n times. If all the k treatments are allocated to the plots at random, then the resultant design is known as completely randomized design (CRD).

Let x_{ij} be the observation if j-th replication ($j = 1, 2, \dots, n$) of i-th treatment ($i = 1, 2, \dots, k$). The observation follows the linear model:

$$x_{ij} = \mu + \alpha_i + e_{ij}, \text{ where}$$
$$\mu = \text{general mean, } \alpha_i = \text{the effect of } i\text{-th treatment}$$
$$\text{and } e_{ij} = \text{random error.}$$

The total variation of x_{ij} is mainly due to treatment except the variation of error. Here the components of variations of treatment and error are to be calculated as follows:

$$S_1 = \text{sum of squares (S.S.) of treatment}$$
$$= \sum \frac{x_i^2}{n} - \text{C.T., S.S. (total)} = \sum\sum x_{ij}^2 - \text{C.T.}$$
$$S_2 = \text{S.S. (error)} = \text{S.S. (total)} - \text{S.S. (treatment), where}$$
$$\text{C.T.} = \frac{G^2}{nk}, \; G = \text{grand total of } x_{ij}$$
$$x_i = \sum_j x_{ij}.$$

The analytical results are shown in the analysis of variance [ANOVA] table as follows:

ANOVA Table

Sources of variation	*d.f.*	*S.S.*	*M.S.* $= \frac{S.S.}{d.f.}$	*Calculated F*	*5% value of F*
Treatment	$k-1$	S_1	$s_1 = S_1/k-1$	$F = s_1/s_2$	
Error	$k(n-1)$	S_2	$s_2 = S_2/k(n-1)$		
Total	$nk-1$				

EXAMPLE 8.1 To study the impact of education on a number of ever born children, 20 mothers of age 50 years and above are observed. These mothers are classified into 4 levels of education, viz., illiterate, primary educated, secondary educated and higher educated. For each level of education 5 mothers are observed and their number of ever born children are recorded. The data are given below:

Number of ever born children (x_{ij}) of different mothers having different levels of education

Level of education	*Number of ever born children* (x_{ij})					*Total* x_i	*Mean* $\bar{x}_i$
Illiterate	7	3	5	5	4	24	4.8
Primary	6	8	3	2	4	23	4.6
Secondary	4	5	2	2	3	16	3.2
Higher	2	1	3	2	1	9	1.8

$$G = 24 + 23 + 16 + 9 = 72$$

(i) Are the mothers of different levels of education similar in respect of number of ever born children?

(ii) Is there any difference between illiterate and primary educated mothers?

(iii) Are the mothers of other levels of education different from illiterate mothers?

Solution. Here $k = 4$, $n = 5$, $G = 72$,

$$\text{C.T.} = \frac{G^2}{nk} = \frac{72^2}{5 \times 4} = 259.2.$$

$$\text{Total S.S.} = \sum\sum x_{ij}^2 - \text{C.T.} = 300 - 259.2 = 70.8.$$

$$S_1 = \text{S.S. (education)} = \sum \frac{x_i^2}{n} - \text{C.T.} = \frac{1442}{5} - 259.2 = 29.2.$$

$$S_2 = \text{S.S. (error)} = \text{Total S.S.} - \text{S.S. (education)} = 70.8 - 29.2 = 41.6.$$

ANOVA Table

Sources of variation	*d.f.*	*S.S.*	*M.S.* $= \frac{S.S.}{d.f.}$	*Calculated F*	*5% value of F*
Education	3	$S_1 = 29.2$	$s_1 = 9.73$	$F = s_1/s_2 = 3.74$	3.2
Error	16	$S_2 = 41.6$	$s_2 = 2.6$		
Total	19				

(i) H_0 : Mothers of different levels of education are similar.

H_A : Mothers of different levels of education are not similar.

Comment. As calculated $F > 5\%$ value of F with 3, 16 d.f. $= 3.2$, H_0 is rejected. Mothers of different levels of education are not similar.

(ii) H_0 : Illiterate mothers and primary educated mothers are similar.

H_A : Illiterate mothers and primary educated mothers are not similar.

The test statistic is

$$t = \frac{\bar{x}_1 - \bar{x}_2}{\sqrt{\frac{2\text{ M.S. (error)}}{n}}} = \frac{4.8 - 4.6}{\sqrt{\frac{2 \times 2.6}{5}}} = 0.20.$$

Comment. As calculated $t <$ 5% value of t with 16 d.f. $= 2.12$, H_0 is accepted. The above mentioned two groups of mothers are similar.

(iii) We need to compare illiterate mothers with other groups of mothers. This illiterate group of mothers can be considered as control level of treatment. Here treatment is the level of education. It has 4 levels.

The comparison of control level of treatment with other levels can be done by Dunnett's test, where the test statistic is

$$D = d_{0.05,4,16}\sqrt{\frac{2\text{ M.S. (error)}}{n}}, \text{ where}$$

$d_{0.05,4,16}$ = tabulated value at 5% level of significance for $k = 4$ means with error d.f. $= 16$ from Dunnett's table $= 2.59$.

If $|\bar{x}_i - \bar{x}_j| \geq D$, there is difference between control level with other levels.

Here $\bar{x}_i$ = mean of control level of treatment.

$\bar{x}_j$ = mean of other level of treatment, $j \neq i = 1, 2, 3, 4$.

In this case, $D = 2.59\sqrt{\frac{2 \times 2.6}{5}} = 2.64.$

$|\bar{x}_1 - \bar{x}_2| = |4.8 - 4.6| = 0.2 < D$, there is no difference between illiterate and primary educated mothers.

$|\bar{x}_1 - \bar{x}_3| = |4.8 - 3.2| = 1.6 < D$, there is no difference between illiterate and secondary educated mothers.

$|\bar{x}_1 - \bar{x}_4| = |4.8 - 1.8| = 3.0 > D$, there is difference between illiterate and higher educated mothers.

8.4 Randomized Block Design

Let there be pq experimental plots to verify the effects of q treatments (B), where each treatment is to be replicated in p blocks (A) of q homogeneous plots. Here q homogeneous plots constitute a block. Now, in each block, if q treatments are randomly allocated to q plots of a block, then the resultant design is known as randomized block design. For example, a company claims that it is producing Horlicks of different types suitable for children of different ages. A researcher may be interested in type of Horlicks which is better for a particular group of children. Here Horlicks can be given to children of all age groups. A group of children of same age group and same body condition score will make a block. If there are 3 types of Horlicks, then at least 3 children of same age group and same body condition score are needed for the experiment and each type of Horlicks is to be given to the children of same age group and of same body condition score at random. The process is to be repeated for children of different age groups. The result, growth of children after use of Horlicks, are to be recorded for analysis. The resultant analysis is the analysis of data of randomized block design. Here different types of Horlicks are treatment (B) and children of different age groups are blocks (A).

The data of randomized block design follow the linear model

$$x_{ij} = \mu + \alpha_i + \beta_j + e_{ij};\ i = 1, 2, \ldots, p;\ j = 1, 2, \ldots, q.$$

Here x_{ij} = observation of j-th treatment (j-th level of B) in i-th block (i-th level of A),

μ = general mean,

α_i = effect of i-th block (A),

β_j = effect of jth treatment (B)

e_{ij} = random error.

Total variation of the experimental results is mainly due to two factors, viz., block (A) and treatment (B). The sum of squares (S.S.) is calculated by the formula

$$\text{S.S. (total)} = \sum\sum x_{ij}^2 - \text{C.T., C.T.} = \frac{G^2}{pq},\ G = \text{grand total.}$$

$$S_1 = \text{S.S. (block)} = \text{S.S. }(A) = \sum \frac{A_i^2}{q} - \text{C.T.}$$

$$S_2 = \text{S.S. (treatment)} = \text{S.S. }(B) = \sum \frac{T_j^2}{p} - \text{C.T.}$$

Here A_i = total of i-th block,

T_j = total of j-th treatment.

S_3 = S.S. (error) = S.S. (total) − S.S. (block) − S.S. (treatment).

The analytical results are shown in the ANOVA Table as follows:

ANOVA Table

Sources of variation	*d.f.*	*S.S.*	$M.S. = \frac{S.S.}{d.f.}$	*Calculated F*	*5% value of F*
Block (A)	$p-1$	S_1	$s_1 = S_1/(p-1)$	$F_1 = s_1/s_3$	
Treatment (B)	$q-1$	S_2	$s_2 = S_2/(q-1)$	$F_2 = s_2/s_3$	
Error	$(p-1)(q-1)$	S_3	$s_3 = S_3/(p-1)(q-1)$	—	
Total	$pq-1$				

If $F_1 \geq 5\%$ value of F with $(p-1)$ and $(p-1)(q-1)$ d.f. from F-table, null hypothesis of homogeneity of blocks (levels of A) is rejected. If $F_2 \geq 5\%$ value of F with $(q-1)$ and $(p-1)(q-1)$ d.f. from F-table, null hypothesis of homogeneity of treatments (levels of B) is rejected.

This analysis is known as two-way classification.

EXAMPLE 8.2 The following are the amounts of s. uric acid (mg/dl) of some males and females after the use of different medicines:

Amount of s. uric acid (x_{ij} mg/dl) of patients

Sex	*Medicines*						*Total A_i*
	No medicine	M_1	M_2	M_3	M_4	M_5	
Male	8.8	5.2	6.5	7.5	4.5	5.0	37.5
Female	7.5	4.8	5.5	6.8	3.5	4.0	32.1
Total T_i	16.3	10.0	12.0	14.3	8.0	9.0	69.6
Mean $\bar{T}_j$	8.15	5.0	6.0	7.15	4.0	4.5	

(i) Are medicines similarly effective to reduce level of s. uric acid?

(ii) Is there any difference between the use of M_3 and M_2?

(iii) Is there any use of medicine compared to the no-use of medicine?

Solution. Here $p = 2,\ q = 6,\ G = 69.6$.

$$\text{C.T.} = \frac{G^2}{pq} = \frac{(69.6)^2}{2 \times 6} = 403.68.$$

$$\text{S.S. (total)} = \sum\sum x_{ij}^2 - \text{C.T.} = 432.26 - 403.68 = 28.58.$$

$$\text{S.S. (sex)} = \text{S.S. }(A) = \sum \frac{A_i^2}{q} - \text{C.T.} = \frac{2436.66}{6} - 403.68 = 2.43.$$

$$\text{S.S. (medicines)} = \text{S.S. }(B) = \sum \frac{T_j^2}{p} - \text{C.T.} = \frac{859.18}{2} - 403.68 = 25.91.$$

$$\text{S.S. (error)} = \text{S.S. (total)} - \text{S.S. }(A) - \text{S.S. }(B)$$

$$= 28.58 - 2.43 - 25.91 = 0.24.$$

ANOVA Table

Sources of variation	*d.f.*	*S.S.*	$M.S. = \frac{S.S.}{d.f.}$	*Calculated F*	*5% value of F*
Block (A)	1	2.43	2.43	$60.75 = F_1$	6.0
Treatment (B)	5	25.91	5.182	$122.55 = F_2$	4.4
Error	6	0.24	0.04		
Total	11				

(i) H_0 : All levels of medicine are similarly effective,

H_A : Levels of medicine are different.

Comment. As $F_2 >$ 5% value of F with 1 and 6 d.f. = 4.4, H_0 is rejected. The levels of medicine are not similarly effective to reduce s. uric acid.

(ii) H_0 : The medicine M_2 and M_3 are similar

H_A : The medicine M_2 and M_3 are different.

The test statistic is

$$t = \frac{\bar{T}_2 - \bar{T}_3}{\sqrt{\frac{2\text{ M.S. (error)}}{p}}} = \frac{6.0 - 7.15}{\sqrt{\frac{2\times 0.04}{2}}} = -5.75.$$

Comment. As $|t| >$ 5% value of t with 6 d.f. = 2.447, H_0 is rejected. Medicine M_2 and M_3 are different. To reduce s. uric acid M_2 is better than M_3.

(iii) We need to compare the results of other medicines with the results of no-use of medicine. Here no-use of medicine is considered as controlled treatment. Comparison of controlled treatment with other treatments is done by Dunnett's test, where the test statistic is

$$D = d_{0.05,q,\text{error d.f.}}\sqrt{\frac{2\times\text{M.S. (error)}}{6}},$$

where $d_{0.05,q,\text{error d.f.}}$ = 5% value from Dunnett's table for $q = 6$ means with error d.f. = $6 = 3.39$.

$$D = 3.39\sqrt{\frac{2\times 0.24}{2}} = 0.678.$$

If $|\bar{T}_1 - \bar{T}_j| \geq D$, $\bar{T}_1$ is different from $\bar{T}_j$, $j = 2, 3, 4, 5$. In this case,

$|\bar{T}_1 - \bar{T}_2| = |8.15 - 5| = 3.15 > D$, M_1 is different from no-use of medicine.

$|\bar{T}_1 - \bar{T}_3| = |8.15 - 6| = 2.15 > D$, M_2 is different from no-use of medicine.

$|\bar{T}_1 - \bar{T}_4| = |8.15 - 7.15| = 1.0 > D$, M_3 is different from no-use of medicine.

$|\bar{T}_1 - \bar{T}_5| = |8.15 - 4.0| = 4.15 > D$, M_4 is different from no-use of medicine.

$|\bar{T}_1 - \bar{T}_6| = |8.15 - 4.5| = 3.65 > D$, M_5 is different from no-use of medicine.

In the above analysis, male and female patients are considered patients of two blocks and usually block to block variation is observed. So, we are less interested in testing the differences in blocks. However, instead of block, if we use sex as another factor having two levels, we may need to test the significance of the difference between sex. The F-test in ANOVA Table shows that the amount of s. uric acid is significantly different for males and females. Thus, this analysis of the data obtained from randomized block design may be considered as the analysis of cross classification, where medicine is crossed with sex.

8.5 Two-way Classification with Several Observations (equal) per cell

Let there be pqr experimental plots to verify the effects of q levels of a factor B and the effects of p levels of another factor A allocating i-th level of $A(i = 1, 2, \ldots, p)$ and j-th level of $B(j = 1, 2, \ldots, q)$ in r homogeneous plots. For example, let us consider that 40 diabetic patients of same age group visited a diabetic centre, where there are $5(p = 5)$ doctors in a shift to investigate the patients. Each doctor investigates 8 patients and he/she usually prescribes $4(q = 4)$ medicines to the patients depending on the level of blood sugar. Each doctor prescribes same medicine for $2(r = 2)$ patients. After using the medicines for 15 days the blood sugar level of each patient is recorded. The recorded data are the data of two-way classification with several observations per cell.

The data of such classification can be observed by randomized block design, where p doctors are considered as p blocks and q medicines are considered as q treatments. Each treatmentis replicated in r plots of any block.

The data of such a design follow linear model

$$x_{ijl} = \mu + \alpha_i + \beta_j + (\alpha\beta)_{ij} + e_{ijl}$$
$$i = 1, 2, \ldots, p;\ j = 1, 2, \ldots, q;\ l = 1, 2, \ldots, r.$$

Here x_{ijl} = observation of j-th level of B with i-th level of A in l-th plot
μ = general mean
α_i = effect of i-th level of A
β_j = effect of j-th level of B
$(\alpha\beta)_{ij}$ = interaction of i-th level of A with j-th level of B
e_{ijl} = random error.

Total variation of the experimental results is mainly due to two factors, viz., A and B and their interaction AB. The sum of squares are calculated by the formula

$$\text{S.S. (total)} = \sum\sum\sum x_{ijl}^2 - \text{C.T.},\ \text{C.T.} = \frac{G^2}{pqr},\ G = \text{grand total}$$

$$S_1 = \text{S.S. }(A) = \sum \frac{A_i^2}{qr} - \text{C.T.},\ S_2 = \text{S.S. }(B) = \sum \frac{B_j^2}{pr} - \text{C.T.}$$

$$S_3 = \text{S.S. }(AB) = \sum\sum \frac{x_{ij}^2}{r} - \text{C.T.} - \text{S.S. }(A) - \text{S.S. }(B)$$

$$S_4 = \text{S.S. (error)} = \text{S.S. (total)} - \text{S.S. }(A) - \text{S.S. }(B) - \text{S.S. }(AB).$$

Here A_i = total of i-th level of A
B_j = total of j-th level of B
x_{ij} = total corresponding to i-th level of A and j-th level of B.

The analytical results are shown in the ANOVA Table as follows:

ANOVA Table

Sources of variation	*d.f.*	*S.S.*	*M.S.* $= \frac{S.S.}{d.f.}$	*Calculated F*	*5% value of F*
A	$p-1$	S_1	$s_1 = S_1/(p-1)$	$F_1 = s_1/s_4$	
B	$q-1$	S_2	$s_2 = S_2/(q-1)$	$F_2 = s_2/s_4$	
AB	$(p-1)(q-1)$	S_3	$s_3 = S_3/(p-1)(q-1)$	$F_3 = s_3/s_4$	
Error	$pq(r-1)$	S_4	$s_4 = S_4/pq(r-1)$	—	
Total	$pqr-1$				

If $F_1 \geq 5\%$ value of F with$(p-1)$ and $pq(r-1)$ d.f. from F-table, H_0 of homogeneity of levels of A is rejected. If $F_2 \geq 5\%$ value of F with $(q-1)$ and $pq(r-1)$ d.f. from F-table, H_0 of homogeneity of level of B is rejected. If $F_3 \geq 5\%$ value of F with $(p-1)(q-1)$ and $pq(r-1)$ d.f. from F-table, H_0 of insignificance of interaction is rejected.

EXAMPLE 8.3 The following are the heartbeats/minute (x_{ijl}) of each of 4 animals of two types after giving 5 different chemicals. One group of 4 animals of each type are observed without chemical (control treatment).

Animal	*Chemical*						*Total A_i*	*Mean $\bar{A}_i$*
	Control	C_1	C_2	C_3	C_4	C_5		
A_1	14, 11 13, 13	9, 8 8, 9	9, 10 9, 9	9, 9 9, 9	7, 8 7, 8	9, 10 9, 8	224	9.33
A_2	12, 14 12, 14	8, 8 9, 8	11, 10 11, 11	10, 10 10, 9	11, 12 12, 12	13, 13 12, 13	265	11.04
Total B_j	103	67	80	75	77	87	489	
Mean $\bar{B}_j$	12.875	8.375	10.00	9.375	9.625	10.875		

(i) Are chemicals similarly effective to reduce heartbeat?

(ii) Is there any difference between C_4 and C_5?

(iii) Are the levels of chemical different from control?

Solution. Here $p = 2, q = 6, r = 4, G = 489$

$$\text{C.T.} = \frac{G^2}{pqr} = \frac{489^2}{2 \times 6 \times 4} = 4981.6875$$

$$\text{S.S. (total)} = \sum\sum\sum x_{ijl}^2 - \text{C.T.} = 5165 - 4981.6875 = 183.3125$$

$$\text{S.S. (animal)} = \text{S.S. }(A) = \sum \frac{A_i^2}{qr} - \text{C.T.} = \frac{120401}{6 \times 4} - 4981.6875 = 35.0208$$

$$\text{S.S. (chemical)} = \text{S.S. }(B) = \sum \frac{B_j^2}{pr} - \text{C.T.} = \frac{40621}{2 \times 4} - 4981.6875 = 95.9375.$$

The observations of animals and levels of chemical (x_{ij}) including control are as follows:

Animal	*Chemical*						*Total A_i*
	Control	C_1	C_2	C_3	C_4	C_5	
A_1	51	34	37	36	30	36	224
A_2	52	33	43	39	47	51	265

$$\text{S.S. (animal}\times\text{chemical)} = \text{S.S. }(AB) = \sum\sum \frac{x_{ij}^2}{r} - \text{C.T.} - \text{S.S. }(A) - \text{S.S. }(B)$$

$$= \frac{20591}{A} - 4981.6875 - 35.0208 - 95.9375$$

$$= 35.1042$$

$$\text{S.S. (error)} = \text{S.S. (total)} - \text{S.S. }(A) - \text{S.S. }(B) - \text{S.S. }(AB)$$

$$= 183.3125 - 35.0208 - 95.9375 - 35.1042$$

$$= 17.25.$$

ANOVA Table

Sources of variation	*d.f.*	*S.S.*	*M.S.* $= \dfrac{S.S.}{d.f.}$	*Calculated* F	*5% value of* F	*p-value*
A	1	35.0208	35.0208	73.08	4.12	0.00
B	5	95.9375	19.1875	40.04	2.46	0.00
AB	5	35.1042	7.0208	14.65	2.46	0.00
Error	36	17.25	0.4792	—	—	—
Total	47					

H_0 : All types of animal are similar.

H_A : Animals are different.

Comment. As $F_1 > 5\%$ value of F with 1 and 36 d.f. $= 4.12$, H_0 is rejected. Animals are different.

(i) H_0 : All levels of chemical are similar.

H_A : Levels of chemical are different.

Comment. As $F_2 > 5\%$ value of F with 5 and 36 d.f. $= 2.46$, H_0 is rejected. Levels of chemical are not similarly effective to reduce heartbeat.

(ii) We need to test $H_0 : \beta_A = \beta_5$, against $H_A : \beta_4 \neq \beta_5$.

The test statistic is $t = \dfrac{\bar{x}_4 - \bar{x}_5}{\sqrt{\frac{2\text{ M.S. (error)}}{pr}}} = \dfrac{9.625 - 10.875}{\sqrt{\frac{2\times 0.4792}{2\times 4}}} = -3.61.$

Comment. As $|t| > 5\%$ value of t with 36 d.f. $= 2.03$, H_0 is rejected. C_4 and C_5 are different.

(iii) We need to compare control treatment with other levels of chemical. This is done by Dunnett's test, where the test statistic is

$$D = d_{0.05,q,\text{error d.f.}}\sqrt{\frac{2\text{ M.S. (error)}}{pr}} = 2.636\sqrt{\frac{2\times 0.4792}{2\times 4}} = 0.912.$$

$|\bar{B}_1 - \bar{B}_2| = |12.875 - 8.375| = 4.5 > D$, C_1 is different from control.

$|\bar{B}_1 - \bar{B}_3| = |12.875 - 10.00| = 2.875 > D$, C_2 is different from control.

$|\bar{B}_1 - \bar{B}_4| = |12.875 - 9.375| = 3.5 > D$, C_3 is different from control.

$|\bar{B}_1 - \bar{B}_5| = |12.875 - 9.625| = 3.25 > D$, C_4 is different from control.

$|\bar{B}_1 - \bar{B}_6| = |12.875 - 10.875| = 2.0 > D$, C_5 is different from control.

8.6 Three-way Classification

Let there be 3 factors A, B and C. A has p levels, B has q levels and C has r levels. Let x_{ijl} be the observation of l-th level of $C\,(l = 1, 2, \ldots, r)$ in presence of j-th level of $B\,(j = 1, 2, \ldots, q)$ and i-th level of $A\,(i = 1, 2, \ldots, p)$. The linear model for the observation of x_{ijl} is

$$x_{ijl} = \mu + \alpha_i + \beta_j + \gamma_l + (\alpha\beta)_{ij} + (\alpha\gamma)_{il} + (\beta\gamma)_{jl} + e_{ijl},$$
$$i = 1, 2, \ldots, p;\ j = 1, 2, \ldots, q;\ l = 1, 2, \ldots, r.$$

Here μ = general mean,
α_i = effect of i-th level of A,
β_j = effect of j-th level of B,
γ_l = effect of l-th level of C,
$(\alpha\beta)_{ij}$ = interaction of i-th level of A with j-th level of B,
$(\alpha\gamma)_{il}$ = interaction of i-th level of A with l-th level of C,
$(\beta\gamma)_{jl}$ = interaction of j-th level of B with l-th level of C,
e_{ijl} = random error.

Total variation of x_{ijl} is mainly for variations of 3 factors A, B and C and for variations of their interactions. The sum of squares (S.S.) is calculated by the formula:

$$\text{S.S. (total)} = \sum\sum\sum x_{ijl}^2 - \text{C.T.},\ \text{C.T.} = \frac{G^2}{pqr},\ G = \text{grand total}$$

$$S_1 = \text{S.S. }(A) = \sum \frac{A_i^2}{qr} - \text{C.T.},\ A_i = \text{total of } i\text{-th level of } A.$$

$$S_2 = \text{S.S. }(B) = \sum \frac{B_j^2}{pr} - \text{C.T.},\ B_j = \text{total of } j\text{-th level of } B.$$

$$S_3 = \text{S.S. }(C) = \sum \frac{C_l^2}{pr} - \text{C.T.},\ C_l = \text{total of } l\text{-th level of } C.$$

$$S_4 = \text{S.S. }(AB) = \sum\sum \frac{x_{ij}^2}{r} - \text{C.T.} - \text{S.S. }(A) - \text{S.S. }(B).$$

$$S_5 = \text{S.S. }(AC) = \sum\sum \frac{x_{il}^2}{q} - \text{C.T.} - \text{S.S. }(A) - \text{S.S. }(C).$$

$$S_6 = \text{S.S. }(BC) = \sum\sum \frac{x_{jl}^2}{p} - \text{C.T.} - \text{S.S. }(B) - \text{S.S. }(C).$$

$$S_7 = \text{S.S. (error)} = \text{S.S. (total)} - \text{S.S. }(A) - \text{S.S. }(B) - \text{S.S. }(C) - \text{S.S. }(AB) - \text{S.S. }(AC) - \text{S.S. }(BC).$$

Here x_{ij} = observation of i-th level of A with j-th level of B,
x_{il} = observation of i-th level of A with l-th level of C,
x_{jl} = observation of j-th level of B with l-th level of C.

The analytical results are shown in the ANOVA Table as follows:

ANOVA Table

Sources of variation	*d.f.*	*S.S.*	$M.S. = \frac{S.S.}{d.f.}$	*Calculated F*	*5% value of F*
A	$p-1$	S_1	s_1	$F_1 = s_1/s_7$	
B	$q-1$	S_2	s_2	$F_2 = s_2/s_7$	
C	$r-1$	S_3	s_3	$F_3 = s_3/s_7$	
AB	$(p-1)(q-1)$	S_4	s_4	$F_4 = s_4/s_7$	
AC	$(p-1)(r-1)$	S_5	s_5	$F_5 = s_5/s_7$	
BC	$(q-1)(r-1)$	S_6	s_6	$F_6 = s_6/s_7$	
Error	$(p-1)(q-1)(r-1)$	S_7	s_7	—	
Total	$pqr-1$				

The main objective of the analysis is to test the following hypotheses:

(i) H_0 : All levels of A are similar,
H_A : Levels of A are not similar.
Comment. If $F_1 \geq 5\%$ value of F with $(p-1)$ and $(p-1)(q-1)(r-1)$ d.f. from F-table, H_0 is rejected.

(ii) H_0 : All levels of B are similar,
H_A : Levels of B are not similar.
Comment. If $F_2 \geq 5\%$ value of F with $(q-1)$ and $(p-1)(q-1)(r-1)$ d.f. from F-table, H_0 is rejected.

(iii) H_0 : All levels of C are similar,
H_A : Levels of C are not similar.
Comment. If $F_3 \geq 5\%$ value of F with $(r-1)$ and $(p-1)(q-1)(r-1)$ d.f. from F-table, H_0 is rejected.

(iv) H_0 : The interaction of i-th level of A with j-th level of B is insignificant,
H_A : The interaction of i-th level of A with j-th level of B is significant.
Comment. If $F_4 \geq 5\%$ value of F with $(p-1)(q-1)$ and $(p-1)(q-1)(r-1)$ d.f. from F-table, H_0 is rejected.

(v) H_0 : The interaction of i-th level of A and l-th levelof C is insignificant,
H_A : The interaction of i-th level of A and l-th level of C is significant.
Comment. If $F_5 \geq 5\%$ value of F with $(p-1)(r-1)$ and $(p-1)(q-1)(r-1)$ d.f. from F-table, H_0 is rejected.

(vi) H_0 : The interaction of j-th level of B with l-th level of C is insignificant,
H_A : The interaction of j-th level of B with l-th level of C is significant.
Comment. If $F_6 \geq 5\%$ value of F with $(q-1)(r-1)$ and $(p-1)(q-1)(r-1)$ d.f. from F-table, H_0 is rejected.

In case of rejection of hypotheses (i), (ii) and (iii), there may need to test the hypotheses.

(vii) $H_0 : \bar{A}_i = \bar{A}_{i'}$.

(viii) $H_0 : \bar{B}_j = \bar{B}_{j'}$.

(ix) $H_0 : \bar{C}_l = \bar{C}_{l'}$.

The test statistics for these latter hypotheses are

$$t = \frac{\bar{A}_i - \bar{A}_{i'}}{\sqrt{\frac{2 \text{ M.S. (error)}}{qr}}}, \quad t = \frac{\bar{B}_j - \bar{B}_{j'}}{\sqrt{\frac{2 \text{ M.S. (error)}}{pr}}}, \quad t = \frac{\bar{C}_l - \bar{C}_{l'}}{\sqrt{\frac{2 \text{ M.S. (error)}}{pq}}}$$

respectively. Any of these t-statistics is distributed as Student's t with $(p-1)(q-1)(r-1)$ d.f.

If $|t| > 5\%$ value of t with $(p-1)(q-1)(r-1)$ d.f. from t table, H_0 is rejected.

EXAMPLE 8.4 A pharmaceutical company has manufactured 4 different varieties of protein and investigated an experiment to verify the effectiveness of the protein varieties. The varieties of protein are given to a group of guinea pigs, where the guinea pigs are classified by 4 age groups and the guinea pigs of each age group are classified by 3 sub-groups of different weights. The weights (in gm) of guinea pigs before and after giving protein varieties are observed. The increased in body weights are recorded for analysis. The increased body weights (x_{ijl} gm) are shown below:

Increased body weights (x_{ijl} gm) of guinea pigs

Age group	*Weight*	*Varieties of protein B*				*Total*	*Total*
A	C	P_1	P_2	P_3	P_4	x_{il}	A_i
	W_1	5	3	8	2	18	
A_1	W_2	5.5	2.5	9	2.5	19.5	59.3
	W_3	6	3	10	2.8	21.8	
	W_1	8	6.5	9.2	6.8	30.5	
A_2	W_2	7.5	6	9.5	7	30	92.9
	W_3	8.5	6.4	10	7.5	32.4	
	W_1	15	12	16	14	57	
A_3	W_2	14.5	11	17	13	55.5	166.5
	W_3	14	11.5	16.5	12	54	
	W_1	11.5	8.3	12.5	6.5	38.8	
A_4	W_2	11.8	9	12.8	10.2	43.8	126.3
	W_3	10.2	9.5	13	11	43.7	
Total	B_j	117.5	88.7	143.5	95.3	445	

(i) Analyse the data and comment on the variation of the protein varieties.

(ii) Identify the best variety of protein.

Solution. Here $i = 1, 2, \ldots, p = 4;\ j = 1, 2, \ldots, q = 4;\ l = 1, 2, \ldots, r = 3.$

$$G = 445, \text{ C.T.} = \frac{G^2}{pqr} = \frac{(445)^2}{4 \times 4 \times 3} = 4125.52.$$

$$\text{S.S. (total)} = \sum\sum\sum x_{ijl}^2 - \text{C.T.} = 4843.48 - 4125.52 = 717.96.$$

$$\text{S.S. } (A) = \sum \frac{A_i^2}{qr} - \text{C.T.} = \frac{55820.84}{4 \times 3} - 4125.52$$

$$= 4651.7367 - 4125.52 = 526.22.$$

$$\text{S.S. } (B) = \sum \frac{B_j^2}{pr} - \text{C.T.} = \frac{51348.28}{4 \times 3} - 4125.52 = 153.50.$$

Observations of protein and age (x_{ij}) and observations of age and weight (x_{il}) are as follows:

Age	*Protein (x_{ij})*				*Weight (x_{il})*		
	P_1	P_2	P_3	P_4	W_1	W_2	W_3
A_1	16.5	8.5	27	7.3	18	19.5	21.8
A_2	24	18.9	28.7	21.3	30.5	30	32.4
A_3	43.5	34.5	49.5	39	57	55.5	54
A_4	33.5	26.8	38.3	27.7	38.8	43.8	43.7
Total C_l	—	—	—	—	144.3	148.8	151.9

$$\text{S.S. }(AB) = \sum\sum \frac{x_{ij}^2}{r} - \text{C.T.} - \text{S.S. }(A) - \text{S.S. }(B)$$

$$= \frac{14465.8}{3} - 4125.52 - 526.22 - 153.50 = 16.69.$$

$$\text{S.S. }(C) = \sum \frac{c_l^2}{pq} - \text{C.T.} = \frac{66037.54}{4 \times 4} - 4125.52 = 1.83.$$

$$\text{S.S. }(AC) = \sum\sum \frac{x_{il}^2}{q} - \text{C.T.} - \text{S.S. }(A) - \text{S.S. }(C)$$

$$= \frac{18638.32}{4} - 4125.52 - 526.22 - 1.83 = 6.01.$$

Observations of protein and weight (x_{jl}) are as follows:

Age	*Weight*		
	W_1	W_2	W_3
A_1	18	19.5	21.8
A_2	30.5	38	32.4
A_3	57	55.5	54
A_4	38.8	43.8	43.7

$$\text{S.S. }(BC) = \sum\sum \frac{x_{jl}^2}{p} - \text{C.T.} - \text{S.S. }(B) - \text{S.S. }(C)$$

$$= \frac{17135.18}{4} - 4125.52 - 153.5 - 1.83 = 2.94.$$

$$\text{S.S. (error)} = \text{S.S. (total)} - \text{S.S. }(A) - \text{S.S. }(B) - \text{S.S. }(C)$$
$$-\text{S.S. }(AB) - \text{S.S. }(AC) - \text{S.S. }(BC)$$
$$= 717.96 - 526.22 - 153.50 - 1.83 - 16.69 - 6.01 - 2.94$$
$$= 10.77.$$

ANOVA Table

Sources of variation	*d.f.*	*S.S.*	*M.S.* $= \frac{S.S.}{d.f.}$	*Calculated F*	*5% value of F*
A	3	526.22	175.41	$F_1 = 293.18$	3.16
B	3	153.50	51.57	$F_2 = 86.19$	3.16
C	2	1.83	0.91	$F_3 = 1.52$	3.55
AB	9	16.69	1.85	$F_4 = 3.09$	2,50
AC	6	6.01	1.002	$F_5 = 1.67$	2.66
BC	6	2.94	0.49	$F_6 = 0.82$	2.66
Error	18	10.77	0.5983	—	—
Total	47				

(i) H_0 : The protein varieties are similar,

H_A : Protein varieties are not similar.

Comment. As $F_2 >$ 5% value of F with 3 and 18 d.f. = 3.16, H_0 is rejected.

(ii) The selection of the best variety of protein can be done by performing t-test or Duncan's multiple range test. Here t-test is used to compare all pairs of means for protein. The least significant difference (L.S.D.) using t-distribution is

$$\text{L.S.D.} = 5\% \text{ value of } t \text{ with error d.f.} = 18\sqrt{\frac{2\text{ M.S. (error)}}{pr}}$$

$$= 2.10\sqrt{\frac{2 \times 0.5983}{4 \times 3}} = 0.663.$$

The mean increased in body weights due to varieties of protein are

$$\bar{P}_1 = 9.79,\ \bar{P}_2 = 7.39,\ \bar{P}_3 = 11.96,\ \bar{P}_4 = 7.94$$

$|\bar{P}_1 - \bar{P}_2| = |9.79 - 7.39| = 2.4 >$ L.S.D., P_1 and P_2 are different.

$|\bar{P}_1 - \bar{P}_3| = |9.79 - 11.96| = 2.17 >$ L.S.D., P_1 and P_3 are different.

$|\bar{P}_1 - \bar{P}_4| = |9.79 - 7.94| = 1.85 >$ L.S.D., P_1 and P_4 are different.

$|\bar{P}_2 - \bar{P}_3| = |7.39 - 11.96| = 4.57 >$ L.S.D., P_2 and P_3 are different.

$|\bar{P}_2 - \bar{P}_4| = |7.39 - 7.94| = 0.55 <$ L.S.D., P_2 and P_4 are similar.

$|\bar{P}_3 - \bar{P}_4| = |11.96 - 7.94| = 4.02 >$ L.S.D., P_3 and P_4 are different.

It is observed that P_1, P_2 and P_3 are different. But more increased body weight is observed due to the use of P_3. So, P_3 is the best variety of protein.

8.7 Latin Square Design

In randomized block design, we control one external source of variation in the experimental plots by arranging homogeneous group of plots as blocks, where blocks are taken to the perpendicular direction of the variation in the plots. In practice, there may be two external sources of variation in the experimental plots. For example, in selecting the best type of protein for children, the types of protein are to be given to the children. But, children may vary by ages and by body weights. Due to change in age and body weight, the impact of protein may change. In such a case, impact of protein is to be measured by eliminating the impacts of age and body weight or by controlling the impact of age and body weight. This control of effects of age and body weight can be done, if children of different ages and of different body weights are selected for the experiment. The number of age groups of children and the number of groups of body weights must be equal to the number of proteins.

Let there be k proteins for the experiment. There must be k^2 children of k ages and k body weights to control the age effect and effect of body weight. Here row is taken perpendicular to the source of variation of age, and column can be taken perpendicular to the source of variation of body weight. k proteins are given to the children in such a way that each protein is allocated once and only once to a child (children) of any age and any body weight. The resultant design is known as Latin square design. For example, the proteins are, say, P_1, P_2 and P_3; ages of children are A_1, A_2 and A_3 and body weights are W_1, W_2 and W_3. Then, arrangement of protein to the children will be as follows:

Age	*Body weight*		
	W_1	W_2	W_3
A_1	P_1	P_2	P_3
A_2	P_2	P_3	P_1
A_3	P_3	P_1	P_2

This above design is 3×3 Latin square design, where there are $k = 3$ rows, $k = 3$ columns and $k = 3$ treatments. The data of Latin square design follow the linear model:

$$x_{ijl} = \mu + \alpha_i + \beta_j + \gamma_l + e_{ijl}; \ i = j = l = 1, 2, \ldots, k.$$

Here x_{ijl} = observation of l-th treatment in j-th column of i-th row,
μ = general mean, α_i = effect of i-th row,
β_j = effect of j-th column,
γ_l = effect of l-th treatment,
e_{ijl} = random error.

The sum of squares are calculated as follows:

$$\text{S.S. (total)} = \sum\sum\sum x_{ijl}^2 - \text{C.T.}, \ \text{C.T.} = \frac{G^2}{k^2}$$

$$S_1 = \text{S.S. (row)} = \sum \frac{R_i^2}{k} - \text{C.T.}$$

$$S_2 = \text{S.S. (column)} = \sum \frac{C_j^2}{k} - \text{C.T.}$$

$$S_3 = \text{S.S. (treatment)} = \sum \frac{T_l^2}{k} - \text{C.T.}$$

where C_j = total of j-th column,
R_i = total of i-th row,
T_l = total of l-th treatment,
G = grand total,
S_4 = S.S. (error) = S.S. (total) – S.S. (row) – S.S. (column) – S.S. (treatment).

The analytical results are shown in the **ANOVA Table** as follows:

ANOVA Table

Sources of variation	*d.f.*	*S.S.*	*M.S.* = $\frac{S.S.}{d.f.}$	*Calculated* F	*5% value of* F
Row	$k-1$	S_1	s_1	$F_1 = s_1/s_4$	
Column	$k-1$	S_2	s_2	$F_2 = s_2/s_4$	
Treatment	$k-1$	S_3	s_3	$F_3 = s_3/s_4$	
Error	$(k-1)(k-2)$	S_4	s_4	—	
Total	k^2-1				

If $F_3 \geq 5\%$ value of F with $(k-1)$ and $(k-1)(k-2)$ d.f., null hypothesis of similarity of treatments is rejected.

EXAMPLE 8.5 The following are the body condition scores of 4 differrent persons suffering from similar diseases and treated by 4 doctors using 4 different medicines:

Body condition scores (x_{ijl})

Doctors	*Persons*				*Total* R_i
	P_1	P_2	P_3	P_4	
D_1	$M_1 - 80$	$M_2 - 77$	$M_3 - 70$	$M_4 - 85$	312
D_2	$M_2 - 75$	$M_1 - 75$	$M_4 - 80$	$M_3 - 70$	300
D_3	$M_3 - 64$	$M_4 - 80$	$M_2 - 78$	$M_1 - 75$	297
D_4	$M_4 - 90$	$M_3 - 70$	$M_1 - 70$	$M_2 - 70$	300
Total C_j	309	302	298	300	1209

(i) Are the medicines similar in changing the body condition scores of persons?
(ii) Is there any difference between M_2 and M_3?
(iii) Is M_1 better than other medicines?

Here M_1, M_2, M_3 and M_4 are 4 medicines.

Solution. Here $k = 4$, $G = 1209$.

$$\text{C.T.} = \frac{G^2}{k^2} = \frac{(1209)^2}{4 \times 4} = 91355.0625.$$

$$\text{S.S. (total)} = \sum\sum\sum x_{ijl}^2 - \text{C.T.} = 92009 - 91355.0625 = 653.9375.$$

$$\text{S.S. (row)} = \sum \frac{R_i^2}{k} - \text{C.T.} = \frac{365553}{4} - 91355.0625 = 33.1875.$$

$$\text{S.S. (column)} = \sum \frac{C_j^2}{k} - \text{C.T.} = \frac{365489}{4} - 91355.0625 = 17.1875.$$

Total of treatment (T_l) : $T_1(M_1) = 300$, $T_2(M_2) = 300$,
$T_3(M_3) = 274$, $T_4(M_4) = 335$.

Mean of treatment : $\bar{T}_1 = 75$, $\bar{T}_2 = 75$, $\bar{T}_3 = 68.5$, $\bar{T}_4 = 83.75$.

$$\text{S.S. (treatment)} = \sum \frac{T_l^2}{k} - \text{C.T.} = \frac{367301}{4} - 91355.0625 = 470.1875.$$

$$\text{S.S. (error)} = \text{S.S. (total)} - \text{S.S. (row)} - \text{S.S. (column)} - \text{S.S. (treatment)}$$
$$= 653.9375 - 33.1875 - 17.1875 - 470.1875 = 133.375.$$

ANOVA Table

Sources of variation	*d.f.*	*S.S.*	*M.S.* = $\frac{S.S.}{d.f.}$	*Calculated F*	*5% value of F*
Row	3	33.1875	11.0625	$F_1 = 0.50$	4.8
Column	3	17.1875	5.7292	$F_2 = 0.26$	4.8
Treatment	3	470.1875	156.7292	$F_3 = 7.05$	4.8
Error	6	133.375	22.2292	—	—
Total	15				

(i) H_0 : All medicines are similar,
H_A : Medicines are not similar.

Comment. As $F_3 > 5\%$ value of F with 3 and 6 d.f. = 4.8, H_0 is rejected. Medicines are different.

(ii) H_0 : M_2 and M_3 are similar,
H_A : M_2 and M_3 are not similar.

The test statistic is

$$t = \frac{\bar{T}_2 - \bar{T}_3}{\sqrt{\frac{2\text{ M.S. (error)}}{k}}} = \frac{75.0 - 68.5}{\sqrt{\frac{2 \times 22.2292}{4}}} = \frac{6.5}{3.3339} = 1.95.$$

Comment. As $|t| < 5\%$ value of t with 6 d.f. = 2.447, H_0 is accepted. M_2 and M_3 are not different.

(iii) We need to compare M_1 with other medicines. M_1 is considered as control treatment. The comparison is done by Dunnett's test, where the test statistic is

$$D = d_{0.05,k,\text{error d.f.}}\sqrt{\frac{2\text{ M.S. (error)}}{k}},$$

where $d_{0.05,k,\text{error d.f.}}$ = 5% value from Dunnett's table for $k = 4$ means with error d.f. = 6 = 3.1.

$$D = 3.1\sqrt{\frac{2 \times 22.2292}{4}} = 10.34.$$

$|\bar{M}_1 - \bar{M}_2| = |75 - 75| = 0 < D$, M_1 and M_2 are similar.
$|\bar{M}_1 - \bar{M}_3| = |75 - 65.5| = 6.5 < D$, M_1 and M_3 are similar.
$|\bar{M}_1 - \bar{M}_4| = |75 - 83.75| = 8.75 < D$, M_1 and M_4 are similar.

8.8 Nested Classification

In randomized block design the factor block has p levels and treatment has q levels. All q levels of treatments are randomly allocated in a block separately and the process of random allocation of treatments is repeated in p blocks. Thus, the factor treatment and blocks are crossed. But, there are some factors, where levels of one factor cannot be crossed with levels of another factor. For example, let us consider that there are two doctors, one is a heart specialist and another is a doctor of medicine. Doctor is a factor, it has two levels. Patients of different ages are treated by doctors. So, age is another factor having different levels. But, patients of all ages may not visit the heart specialist or medicine specialist. Thus, levels of ages are not crossed with levels of doctors. The patients of different ages are nested within the levels of doctors. This type of experimental plan is known as nested design and the analysis of the data obtained from such experiment is known as nested classification.

The example what is mentioned above is a two-stage nested design. The linear model for this two-stage nested classification is

$$x_{ijl} = \mu + \alpha_i + \beta_{j(i)} + e_{ijl}, \text{ where}$$
$$x_{ijl} = l\text{-th observation of } j\text{-th level of } B \text{ in } i\text{-th level of } A,$$
$$i = 1, 2, \ldots, p;\ j = 1, 2, \ldots, q;\ l = 1, 2, \ldots, r$$
$$\mu = \text{general mean},$$
$$\alpha_i = \text{effect of } i\text{-th level of } A,$$
$$\beta_{j(i)} = \text{effect of } j\text{-th level of } B \text{ within } i\text{-th level of } A,$$
$$e_{ijl} = \text{random error}.$$

Here $\text{S.S. (total)} = \sum\sum\sum x_{ijl}^2 - \text{C.T., C.T.} = \frac{G^2}{pqr}$.

$$S_1 = \text{S.S. } (A) = \sum \frac{x_i^2}{qr} - \text{C.T.}$$

$$S_2 = \text{S.S. } (B \text{ within } A) = \sum_i \left[\frac{\sum x_{ij}^2}{r} - \frac{x_i^2}{qr}\right].$$

$$x_i = \sum_j \sum_l x_{ijl},\ x_{ij} = \sum_l x_{ijl}.$$

$$S_3 = \text{S.S. (error)} = \text{S.S. (total)} - \text{S.S. } (A) - \text{S.S. } (B \text{ within } A).$$

ANOVA Table

Sources of variation	*d.f.*	*S.S.*	*M.S.* = $\frac{S.S.}{d.f.}$	*Calculated F*	*5% value of F*
A	$p-1$	S_1	s_1	$F_1 = s_1/s_3$	
B within A	$p(q-1)$	S_2	s_2	$F_2 = s_2/s_3$	
Error	$pq(r-1)$	S_3	s_3	—	
Total	$pqr-1$				

If $F_1 \geq 5\%$ value of F with $(p-1)$ and $pq(r-1)$ d.f., levels of A are different. If $F_2 \geq 5\%$ value of F with $p(q-1)$ and $pq(r-1)$ d.f., levels of B within A are different.

EXAMPLE 8.6 The following are the amount of blood sugar (fasting) levels (x_{ijl} mmol/l) of some patients treated by 4 doctors using 3 different medicines.

Patients	*Amount of F.B.S. (x_{ijl} mmol/l)*											
	Doctors											
	D_1			D_2			D_3			D_4		
	Medicines											
	M_1	M_2	M_3	M_1	M_2	M_3	M_1	M_2	M_3	M_1	M_2	M_3
1	4.5	5.0	5.8	5.5	5.8	7.0	5.5	6.5	7.2	6.0	4.8	4.8
2	5.8	5.6	6.8	6.4	6.4	6.8	6.3	6.5	7.5	6.0	5.6	5.9
3	5.5	6.3	6.6	6.5	6.0	7.5	6.3	6.8	6.9	6.2	4.9	6.0
4	4.4	6.4	6.0	6.0	6.5	6.5	6.0	6.2	6.5	6.3	5.0	6.3
Total x_{ij}	20.2	23.3	25.2	24.4	24.7	27.8	24.1	26.0	28.1	24.5	20.3	23.0
Total x_i	68.7			76.9			78.2			67.8		

$$G = 68.7 + 76.9 + 78.2 + 67.8 = 291.6$$

(i) Are the doctors similar in treating the patients?

(ii) Are the medicines similarly used by doctors?

(iii) Is doctor D_4 better than other doctors?

Solution. Here $p = 4,\ q = 3,\ r = 4,$

$$\text{C.T.} = \frac{G^2}{pqr} = \frac{(291.6)^2}{4 \times 3 \times 4} = 1771.47.$$

$$\text{S.S. (total)} = \sum\sum\sum x_{ijl}^2 - \text{C.T.} = 1795.84 - 1771.47 = 24.37.$$

$$\text{S.S. (medicines within doctors)} = \sum_i \left[\frac{x_{ij}^2}{r} - \frac{x_i^2}{qr}\right]$$

$$= \left[\frac{20.2^2 + 23.3^2 + 25.2^2}{4} - \frac{(68.7)^2}{3 \times 4}\right] + \left[\frac{24.4^2 + 24.7^2 + 27.8^2}{4} - \frac{(76.9)^2}{3 \times 4}\right]$$

$$+ \left[\frac{24.1^2 + 26.0^2 + 28.1^2}{4} - \frac{(78.2)^2}{3 \times 4}\right] + \left[\frac{24.5^2 + 20.3^2 + 23.0^2}{4} - \frac{(67.8)^2}{3 \times 4}\right]$$

$$= 3.185 + 1.772 + 2.002 + 2.265 = 9.224.$$

$$\text{S.S. (doctor)} = \sum \frac{x_i^2}{qr} - \text{C.T.} = \frac{68.7^2 + 76.9^2 + 78.2^2 + 67.8^2}{3 \times 4} - 1771.47$$
$$= 1778.782 - 1771.47 = 7.312.$$

$$\text{S.S. (error)} = \text{S.S. (total)} - \text{S.S. (doctors)} - \text{S.S. (medicines within doctors)}$$
$$= 24.37 - 7.312 - 9.224 = 7.834.$$

ANOVA Table

Sources of variation	*d.f.*	*S.S.*	*M.S.* = $\frac{S.S.}{d.f.}$	*Calculated F*	*5% value of F*
Doctors	3	7.312	2.437	11.20	2.86
Medicines within doctors	8	9.224	1.153	5.30	2.21
Error	36	7.834	0.2176	—	—
Total	47				

(i) H_0 : All doctors are similar in treating patients.

H_A : All doctors are not similar.

Comment. As calculated $F >$ 5% value of $F = 2.86$ with 3 and 36 d.f., H_0 is rejected. Medicines used by doctors are not similar.

(ii) H_0 : All medicines used by doctors are similar.

H_A : All medicines used by doctors are not similar.

Comment. As calculated $F >$ 5% value of $F = 2.21$ with 8 and 36 d.f., H_0 is rejected. Medicines used by doctors are not similar.

(iii) We need to compare doctor D_4 with other doctors. So, D_4 is considered as control treatment. The comparison is done by Dunnett's test, where the test statistic is

$$D = d_{0.05,p,\text{error d.f.}} \sqrt{\frac{2\ \text{M.S. (error)}}{qr}},$$

where $d_{0.05,p,\text{error d.f.}}$ = 5% value of d from Dunnett's table for $p(=4)$ means with error d.f.$(=36)$

$= 2.452.$

$$\therefore D = 2.452\sqrt{\frac{2 \times 0.2176}{3 \times 4}} = 0.467.$$

Means of doctors: $\bar{x}_1 = 5.725, \bar{x}_2 = 6.408, \bar{x}_3 = 6.517, \bar{x}_4 = 5.65$

$|\bar{x}_4 - \bar{x}_1| = |5.65 - 5.725| = 0.075 < D$; D_4 and D_1 are similar.

$|\bar{x}_4 - \bar{x}_2| = |5.65 - 6.408| = 0.758 > D$; D_4 is different from D_2.

$|\bar{x}_4 - \bar{x}_3| = |5.65 - 6.517| = 0.867 > D$; D_4 is different from D_3.

8.9 Factorial Experiment

The main objective of the design of experiment is to study the impacts of treatments through controlled experiment. Example of treatments are: (i) different medicines against a disease, (ii) treatment of disease/diseases by different doctors, (iii) doses of medicine against a disease, (iv) impacts of protein on health index, (v) child health care systems, (vi) use of combination of medicines against a disease, etc. For example, some doctors may suggest the use of Nifedifine SR 20 in the morning and Atenolol-50 in the night against high blood pressure. Again, some doctors may suggest the use of a drug which is a mixture of the above two components. These two components may be combined in different doses. The combination of different components having different levels are used as treatment. If the treatments are decided by the combination of components with different levels, and treatment effects are observed by design of experiment, then the experiment is called factorial experiment. The objective of this factorial experiment is to study the main effect of each factor and the interaction effect of the factors.

Let there be two factors A and B having two levels of each of them. The levels of A are, say, a_0 and a_1 and the levels of B are b_0 and b_1. The level combinations are a_0b_0, a_0b_1, b_0a_1 and a_1b_1. If these 4-level combinations are used as treatment in any experiment, then the experiment is called factorial experiment. Specially, this experiment is called 2^2-factorial experiment. Here a_0 is the first level of A and a_1 is the second level of A. Similarly, b_0 and b_1 are the first and second levels of B, respectively. The level combinations can also be written as $a_0b_0 = (1)$, $a_1b_0 = a$, $a_0b_1 = b$, $a_1b_1 = ab$. Finally, the level combinations of 2^2-factorial experiment are $(1), a, b$ and ab.

In factorial experiment all factors may not have same level. Let the factors are $F_1, F_2, \ldots, F_k$ and the levels of these factors are $l_1, l_2, \ldots, l_k$, respectively. Then the total number of treatments (level combinations) is $l_1 \times l_2 \times \cdots \times l_k$. If $l_1 = l_2 = \cdots = l_k$, then total level combinations are l^k and the experiment is known as symmetrical factorial experiment. If $l_1 \neq l_2 \neq \cdots \neq l_k$, then the experiment is called asymmetrical factorial experiment. The analysis of the data of factorial experiment is to estimate the main effects of the factors and to estimate the interaction effects of factors.

Main effect. For 2^2-factorial experiment with factors A and B the level combinations are $(1), a, b$ and ab. Here

$$a - (1) = \text{simple effect of } A \text{ in absence of } B \text{ (first level of } B)$$
$$ab - b = \text{simple effect of } A \text{ in presence of } B \text{ (second level of } B).$$

So, the total effect of A is

$$[A] = a - (1) + ab - b.$$

If (1) is considered as 1, mathematically it is written as

$$[A] = (a - 1)(b + 1).$$

The average effect of A (main effect of A) is given by

$$A = \frac{1}{2}(a - 1)(b + 1) = \frac{1}{2^{2-1}}(a - 1)(b + 1).$$

In a similar way, we can write:

$$b - (1) = \text{simple effect of } B \text{ in absence of } A.$$
$$ab - a = \text{simple effect of } B \text{ in presence of } A.$$

Total effect of B is

$$[B] = b - (1) + ab - a = (a + 1)(b - 1).$$

The average effect of B (main effect of B) is

$$B = \frac{1}{2^{2-1}}(a + 1)(b - 1).$$

The interaction effect of A and B is calculated by subtracting simple effect of A in absence of B from simple effect of A in presence of B. Thus, we have

$$[AB] = (ab - b) - (a - (1)) = ab - a - b + (1) = (a-1)(b-1).$$

The interaction effect is

$$AB = \frac{1}{2^{2-1}}(a-1)(b-1).$$

If there are n factors $A, B, C, \ldots, N$ having 2 levels each, then main effects of the factors are:

$$A = \frac{1}{2^{n-1}}(a-1)(b+1)(c+1)\cdots(n+1)$$

$$B = \frac{1}{2^{n-1}}(a+1)(b-1)(c+1)\cdots(n+1)$$

$$\cdots\cdots\cdots\cdots\cdots\cdots\cdots\cdots$$

$$N = \frac{1}{2^{n-1}}(a+1)(b+1)(c+1)\cdots(n-1).$$

Two-factor interaction effects are:

$$AB = \frac{1}{2^{n-1}}(a-1)(b-1)(c+1)\cdots(n+1)$$

$$AC = \frac{1}{2^{n-1}}(a-1)(b+1)(c-1)\cdots(n+1)$$

$$\cdots\cdots\cdots\cdots\cdots\cdots\cdots\cdots$$

Three-factor interaction effect, namely ABC, is

$$ABC = \frac{1}{2^{n-1}}(a-1)(b-1)(c-1)(d+1)\cdots(n+1).$$

$$\cdots\cdots\cdots\cdots\cdots\cdots\cdots\cdots$$

The n-factor interaction effect is

$$ABCD\cdots N = \frac{1}{2^{n-1}}(a-1)(b-1)(c-1)\cdots(n-1).$$

In practice, the total effect is calculated easily by Yate's method. The method is explained below:

Let us consider that for 2^3-factorial experiment each treatment is replicated r times and the observations of each replication are shown below:

Level combination	*Observations in r replication* (x_{ij})				*Total*
(1)	x_{11}	x_{12}	......	x_{1r}	X_1
a	x_{21}	x_{22}	......	x_{2r}	X_2
b	x_{31}	x_{32}	......	x_{3r}	X_3
ab	x_{41}	x_{42}	......	x_{4r}	X_4
c	x_{51}	x_{52}	......	x_{5r}	X_5
ac	x_{61}	x_{62}	......	x_{6r}	X_6
bc	x_{71}	x_{72}	......	x_{7r}	X_7
abc	x_{81}	x_{82}	......	x_{8r}	X_8

Here $X_1, X_2, \ldots, X_8$ are the total results of 8 different treatments.

Yate's method

(i) Treatments (level combinations) are to be arranged as shown in the table in column-1.

(ii) Total results of the treatments are to be shown in column-2.

(iii) The first step of Yate's method is to add the total results in pairs and will be shown in column-3. The first half of the observations of column-3 will be available from the addition of results in pairs. The second half of observations of column-3 are calculated from subtraction of r results of column-2 in pairs. In subtracting, the first observation of a pair is to be subtracted from the second observation of the pair.

(iv) The results of column-4 will be available from the results of column-3 in a similar way (step 3).

(v) For 2^n-factorial experiment, Yate's method will be continued up to n steps. The results of n-th step will provide total effects. First value of n-th step is the grand total (G). The main effect and interaction effect are calculated by the formula:

$$\text{Main effect} = \frac{[\ \ ]}{r2^{n-1}}, \text{ where } r = \text{number of replication of a treatment.}$$

The sum of squares of effects and interactions are calculated by the formula

$$\text{S.S. (effects)} = \frac{[\ \ ]^2}{r2^n}.$$

EXAMPLE 8.7 A heart specialist investigates 8 patients of hypertension in a day. For some patients he prescribes Cardipro 50 (C), for some patients he prescribes Nidifine SR 20 (N) and for some others he prescribes Prosan (P) and for some patients he prescribes combination of any two medicines or combination of three medicines. After one month of prescription he records the blood pressure levels of the patients. The blood pressure levels (x_{ij} mmHg) of patients after the use of any medicine or their combination are as follows:

Level combination of medicines	*Blood pressure levels (x_{ij}) of patients of 4 days*				*Total* X_i
	D_1	D_2	D_3	D_4	
(1)	88	90	95	85	358
c	78	85	75	80	318
n	70	78	75	80	303
cn	72	75	70	70	287
p	80	85	85	82	332
pc	77	80	82	78	317
np	75	78	75	70	298
npc	70	70	76	82	298
Total R_i	610	641	633	627	2511

Here (1) indicates no-use of any of the medicine. Test the effectiveness of medicines and their combinations. These above results are from the 2^3-factorial experiment, where each treatment is replicated $r = 4$ times. The estimates of effects and interactions and their sum of squares are calculated by Yate's method. Application of Yate's method is shown below:

Level combinations	Total results X_i	Steps of Yate's method Step-1	Step-2	Step-3 [ ]	Effects and interactions	S.S. $= \frac{[\ ]^2}{r2^3}$	Effects and interactions $\frac{[\ ]^2}{r2^{3-1}}$
(1)	358	676	1266	2511	G	197035.03	—
c	318	590	1245	−71	C	157.53	−4.44
n	303	649	−56	−139	N	603.78	−8.69
cn	287	596	−15	39	CN	47.53	2.44
p	332	−40	−86	−21	P	13.78	−1.31
pc	317	−16	−53	41	PC	52.53	2.56
np	298	−15	24	33	NP	34.03	2.06
npc	298	0	15	−9	NPC	2.53	−0.56

$$\text{S.S. (total)} = \sum\sum x_{ij}^2 - \text{C.T.} = 198291 - 197035.03 = 1255.97.$$

$$\text{S.S. (replications)} = \sum \frac{R_i^2}{2^3} - \text{C.T.} = \frac{610^2 + 641^2 + 633^2 + 627^2}{8} - 197035.03 = 64.845.$$

$$\begin{aligned}\text{S.S. (error)} &= \text{S.S. (total)} - \text{S.S. (C)} - \text{S.S. (N)} - \text{S.S. (CN)} - \text{S.S. (P)} - \text{S.S. (PC)} \\ &\quad - \text{S.S. (NP)} - \text{S.S. (NPC)} - \text{S.S. (replications)} \\ &= 1255.97 - 157.53 - 603.78 - 47.53 - 13.78 - 52.53 \\ &\quad - 34.03 - 2.53 - 64.845 \\ &= 279.415.\end{aligned}$$

ANOVA Table

Sources of variation	d.f.	S.S.	M.S. $= \frac{S.S.}{d.f.}$	Calculated F	5% value of F
Replications	3	64.845	21.615	1.62	3.07
C	1	157.53	157.53	11.84	4.32
N	1	603.78	603.78	45.38	"
CN	1	47.53	47.53	3.57	"
P	1	13.78	13.78	1.04	"
PC	1	52.53	52.53	3.95	"
NP	1	34.03	34.03	2.56	"
NPC	1	2.53	2.53	0.19	"
Error	21	279.415	13.305	—	
Total	31				

The analysis indicates that use of Cardipro-50 and Nidifine SR-20 are effective in reducing blood pressure level [Effects of these two medicines are significant as calculated $F > 5\%$ value of F].

8.10 Split-plot Design

It is an asymmetrical factorial experiment, where there are two factors, say, A and B. Consider that A has p levels $A_1, A_2, \ldots, A_p$ and B has q levels $B_1, B_2, \ldots, B_q$. The total number of level combinations are pq. This pq level combinations are treated as treatment for the experiment.

The levels of A are such that each one can be allocated to larger experimental plots. For example, a researcher may be interested to study the birth-weight of babies of currently married mothers who are under medical check-up. The mothers are of different age groups, say, ≤ 20, 21–25, 26–30, 31–35 and 35+. In each group, there may be more pregnant mothers. Assume that all groups of mothers suffer from anaemia. So, they need iron to get rid of the disease. For each group of mothers the doctor may prescribe folic acid, injection and improved diet to increase their haemoglobin level. For each group of mothers three levels of treatment can be applied. Here a group of mothers of ages ≤ 20 or 21–25 or 26–30 or 31–35 or 35+ can be considered as whole plot. And to apply three levels of treatment a whole plot can be splitted into 3 sub-plots. Now, each level of treatment can be allocated to sub-plots of a randomly allocated whole plot factor. Then the design is known as split-plot design.

Usually, the whole plot factor is considered as A having p levels and sub-plot factor is considered as B having q levels. The experiment can be replicated r times. The observation of a sub-plot factor (B) in a whole plot (A) of a replication is denoted by x_{ijl} [$i = 1, 2, \ldots, r$; $j = 1, 2, \ldots, p$ and $l = 1, 2, \ldots, q$.

Here $x_{ijl} = \mu + \alpha_i + \beta_j + \gamma_l + (\beta\gamma)_{jl} + e_{ijl}$

$$i = 1, 2, \ldots, r;\ j = 1, 2, \ldots, p;\ l = 1, 2, \ldots, q.$$

x_{ijl} = observation of l-th level of B in j-th level of A in i-th replication

μ = general mean,

α_i = effect of i-th replication,

β_j = effect of j-th level of A,

γ_l = effect of l-th level of B,

$(\beta\gamma)_{jl}$ = interaction of j-th level of A with l-th level of B,

e_{ijl} = random error.

The analysis of the data is done in two steps. One step of analysis is the whole plot analysis and another step of analysis is the sub-plot analysis. So, we have two sources of error variation; one for whole plot analysis and one for sub-plot analysis. The sum of squares are calculated as follows:

$$\text{S.S. (total)} = \sum\sum\sum x_{ijl}^2 - \text{C.T.},\ \text{C.T.} = \frac{G^2}{pqr},\ G = \text{grand total}$$

$$\text{S.S. (replication)} = \sum \frac{R_i^2}{pq} - \text{C.T.},\ \text{S.S. } (A) = \sum \frac{A_j^2}{qr} - \text{C.T.}$$

$$\text{S.S. (error-1)} = \sum\sum \frac{x_{ij}^2}{r} - \text{C.T.} - \text{S.S. (replications)} - \text{S.S. } (A)$$

$$\text{S.S. } (B) = \sum \frac{B_l^2}{pr} - \text{C.T.}$$

$$\text{S.S. } (AB) = \sum\sum \frac{x_{jl}^2}{r} - \text{C.T.} - \text{S.S. } (A) - \text{S.S. } (B)$$

$$\text{S.S. (error-2)} = \text{S.S. (total)} - \text{S.S. (replication)} - \text{S.S. } (A) - \text{S.S. (error-1)} - \text{S.S. } (B) - \text{S.S. } (AB).$$

EXAMPLE 8.8 In a heart foundation hospital, male patients and female patients are admitted in separate wards. The patients of each ward are classified according to the severity of disease. Some patients are in serious conditions and need operation, some do not need operation and some others have mild problem. Accordingly, treatment of patients are not same for all patients. Those who have mild problem they are given only Cardipro-50 once in a day, those who do not need operation, they are given Tenocab-50 twice a day. Those who need operation they are given Nidipro twice a day. After one week of treatment their blood pressure levels are recorded. The results are shown below for different factor-levels combination:

Let A_1 = male ward, A_2 = female ward.

Wards are selected from 3 floors (replication). Patients of a ward given:

B_1 = Cardipro-50

B_2 = Tenocab-50 twice a day

B_3 = Nidipro twice a day.

Blood pressure levels (x_{ijl} mmHg) of patients of different wards of different floors after using different medicines:

Replication [Patients of floor]	*Male ward* (A_1)			*Female ward* (A_2)			*Total* R_i
	Medicine used			*Medicine used*			
	B_1	B_2	B_3	B_1	B_2	B_3	
1	80	85	78	75	79	80	477
2	90	88	85	82	88	78	511
3	75	70	77	70	74	78	444
Total A_j		728			704		1432 = G

(i) Are the responses of medicine similar for male and female patients?

(ii) Are the medicines similar?

Solution. Here $r = 3, p = 2, q = 3$;

$$\text{C.T.} = \frac{G^2}{pqr} = \frac{(1432)^2}{2 \times 3 \times 3} = 113923.56.$$

$$\text{S.S. (total)} = \sum\sum\sum x_{ijl}^2 - \text{C.T.} = 114510 - 113923.56 = 586.44.$$

$$\text{S.S. (replications)} = \sum \frac{R_i^2}{pq} - \text{C.T.} = \frac{477^2 + 511^2 + 444^2}{2 \times 3} - 113923.56 = 374.11.$$

$$\text{S.S. (ward)} = \sum \frac{A_j^2}{qr} - \text{C.T.} = \frac{728^2 + 704^2}{3 \times 3} - 113923.56 = 32.00.$$

Observations of ward and medicine (x_{jl})

Ward	*Medicine*			*Total* A_j
	B_1	B_2	B_3	
A_1	245	243	240	728
A_2	227	241	236	704
Total B_l	472	484	476	1432

$$\text{S.S. (medicine)} = \sum \frac{B_l^2}{p \times r} - \text{C.T.} = \frac{472^2 + 484^2 + 476^2}{2 \times 3} - 113923.56 = 12.44.$$

$$\text{S.S. (ward}\times\text{medicine)} = \sum\sum \frac{x_{jl}^2}{r} - \text{C.T.} - \text{S.S. (ward)} - \text{S.S. (medicine)}$$

$$= \frac{245^2 + 243^2 + 240^2 + 227^2 + 241^2 + 236^2}{3}$$

$$-113923.56 - 32.00 - 12.44$$

$$= 25.33.$$

$$\text{S.S. (error-2)} = \text{S.S. (total)} - \text{S.S. (replication)} - \text{S.S. (ward)} - \text{S.S. (error-1)}$$

$$-\text{S.S. (medicine)} - \text{S.S. (ward}\times\text{medicine)}$$

$$= 586.44 - 374.11 - 32.00 - 19.00 - 12.44 - 25.33 = 123.56.$$

The observations of i-th replication and j-th whole plot (x_{ij}) are:

Replication	*Whole plot*		Total R_i
	A_1	A_2	
1	243	234	477
2	263	248	511
3	222	222	444
Total A_j	728	704	1432

$$\text{S.S. (error-1)} = \frac{\sum\sum x_{ij}^2}{q} - \text{C.T.} - \text{S.S. (replication)} - \text{S.S. (ward)}$$

$$= \frac{1}{3}\left[243^2 + 234^2 + 263^2 + 248^2 + 222^2 + 222^2 - 113923.56\right] - 374.11 - 32.00$$

$$= 19.00.$$

ANOVA Table

Sources of variation	*d.f.*	*S.S.*	*M.S.* $= \frac{S.S.}{d.f.}$	*Calculated* F	*5% value of* F
Replication	2	374.11	$s_1 = 187.055$	$F_1 = s_1/s_3 = 19.69$	19.0
Ward (A)	1	32.00	$s_2 = 32.00$	$F_2 = s_2/s_3 = 3.37$	18.5
Error-1	2	19.00	$s_3 = 9.5$		
Medicine (B)	2	12.44	$s_4 = 6.22$	$F_3 = s_4/s_6 = 0.43$	4.46
AB	2	25.33	$s_5 = 12.665$	$F_4 = s_5/s_6 = 0.88$	4.46
Error-2	8	123.56	$s_6^2 = 14.445$		
Total	17				

(i) H_0 : The male and female patients are similarly responsive to medicines.

H_A : The male and female patients are not similarly responsive.

Comment. As calculated $F_2 <$ 5% value of F, H_0 is accepted. Male and female patients are similarly responsive to medicines.

(ii) H_0 : The medicines are similar.

H_A : The medicines are not similar.

Comment. As calculated $F_3 <$ 5% value of F, H_0 is accepted. All medicines are similar.

EXERCISE – 8

1. The following are the breast feeding period in months (x_{ij}) of mothers of different levels of education to their babies:

Levels of education	Breast feeding period (x_{ij}) to babies					Total x_i
Illiterate	12	15	18	12	10	67
Primary	12	10	15	12	12	61
Secondary	6	6	8	9	6	35
Higher	6	6	10	12	6	40
Total						203

(i) Are the mothers of different levels of education similar in respect of duration of breast feeding period?

(ii) Is there any difference between illiterate mothers and mothers having primary education in respect of breast feeding period?

(iii) Compare higher educated mothers with mothers of other levels of education.

2. The following are the body weights (x_{ij} in kg) of some persons having different levels of blood sugar and different levels of blood pressure.

Levels of blood pressure	Levels of blood sugar ≤ 6.3	6.4–7.5	7.5–9.0	9.0+	Total R_i
< 70	58	60	50	50	218
70–80	65	62	58	52	237
80–90	60	55	54	55	224
90–100	55	55	50	54	214
100+	52	53	58	50	213
Total C_j	290	285	270	261	1106

(i) Is there any difference in body weights due to the changes in levels of blood pressure?

(ii) Are the body weights of persons of different levels of blood sugar similar?

(iii) Compare body weights of persons of different blood sugar levels with body weights of normal blood sugar levels.

(iv) Compare body weights of persons of different levels of blood pressure with body weights of persons who are not suffering from hypertension.

3. The following are the change in body weights (x_{ij} kg) of babies given different types of protein. The experiment is conducted through a randomized block design, where in each block there are babies of similar ages

Block Age of babies in years	Change in body weights (x_{ij}) due to protein P_1	P_2	P_3	P_4	P_5	Total R_i
$A_1 < 3$	1.8	2.5	3.0	2.5	3.4	13.2
A_2 (3–5)	2.0	2.6	3.5	2.5	3.5	14.1
A_3 (5–6)	1.5	2.0	3.0	3.0	4.0	13.5
A_4 (6–7)	1.5	2.8	3.0	3.0	3.0	13.3
Total C_j	6.8	9.9	12.5	11.0	13.9	54.1

(i) Are the proteins similar in changing body weights?

(ii) Compare P_1 with other proteins.

4. The following are the number of ever born children to mothers of different levels of education, different occupation and different duration of marriage. For each level of education, occupation and duration of marriage, one mother is selected following a Latin square design.

Levels of occupation	*Levels of education*				*Total*
	Illiterate	*Primary*	*Secondary*	*Higher*	R_i
Housewife	A–1	B–3	C–3	D–5	12
Service	B–2	C–4	D–3	A–0	9
Business	C–3	D–4	A–1	B–2	10
Others	D–6	A–0	B–2	C–2	10
Total C_j	12	11	9	9	41

Duration of marriage A, B, C, D.

A = less than 2 years, B = 2 to 8 years

C = 9 to 15 years, D = more than 15 years

(i) Are the levels of occupation similar in respect of number of ever born children?

(ii) Are the levels of education similar in respect of number of ever born children?

(iii) Are the levels of duration of marriage similar in respect of number of ever born children?

5. Two researchers separately investigated the birth-weights of babies depending on gestation period of mothers. The investigated results are as follows:

Birth-weight recorded by Researcher-1

Gestation period in days	*Birth-weight of babies*
< 210	2.5, 2.4, 2.0, 2.2
210–230	3.0, 3.5, 3.0, 3.2
230–250	4.0, 3.8, 3.5, 3.5
250–270	3.8, 3.5, 2.8, 4.0

Birth-weight recorded by Researcher-2

Gestation period in days	*Birth-weight of babies*
< 210	3.0, 3.2, 2.0, 2.5
210–230	3.0, 3.5, 3.0, 2.8
230–250	2.8, 2.9, 3.5, 3.7
250–270	4.0, 4.2, 3.0, 3.0

Do you think that the birth-weights are changing with the change in gestation period?

Chapter 9

Correlation and Regression

9.1 Regression

The statistical methods so far discussed are based on uni-variate information, viz., height, weight, age, body temperature, body mass index, blood pressure, blood sugar, etc. of human beings. These variables simultaneously can be observed from each sample point of a sample units or can be observed from all population units. If these variables are recorded from randomly selected sample units, these are termed as multivariate data. If any two of these are recorded from sample units, these are called bivariate data. As the variables are observed from one sample point, these are expected to be correlated. For example, the blood pressure (y mmHg) of any person is expected to be correlated with his age (x years), the birth-weight (y kg) of a new born baby is expected to be correlated with gestation period (x months) of mother. Here blood pressure (x) depends on age (y), birth-weight (y) depends on gestation period (x). Similarly, weight (y kg) of a person depends on height (x meter), body weight (y kg) of children depends on amount of protein (x gm) taken with food, etc. Here the variables denoted by y are termed as dependent variables and the variables x are independent or explanatory variables. We need to study how an explanatory variable influences the dependent variable and what is the strength of relationship. Regression is a technique to analyse and estimate the influence of explanatory variable on dependent variable. On the other hand, correlation is a method to study the degree of linear relationship of dependent and independent variables.

Let us explain the regression analysis by an example. Consider the values of birth-weight (y kg) of different babies and the gestation period (x months) of some mothers:

Gestation period (x *months*)	*Birth-weight* (y *kg*)	*Mean of* y *for given* x
8.0	2.5, 2.0, 2.6, 2.2	2.325
8.2	2.5, 2.4, 2.8, 2.5	2.55
8.4	3.0, 2.8, 3.2, 3.0	3.00
8.6	3.0, 3.4, 2.9, 3.2	3.125
8.8	3.5, 3.5, 3.2, 3.0	3.30
9.0	3.8, 3.0, 4.0, 4.0	3.70
9.2	4.0, 4.2, 2.8, 3.8	3.70

It is observed that the birth-weights of 4 babies are recorded against different gestation periods of mothers. The mean birth-weight (y) for a given gestation period (x) is calculated and found that with the increase in gestation period there is an increase in birth-weight. The process of estimating such a mean value of dependent variable (y) for a given value of independent variable (x) is an aspect of regression analysis. The analysis also helps us to estimate the rate of change of dependent variable (y) for a unit change in independent variable (x) and to formulate a mathematical function of mean value of dependent variable for any value of independent variable(s) so that the mean value against a given independent variable can be estimated. The mean value of dependent variable indicates that without mentioning or explaining the other causes, the cause of increase of weight is due to increase of gestation period.

Definition of regression analysis. It is a technique to study the cause and effect relationship of two or more variables when variables are recorded from each sample point of a sample. The analysis throws light on the nature of relationship of dependent and independent variables.

The mean birth-weight (y) calculated as above against a gestation period (x) can be shown by a diagram as below:

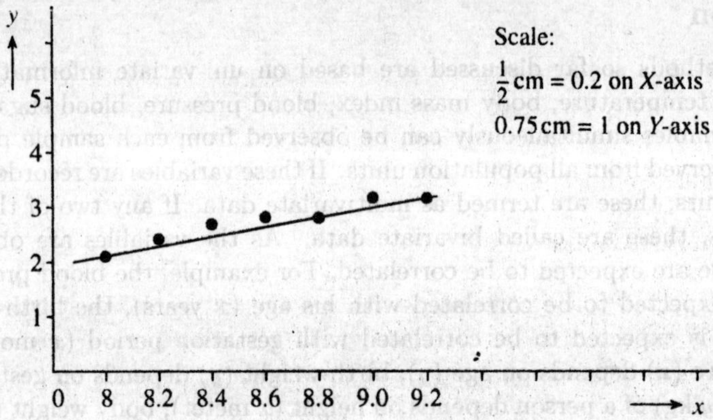

Fig. 9.1 : Scatter diagram showing the birth-weight (y) against gestation period (x).

It is seen that the mean birth-weight forms a straight line indicating a linear relationship of y and x. Such a straight line is called *simple linear regression line* and the method of fitting such a line is called *simple linear regression analysis.* However, the regression may be non-linear also and may involve more than one independent variables. When there are more than one independent variables, the regression is called *multiple regression.*

The British biometrician Sir Francis Galton (1822–1911) was the first scientist to introduce the concept of regression in connection with the inheritance of stature. He found that the offsprings of abnormally tall or short parents tend to 'regress' or step back to the average population height. The same scientist in 1888 also introduced the concept of correlation, where correlation analysis is a technique to study the degree of relationship of two variables. The amount or degree of relationship is measured by a statistic known as *correlation coefficient.* The correlation analysis does not need to identify which is dependent and which is independent variable.

9.2 Regression Model

Let $(x_1, y_1), (x_2, y_2), \ldots, (x_n, y_n)$ be n pairs of values observed from n sample points selected from a bivariate population of the variables X and Y. It is assumed that the variables X and Y are linearly related by a model $y = A + Bx + e$, where y depends on x.

Here $B = B_{yx}$ = the regression coefficient of y on x. It measures the rate of change of y for unit change in x,

A = the regression parameter. It is used to measure the value of y in absence of $x(x \doteq 0)$,

e = random component. It is used to study the influences of other independent variables, which are not included in the model.

The function of y shown above is called simple linear regression model of y on x.

In practice, y does not depend on only one explanatory variable x. For example, blood pressure (y), in general, changes with the change in age (x). But, blood pressure is also influenced by protein intake, s. cholesterol, work habit, etc. Assume that y depends on k explanatory (independent) variables x_i $(i = 1, 2, \ldots, k)$. The study of linear relationship of y and x_i's is known as multiple regression analysis and the relationship of y and x_i's is shown by a model:

$$y = B_0 + B_1x_1 + B_2x_2 + \cdots + B_xx_k + e,$$

where B_i's $(i = 0, 1, 2, \ldots, k)$ are the regression parameters and on some assumption the B_i measures the rate of change of y for unit change in x_i $(i = 1, 2, \ldots, k)$; e has the similar meaning as discussed in case of simple linear regression model.

The above function of y on x_i's is known as *multiple linear regression model.*

9.3 Regression Analysis

Regression analysis involves the estimation of parameter(s) and test the significance of parameter(s). The analysis is done on the basis of some assumptions: These are:

(i) The dependent variable y is assumed to be normally distributed. It is seen [Section 9.1] that for any given value of x there is a set of y values. It is assumed that of all sets of y values have similar variance σ^2.

(ii) The explanatory (dependent) variable(s) is/are non-random variable(s).

(iii) The explanatory variabls are uncorrelated.

(iv) There is no measurement error in measuring the values of explanatory variable(s).

(v) The error term 'e' is normally and independently distributed with mean zero and variance σ^2.

The estimation of parameter in simple linear regression analysis or in multiple linear regression analysis is done by method of least squares [The method of estimation is not shown here].

For simple linear regression model, the estimate of A is a and the estimate of B is b, where

$$a = \bar{y} - b\bar{x}, \; b = \frac{\text{S.P. } (xy)}{\text{S.S. } (x)}.$$

$$\text{S.P. } (xy) = \Sigma xy - \frac{\Sigma x \Sigma y}{n}, \; \text{S.S. } (x) = \Sigma x^2 - \frac{(\Sigma x)^2}{n}.$$

The fitted regression line is $\hat{y} = a + bx$. After fitting this line one can estimate value of y for any given value of x.

As an important step of the analysis, we can test the hypothesis:

$$H_0 : B = 0, \text{ against } H_A : B \neq 0.$$

The test statistic is $t = \dfrac{b}{\sqrt{\dfrac{s^2}{\text{S.S. }(x)}}}$, where $s^2 = \dfrac{1}{n-2}[\text{S.S. }(y) - b\,\text{S.P. }(xy)]$.

Here $\text{S.S. }(y) = \Sigma y^2 - \dfrac{(\Sigma y)^2}{n}$.

This t has $(n-2)$ d.f. The hypothesis will be rejected if $|t| \geq 5\%$ value of t with $(n-2)$ d.f. or if P-value of $t \leq 0.05$.

For multiple regression analysis, let us consider that $k = 2$. Then the multiple linear regression model is $y = B_0 + B_1x_1 + B_2x_2 + e$.

Let b_0, b_1 and b_2 be the estimate of B_0, B_1 and B_2 respectively. Then the fitted regression equation is $\hat{y} = b_0 + b_1x_1 + b_2x_2$, where $b_0 = \bar{y} - b_1\bar{x}_1 - b_2\bar{x}_2$.

$$\begin{bmatrix} b_1 \\ b_2 \end{bmatrix} = \begin{bmatrix} \text{S.S. }(x_1) & \text{S.P. }(x_1x_2) \\ \text{S.P. }(x_1x_2) & \text{S.S. }(x_2) \end{bmatrix}^{-1} \begin{bmatrix} \text{S.P. }(x_1y) \\ \text{S.P. }(x_2y) \end{bmatrix}.$$

$$\text{S.S. }(x_1) = \Sigma x_1^2 - \frac{(\Sigma x_1)^2}{n},\ \text{S.S. }(x_2) = \Sigma x_2^2 - \frac{(\Sigma x_2)^2}{n}.$$

$$\text{S.P. }(x_1x_2) = \Sigma x_1x_2 - \frac{\Sigma x_1 \Sigma x_2}{n},\ \text{S.P. }(x_1y) = \Sigma x_1y - \frac{\Sigma x_1 \Sigma y}{n}$$

$$\text{S.P. }(x_2y) = \Sigma x_2y - \frac{\Sigma x_2 \Sigma y}{n},\ \bar{x}_1 = \frac{1}{n}\Sigma x_1,\ \bar{x}_2 = \frac{1}{n}\Sigma x_2,\ \bar{y} = \frac{1}{n}\Sigma y.$$

On calculation, we have

$$b_1 = \frac{\text{S.P. }(x_1y)\ \text{S.S. }(x_2) - \text{S.P. }(x_2y)\ \text{S.P. }(x_1x_2)}{\text{S.S. }(x_1)\ \text{S.S. }(x_2) - [\text{S.P. }(x_1x_2)]^2}$$

$$b_2 = \frac{\text{S.P. }(x_2y)\ \text{S.S. }(x_1) - \text{S.P. }(x_1x_2)\ \text{S.P. }(x_1y)}{\text{S.S. }(x_1)\ \text{S.S. }(x_2) - [\text{S.P. }(x_1x_2)]^2}$$

Here we need to test the hypothesis:

H_0 : The regression is insignificant.

H_A : The regression is significant.

The regression sum of squares is

$$S_1 = \text{S.S. (regression)} = b_1\text{S.P. }(x_1y) + b_2\text{S.P. }(x_2y).$$

Sum of squares of error is

$$S_2 = \text{S.S. (error)} = \text{S.S. }(y) - \text{S.S. (regression)},\ \text{S.S. }(y) = \Sigma y^2 - \frac{(\Sigma y)^2}{n}.$$

ANOVA Table

Sources of variation	*d.f.*	*S.S.*	*M.S.* $= \dfrac{S.S.}{d.f.}$	*Calculated F*	*5% value of F*
Regression	k	S_1	$s_1 = S_1/k$	$F = s_1/s_2$	
Error	$n-k-1$	S_2	$s_2 = S_2/(n-k-1)$		
Total	$n-1$				

Null hypothesis is rejected if calculated $F \geq 5\%$ value of F or if P-value of $F \leq 0.05$.

If null hypothesis is rejected, we need to test the hypothesis:

(a) $H_0 : B_1 = 0$, against $H_A : B_1 \neq 0$.

(b) $H_0 : B_2 = 0$, against $H_A : B_2 \neq 0$.

The test statistic for (a) is

$$t = \frac{b_1}{\sqrt{\text{s.e. } (b_1)}}, \text{ s.e. } (b_1) = \sqrt{v(b_1)}, \ v(b_1) = \frac{\text{S.S. } (x_2) s_2}{\text{S.S. } (x_1) \text{ S.S. } (x_2) - [\text{S.P. } (x_1 x_2)]^2}.$$

Null hypothesis is rejected if $|t| \geq 5\%$ value of t with $(n-k-1)$ d.f. $[p(t) \leq 0.05]$

The test statistic for (b) is

$$t = \frac{b_2}{\sqrt{\text{s.e. } (b_2)}}, \text{ s.e. } (b_2) = \sqrt{v(b_2)}, \ v(b_2) = \frac{\text{S.S. } (x_1) s_2}{\text{S.S. } (x_1) \text{ S.S. } (x_2) - [\text{S.P. } (x_1 x_2)]^2}$$

Conclusion will be drawn similarly.

At this stage, one may be interested to estimate the percentage of variation of y which is explained by the explanatory variables. This percentage is calculated by

$$R^2 = \frac{\text{S.S. (regression)}}{\text{S.S. } (y)} \times 100\%.$$

Here $R = \sqrt{R^2}$ is called the multiple correlation coefficient of y with explanatory variables $x_1, x_2, \ldots, x_k$.

9.4 Interpretation of Simple Regression Coefficient

Let us consider 4 pairs of observations of height (x metre) and weight (y kg) observed from 4 persons.

x	152	153	154	155	$\Sigma x = 614$	$\bar{x} = 153.5$
y	58	60	62	64	$\Sigma y = 244$	$\bar{y} = 61$

x^2	23104	23409	23716	24025	$\Sigma x^2 = 94254$
y^2	3364	3600	3844	4096	$\Sigma y^2 = 14904$
xy	8816	9180	9548	9920	$\Sigma xy = 37464$

$$\text{S.P. } (xy) = \Sigma xy - \frac{\Sigma x \Sigma y}{n} = 37464 - \frac{614 \times 244}{4} = 10.$$

$$\text{S.S. } (x) = \Sigma x^2 - \frac{(\Sigma x)^2}{n} = 94254 - \frac{(614)^2}{4} = 5.$$

$$b = \frac{\text{S.P. } (xy)}{\text{S.S. } (x)} = \frac{10}{5} = 2, \ a = \bar{y} - b\bar{x} = 61 - 2 \times 153.5 = -246.$$

The fitted regression line is $\hat{y} = -246 + 2x$ and $\hat{y} = 70$, if $x = 158$.

Comment. $b = 2$: if x increases by 1 unit, y increases by 2 units.

Let us consider another set of observations of x and y as follows:

x	152	153	154	155	$\Sigma x = 614, \Sigma x^2 = 94254$
y	64	62	60	58	$\Sigma y = 244, \Sigma y^2 = 14904$
xy	9728	9486	9240	8990	$\Sigma xy = 37444$

$$\text{S.P. }(xy) = \Sigma xy - \frac{\Sigma x \Sigma y}{n} = 37444 - \frac{614 \times 244}{4} = -10.$$

$$b = \frac{\text{S.P. }(xy)}{\text{S.S. }(x)} = \frac{-10}{5} = -2,\ a = \bar{y} - b\bar{x} = 61 - (-2)153.5 = 368.$$

The fitted regression line is $\hat{y} = 368 - 2x, \hat{y} = 52$, if $x = 158$.

Comment. $b = -2$: if x increases by 1 unit, y decreases by 2 units.

Again, let us consider the following set of values of x and y.

x	152	153	154	155	156	157	158	159	$\Sigma x = 1244$
y	58	60	62	64	64	62	60	58	$\Sigma y = 488, \bar{y} = 61$
xy	8816	9180	9548	9920	9984	9734	9480	9222	$\Sigma xy = 75884$

$$\text{S.P. }(xy) = \Sigma xy - \frac{\Sigma x \Sigma y}{n} = 75884 - \frac{1244 \times 488}{8} = 0.$$

$$b = \frac{\text{S.P. }(xy)}{\text{S.S. }(x)} = \frac{0}{\text{S.S. }(x)} = 0,\ a = \bar{y} - b\bar{x} = 61 \text{ (as } b = 0).$$

The fitted line is $\hat{y} = 61$. It implies that y does not depend on x.

Comment. $b = 0$ implies that y does not depend on x.

This last set of values of x and y does not show any trend in the relationship of x and y. It is seen that for 50% values of x the values of y increase when x values increase. But, for another 50% values of x the values of y decrease with the increase in the values of x. This type of relationship in the values of x and y is not, normally, expected.

EXAMPLE 9.1 The following are the age (x years) and blood pressure level (y mmHg) of some persons:

x	45	65	70	72	70	50	50	40	55	58	60	$\Sigma x = 635$
y	75	85	90	88	100	85	75	70	95	90	80	$\Sigma y = 933$

(i) Fit a regression line of y on x and estimate y, if $x = 72$.

(ii) Test the significance of regression.

Solution.

(i) Let the fitted regression line be $\hat{y} = a + bx$, where

$$a = \bar{y} - b\bar{x},\ b = \frac{\text{S.P. }(xy)}{\text{S.S. }(x)};\ n = 11.$$

y^2	5625	7225	8100	7744	10000	7225	5625	4900	9025	8100	6400
xy	3375	5525	6300	6336	7000	4250	3750	2800	5225	5220	4800
x^2	2025	4225	4900	5184	4900	2500	2500	1600	3025	3364	3600

$$\Sigma y^2 = 79969.$$

$$\Sigma xy = 54581,\ \bar{y} = \frac{1}{n}\Sigma y = \frac{933}{11} = 84.82.$$

$$\Sigma x^2 = 37823,\ \bar{x} = \frac{1}{n}\Sigma x = \frac{635}{11} = 57.73.$$

$$\text{S.P. } (xy) = \Sigma xy - \frac{\Sigma x \Sigma y}{n} = 54581 - \frac{635 \times 933}{11} = 721.45.$$

$$\text{S.S. } (x) = \Sigma x^2 - \frac{(\Sigma x)^2}{n} = 37823 - \frac{(635)^2}{11} = 1166.18.$$

$$b = \frac{\text{S.P. } (xy)}{\text{S.S. } (x)} = \frac{721.45}{1166.18} = 0.619, \ a = \bar{y} - b\bar{x} = 84.82 - 0.619 \times 57.73 = 49.085.$$

Therefore, the fitted regression line is

$$\hat{y} = a + bx = 49.085 + 0.619x.$$

The estimated value of y when $x = 72$ is

$$\hat{y} = 49.085 + 0.619 \times 72 = 93.65.$$

(ii) We need to test the hypothesis H_0: $B = 0$, against $H_A : B \neq 0$.

We have

$$\text{S.S. } (y) = \Sigma y^2 - \frac{(\Sigma y)^2}{n} = 79969 - \frac{(933)^2}{11} = 833.64.$$

$$s^2 = \frac{1}{n-2} [\text{S.S. } (y) - b \text{ S.P. } (xy)] = \frac{1}{11-2} [833.64 - 0.619 \times 721.45] = 43.0065.$$

The test statistic is $t = \dfrac{b}{\sqrt{\dfrac{s^2}{\text{S.S. } (x)}}} = \dfrac{0.619}{\sqrt{\dfrac{43.0065}{1166.18}}} = \dfrac{0.619}{0.192} = 3.22.$

This t has $n - 2 = 11 - 2 = 9$ d.f.

Comment. As $|t| > 5\%$ value of t with 9 d.f. $(= 2.262)(p < 0.05)$, H_0 is rejected. Blood pressure is significantly increased with the increase in age.

9.5 Correlation

It is already been discussed that the dependent variable (y) and the independent variable (x) can be observed from each unit of a sample. As these variables are observed from same sample point, these are expected to be related. The linear relationship of two variables x and y is known as simple 'correlation'. The degree (strength) of linear relationship is measured by a quantity known as r, where

$$r = \frac{\text{S.P. } (xy)}{\sqrt{\text{S.S. } (x) \text{ S.S. } (y)}}.$$

This r is called simple correlation coefficient.

The value of r lies between -1 to $+1$ $(-1 \leq r \leq 1)$. It can be shown by example.

Let us consider the data discussed in Section 9.4.

For the first set of data sum of squares of y is

$$\text{S.S. } (y) = \Sigma y^2 - \frac{(\Sigma y)^2}{n} = 14904 - \frac{(244)^2}{4} = 20.$$

Therefore, $r = \dfrac{\text{S.P. } (xy)}{\sqrt{\text{S.S. } (x) \text{ S.S. } (y)}} = \dfrac{10}{\sqrt{5 \times 20}} = 1.$

Comment. $r = 1$ implies that x and y is perfectly positively correlated. This perfect positive correlation is observed if both x and y change uniformly in the same direction.

For the second set of data,

$$r = \frac{\text{S.P. }(xy)}{\sqrt{\text{S.S. }(x)\text{ S.S. }(y)}} = \frac{-10}{\sqrt{5 \times 20}} = -1.$$

Comment. $r = -1$ implies that x and y is perfectly negatively correlated. This perfect negative correlation is observed if both x and y change uniformly but in opposite directions.

For the third set of data, we have S.P. $(xy) = 0$. Therefore,

$$r = \frac{\text{S.P. }(xy)}{\sqrt{\text{S.S. }(x)\text{ S.S. }(y)}} = \frac{0}{\sqrt{\text{S.S. }(x)\text{ S.S. }(y)}} = 0.$$

Comment. $r = 0$ implies that x and y are not linearly related.

EXAMPLE 9.2 The following are the height (x cm) and weight (y kg) of some persons:

x	155	154	158	156	155	157	156	154	155	157
y	62	58	65	60	58	63	64	60	68	66

(i) Fit a regression line of y on x and estimate y, if $x = 156$.
(ii) Test the significance of regression.
(iii) Calculate correlation coefficient of x and y.
(iv) Test the significance of correlation.

Solution.

(i) We have $n = 10$. Let the fitted regression line be $\hat{y} = a + bx$.

x^2	24025	23716	24964	24336	24025	24649	24336	23716	24025	24649
y^2	3844	3364	4225	3600	3364	3969	4096	3600	4624	4356
xy	9610	8932	10270	9360	8990	9891	9984	9240	10540	10362

$$\Sigma x = 1557,\ \bar{x} = \frac{1}{n}\Sigma x = \frac{1557}{10} = 155.7.$$

$$\Sigma y = 624,\ \bar{y} = \frac{1}{n}\Sigma y = \frac{624}{10} = 62.4.$$

$$\Sigma x^2 = 242441, \Sigma y^2 = 39042,\ \Sigma xy = 97179.$$

$$\text{S.S. }(x) = \Sigma x^2 - \frac{(\Sigma x)^2}{n} = 242441 - \frac{(1557)^2}{10} = 16.1.$$

$$\text{S.S. }(xy) = \Sigma xy - \frac{\Sigma x \Sigma y}{n} = 97179 - \frac{1557 \times 624}{10} = 22.2.$$

$$b = \frac{\text{S.P. }(xy)}{\text{S.S. }(x)} = \frac{22.2}{16.1} = 1.38.$$

$$a = \bar{y} - b\bar{x} = 62.4 - 1.38 \times 155.7 = -152.47.$$

Therefore, the fitted regression line is

$$\hat{y} = -152.47 + 1.38x.$$

Estimate of y, if $x = 156$, is

$$\hat{y} = -152.47 + 1.38 \times 156 = 62.81.$$

(ii) We need to test the hypothesis $H_0 : B = 0$, against $H_A : B \neq 0$.

Sum of squares of regression

$$\text{SSR} = b\,\text{S.P.}\ (xy) = 1.38 \times 22.2 = 30.636.$$

$$\text{S.S.}\ (y) = \Sigma y^2 - \frac{(\Sigma y)^2}{n} = 39042 - \frac{(624)^2}{10} = 104.4.$$

Mean sum of squares of error

$$s^2 = \frac{1}{n-2}\left[\text{S.S.}\ (y) - b\ \text{S.P.}\ (xy)\right]$$

$$= \frac{1}{10-2}[104.4 - 30.636] = 9.2205.$$

The test statistic is

$$t = \frac{b}{\sqrt{\frac{s^2}{\text{S.S.}\ (x)}}} = \frac{1.38}{\sqrt{\frac{9.2205}{16.1}}} = \frac{1.38}{0.757} = 1.82.$$

Comment. As $|t| < 5\%$ value of t with $10 - 2 = 8$ d.f. $= 2.306$, H_0 is accepted. The regression is not significant.

(iii) The correlation coefficient of x and y is

$$r = \frac{\text{S.P.}\ (xy)}{\sqrt{\text{S.S.}\ (x)\ \text{S.S.}\ (y)}} = \frac{22.2}{\sqrt{16.1 \times 104.4}} = \frac{22.2}{40.998} = 0.54.$$

(iv) We need to test the hypothesis

$$H_0 : P = 0, \text{ against } H_A : P \neq 0,$$

where P is the population correlation coefficient.

The test statistic is

$$t = \frac{r\sqrt{n-2}}{\sqrt{1-r^2}} = \frac{0.54\sqrt{10-2}}{\sqrt{1-(0.54)^2}} = \frac{1.527}{0.8417} = 1.81.$$

Comment. As $|t| < 5\%$ value of t with 8 d.f. (= 2.306), H_0 is accepted. x and y are not significantly correlated.

EXAMPLE 9.3 The following are number of ever born children (y) of some mothers according to their duration of marriage (x in completed years) and level of education (z in completed years of schooling):

x	15	20	5	8	22	18	15	10	10	15	4	3	6	10	15	16
z	8	0	10	12	0	0	14	6	0	0	14	12	10	14	0	4
y	4	6	1	2	6	6	2	3	3	5	1	1	2	3	4	5

(i) Fit a linear regression model of y on x and z and estimate y, if $x = 12, z = 16$.

(ii) Test the significance of regression. Do you think that education has significant impact on number of ever born children?

(iii) Find percentage of variation of y which is explained by x and z. Calculate multiple correlation coefficient of y, x and z.

Solution. We have $n = 16$

x^2	225	400	25	64	484	324	225	100	100	225	16	9	36	100	225	256
z^2	64	0	100	144	0	0	196	36	0	0	196	144	100	196	0	16
y^2	16	36	1	4	36	36	4	9	9	25	1	1	4	9	16	25
xz	120	0	50	96	0	0	210	60	0	0	56	36	60	140	0	64
xy	60	120	5	16	132	108	30	30	30	75	4	3	12	30	60	80
yz	32	0	10	24	0	0	28	18	0	0	14	12	20	42	0	20

$$\Sigma x = 192,\ \bar{x} = \frac{1}{n}\Sigma x = \frac{192}{16} = 12,\ \Sigma z = 104,\ \bar{z} = \frac{1}{n}\Sigma z = \frac{104}{16} = 6.5.$$

$$\Sigma y = 54,\ \bar{y} = \frac{1}{n}\Sigma y = \frac{54}{16} = 3.375.$$

$$\Sigma x^2 = 2814,\ \Sigma z^2 = 1092,\ \Sigma y^2 = 232,\ \Sigma xz = 892,\ \Sigma xy = 795,\ \Sigma yz = 220.$$

$$\text{S.S. } (x) = \Sigma x^2 - \frac{(\Sigma x)^2}{n} = 2814 - \frac{(192)^2}{16} = 510.$$

$$\text{S.S. } (y) = \Sigma y^2 - \frac{(\Sigma y)^2}{n} = 232 - \frac{(54)^2}{16} = 49.75.$$

$$\text{S.S. } (z) = \Sigma z^2 - \frac{(\Sigma z)^2}{n} = 1092 - \frac{(104)^2}{16} = 416.$$

$$\text{S.P. } (xz) = \Sigma xz - \frac{\Sigma x \Sigma z}{n} = 892 - \frac{192 \times 104}{16} = -356.$$

$$\text{S.P. } (xy) = \Sigma xy - \frac{(\Sigma x \Sigma y}{n} = 795 - \frac{192 \times 54}{16} = 147.$$

$$\text{S.P. } (yz) = \Sigma yz - \frac{\Sigma y \Sigma z}{n} = 220 - \frac{54 \times 104}{16} = -131.$$

(i) The fitted regression line is $\hat{y} = b_0 + b_1 x + b_2 z$, where

$$b_1 = \frac{\text{S.P. } (xy)\ \text{S.S. } (z) - \text{S.P. } (yz)\ \text{S.P. } (xz)}{\text{S.S. } (x)\ \text{S.S. } (z) - [\text{S.P. } (xz)]^2}$$

$$= \frac{147 \times 416 - (-131)(-356)}{510 \times 416 - (-356)^2} = \frac{14516}{85424} = 0.17.$$

$$b_2 = \frac{\text{S.P. } (yz)\ \text{S.S. } (x) - \text{S.P. } (xz)\ \text{S.P. } (xy)}{\text{S.S. } (x)\ \text{S.S. } (z) - [\text{S.P. } (xz)]^2}$$

$$= \frac{-131 \times 510 - (-356)(147)}{510 \times 416 - (-356)^2} = \frac{-14478}{85424} = -0.17.$$

$$b_0 = \bar{y} - b_1\bar{x} - b_2\bar{z} = 3.375 - 0.17 \times 12 - (-0.17)6.5 = 2.8.$$

Therefore, the fitted line is

$$\hat{y} = 2.8 + 0.17x - 0.17z.$$

The value of $\hat{y}$, when $x = 12$ and $z = 16$, is

$$\hat{y} = 2.8 + 0.17 \times 12 - 0.17 \times 16 = 2.12 \approx 2.$$

Alternatively, the values of b_1 and b_2 are:

$$\begin{bmatrix} b_1 \\ b_2 \end{bmatrix} = \begin{bmatrix} \text{S.S. } (x) & \text{S.P. } (xz) \\ \text{S.P. } (xz) & \text{S.S. } (z) \end{bmatrix}^{-1} \begin{bmatrix} \text{S.P. } (xy) \\ \text{S.P. } (yz) \end{bmatrix}$$

$$= \begin{bmatrix} 510 & -356 \\ -356 & 416 \end{bmatrix}^{-1} \begin{bmatrix} 147 \\ -131 \end{bmatrix} = \frac{1}{85424} \begin{bmatrix} 416 & 356 \\ 356 & 510 \end{bmatrix} \begin{bmatrix} 147 \\ -131 \end{bmatrix}$$

$$= \begin{bmatrix} 0.0048698 & 0.0041674 \\ 0.0041674 & 0.0059702 \end{bmatrix} \begin{bmatrix} 147 \\ -131 \end{bmatrix}$$

$$= \begin{bmatrix} 0.0048698 \times 147 + (0.0041674)(-131) \\ 0.0041674 \times 147 + (0.0059702)(-131) \end{bmatrix} = \begin{bmatrix} 0.17 \\ -0.17 \end{bmatrix}.$$

(ii) Sum of squares of regression is

$$S_1 = b_1 \text{S.P. } (xy) - b_2 \text{S.P. } (zy)$$
$$= 0.17 \times 147 - (-0.17)(-131) = 47.26 = \text{S.S. (regression)}.$$

Sum of squares of error is $S_2 = \text{S.S. } (y) - S_1 = 49.75 - 47.26 = 2.49$.

ANOVA Table

Sources of variation	*d.f.*	*S.S.*	*M.S.* $= \frac{S.S.}{d.f.}$	*Calculated F*	*5% value of F*
Regression	2	$S_1 = 47.26$	$s_2 = 23.63$	$F = s_1/s_2 = 123.39$	3.80
Error	13	$S_2 = 2.49$	$s_2 = 0.1915$		
Total	15				

The null hypothesis to be tested is:

H_0 : The regression is insignificant.

H_A : The regression is significant.

Comment. As calculated $F >$ 5% value of F with 2 and 13 d.f. (= 3.80), H_0 is rejected. The regression is significant.

Second part. We need to test $H_0 : B_2 = 0$, against $H_A : B_2 \neq 0$.

The test statistic is $t = \dfrac{b_2}{\text{s.e. } (b_2)} = \dfrac{-0.17}{0.0305} = -5.57$.

$$v(b_2) = \frac{\text{S.S. } (z) s_2}{\text{S.S. } (x)\text{S.S. } (z) - [\text{S.P. } (xz)]^2} = \frac{416 \times 0.1915}{510 \times 416 - (-356)^2} = \frac{79.664}{85424} = 0.00093$$

s.e. $(b_2) = \sqrt{v(b_2)} = \sqrt{0.00093} = 0.0305$.

Comment. As $|t| >$ 5% value of t with 13 d.f. (= 2.16), H_0 is rejected. This implies **that number of ever born children is significantly decreased with the increase in level of education.**

(iii) The percentage of variation of y explained by x and z is measured by

$$R^2 = \frac{\text{S.S. (regression)}}{\text{S.S. } (y)} = \frac{S_1}{\text{S.S. } (y)} = \frac{47.26}{49.75} = 0.9499 \text{ or } 94.99\%$$

The multiple correlation coefficient is $R = \sqrt{R^2} = \sqrt{0.9499} = 0.97$.

EXAMPLE 9.4 Using the data of Example 9.3 fit a regression line of y on x and test the significance of regression.

Solution. The fitted regression line is

$$\hat{y} = a + bx, \text{ where } \quad a = \bar{y} - b\bar{x}, \; b = \frac{\text{S.P. }(xy)}{\text{S.S. }(x)} = \frac{147}{510} = 0.29$$

$$= -0.105 + 0.29x \qquad = 3.375 - 0.29 \times 12$$

$$= -0.105.$$

We need to test the hypothesis $H_0 : B = 0$, against $H_A : B \neq 0$.

The test statistic is $t = \dfrac{b}{\sqrt{\dfrac{s^2}{\text{S.S. }(x)}}}$,

where $s^2 = \dfrac{1}{n-2}[\text{S.S. }(y) - b\,\text{S.P. }(xy)] = \dfrac{1}{16-2}[49.75 - 0.29 \times 147] = 0.5086.$

Therefore, $t = \dfrac{0.29}{\sqrt{\dfrac{0.5086}{510}}} = \dfrac{0.29}{0.0316} = 9.18.$

Comment. As $|t| > 5\%$ value of t with 14 d.f. $(= 2.145)$, H_0 is rejected. Number of ever born children is significantly increased with the increase in duration of marriage.

EXERCISE – 9

1. (a) Show, by an example, that the correlation coefficient of two variables x and y is (i) $r_{xy} = -1$, (ii) $r_{xy} = 1$.

   (b) The following are the amounts of s. triglyceride (x mg/dl) and s. cholesterol (y mg/dl) of some persons:

x	75	80	82	84	85	88	70	90	95	80
y	130	120	125	130	140	150	120	150	140	110

   Calculate correlation coefficient of x and y and test the significance of correlation.

2. (a) Show, by an example, that the regression coefficient of y on x is -2.

   (b) The following are the gestation period (x months) and birth-weight (y kg) of babies of some mothers

x	8.5	8.0	9.0	9.1	8.8	8.5	8.4	8.6	8.0	9.0
y	2.0	2.5	4.0	3.8	2.9	2.8	3.0	3.5	3.0	4.5

   (i) Fit a regression line of y on x and test the significance of regression.

   (ii) Estimate birth-weight y of a baby who takes birth after 8.9 months gestation period.

3. (a) What does regression coefficient measure? What does correlation coefficient measure? Comment on $r = -1$, $r = 1$, $r = 0$.

   (b) The following are the heights (x cm) and weights (y kg) of some persons:

x	152	155	156	157	155	158	152	150	155	156
y	55	60	65	70	62	68	50	58	64	65

   (i) Fit a regression line of y on x and test the significance of regression.

   (ii) Calculate correlation coefficient of x and y and test the significance of correlation.

Chapter 10

Analysis of Time Series

10.1 Introduction

The statistical data which are recorded at different points of time (day, month, year, etc.) are known as time series data. For example, number of patients admitted in a hospital in different days or in different months or in different seasons or in different years; number of patients died in different months or in different years; number of nurses are trained in different years, number of doctors coming out of college in different years, amount of production of medicines in different years, etc. The values of the variables mentioned here are changed with the change in time.

The time series data are influenced by many causes. These causes are affecting the values of the variable in a regular fashion or in a regular time interval. For example, the death of children occurres in different days due to natural cause and again, it occurres due to seasonal variation. Also, it occurres due to some natural calamity. This means that the time series data are influenced by many causes and data are changed in different degrees over time.

Time series analysis is done to study the influences of the components which affect the time series data. The nature of the influences and the relationship of the causes can be studied by analysing time series data.

10.2 Components of Time Series Data

Time series data are influenced by four components. These are:

(a) Trend (T)

(b) Seasonal variation (S)

(c) Cyclical variation (C) and

(d) Irregular variation (I).

Trend (T). It is called long term trend. Time series data are increased or decreased or they remain similar over a long time. The trend is a change in the data for a long time and is known as long term trend. For example, the child and infant mortality in Bangladesh has been decreasing since long time. The trend in gradual decrease of child loss is known as long term trend. The production of medicine is in increasing trend. The MBBS doctors are increasing gradually for a long time.

The analysis of trend helps us to compare two or more time series data. If long term trend is eliminated from time series data, the influences of short term movement and irregular variation can be studied.

Seasonal variation (S). The number of patients of dysentry increases during rainy season. The child death also increases during this season. The number of patients having cough and cold increases during winter season. During this season the demand of cough syrup is increased. The seasonal change in time series data is known as seasonal variation.

The study of seasonal variation helps the health management authority to take precautionary measures to avoid the child loss in seasons. The pharmaceutical company may produce more medicines for cough and cold during winter season.

Irregular variation (I). If trend, seasonal variation and cyclical fluctuation can be eliminated from time series data, the remaining part of the data is irregular variation. This variation occurs suddenly. For example, the number of death increases due to cyclone or earthquake. Cyclone or earthquake occurres at irregular interval. This type of change in time series data is known as irregular variation.

Cyclical fluctuation (C). The time series data though gradually increase or decrease over time, still all observations of different time may not fall in a straight line. The values may fall below and above the trend line either in a regular fashion or irregularly. If the movement of time series data above and below the trend line in a regular fashion continues for a long time in a year, it is called cyclical fluctuations. For example, the consumption of cough syrup increases during winter and rainy seasons and decreases during summer and spring seasons. The fluctuations are observed with the changes in the seasons. This cyclical fluctuation does not follow any rule. It occurs completely irregularly even after 5–6 years or after 8–10 years.

10.3 Analysis of Time Series

It has already been discussed that the time series data are influenced by 4 components, namely, trend (T), seasonal variation (S), cyclical fluctuation (C) and irregular variation (I). If time series observation is denoted by O, one can formulate two different types of relationships among O, T, S, C and I. Generally, a multiplied relationship as follows, is formulated:

$$O = T \times S \times C \times I.$$

However, additive relationship as follows, is also considered:

$$O = T + S + C + I.$$

The objectives of the time series analysis are:

(i) The components by which time series data are influenced need to be identified and their influences are to be measured.

(ii) The identified components are to be removed one by one independently and prediction is made about the components.

Time series analysis is very much important to the economist, health planners, hospital management authorities, population policy planners, maintenance of law and order authorities, etc. It is used to compare and contrast the phenomenon of different time periods and different places.

In doing time series analysis some precautions are needed. These are (i) adjustment in the data due to the change in time period; for example, if monthly data are recorded, adjustment due to the changes in dates of a month must be made; (ii) adjustment due to the change in the population size; for example, consumption of medicine in the country or in a hospital changes with the change in population size, change in number of beds, etc. After the adjustment the time series analysis is performed mainly to measure the trend and seasonal variation. In this chapter, we shall discuss only the measurement of trend by two methods, namely, (a) moving average method, (b) method of least squares.

10.4 Trend Values by the Method of Moving Average

If time series data are recorded for different years, the moving averages of some years, even or odd number of years, are calculated. The averages of odd number of years are usually preferred so that the calculated average is placed in the central position of the years. Usually, 3-yearly, 5-yearly, 7-yearly $\cdots$ and so on averages are calculated. In case of 3-yearly moving average, the average value is placed against the middle of the three years. Similar process is followed for 5-yearly or 7-yearly averages. When an average of 3-years is calculated, the first value included in the average is discarded and observation of fourth year is included for the second moving average. The process is continued until all the observations are used to calculate the average. The calculated average values are trend values of the time series data.

If the averages of even number of years are calculated, two-yearly moving averages are again to be calculated to place the averages against years.

Let us explain the method by a small example. Let the number of nurses be working in a clinic in different years. The data are as follows:

Year	2001	2002	2003	2004	2005	2006	2007	2008
Number of nurses	15	18	20	22	22	25	25	30

We can calculate 3-yearly moving average to find the trend values in the number of nurses working.

Year	*Number of nurses*, 0	*3-yearly moving average* [*Trend values*, *T*]	*4-yearly moving average*	*2-yearly moving average* (*T*)
2001	15	—	—	—
2002	18	$\frac{1}{3}(15+18+20)=17.7$		—
			18.75	
2003	20	$\frac{1}{3}(18+20+22)=20.0$		$\frac{1}{2}(18.75+20.50)=19.625$
			20.50	
2004	22	$\frac{1}{3}(20+22+22)=21.3$		$\frac{1}{2}(20.5+22.25)=21.375$
			22.25	
2005	22	$\frac{1}{3}(22+22+25)=23.0$		$\frac{1}{2}(22.25+23.5)=22.875$
			23.50	
2006	25	$\frac{1}{3}(22+25+25)=24.0$		$\frac{1}{2}(23.5+25.5)=24.50$
			25.50	
2007	25	$\frac{1}{3}(25+25+30)=26.7$		—
2008	30	—		—

The trend values of the first $\frac{1}{2}(n-1)$ years and last $\frac{1}{2}(n-1)$ years are not found out, if n is odd, where n is the number of years used to calculate moving agerage. The trend values of first and last $\frac{n}{2}$th years are also not available, if n is even. However, these values can be observed by drawing trend line in the graph paper plotting trend values in Y-axis against the time in X-axis. The line is extended in both sides to find the values which are not calculated mathematically.

EXAMPLE 10.1 The following are the number of dead patients in a hospital in different years:

Year	1990	1991	1992	1993	1994	1995	1996	1997	1998	1999
No. of dead patients	125	128	127	122	130	135	128	125	124	127
Year	2000	2001	2002	2003	2004	2005	2006	2007	2008	2009
No. of dead patients	142	141	135	136	130	125	120	118	110	105

Find trend values by 5-yearly moving average method.

Solution.

Year	*5-yearly moving total*	*5-yearly moving average (T)*	*Year*	*5-yearly moving total*	*5-yearly moving average (T)*
1990	—	—	2000	669	133.8
1991	—	—	2001	681	136.2
1992	632	126.4	2002	684	136.8
1993	642	128.4	2003	667	133.4
1994	642	128.4	2004	646	129.2
1995	640	128.0	2005	629	125.8
1996	642	128.4	2006	603	120.6
1997	639	127.8	2007	578	115.6
1998	646	129.2	2008	—	—
1999	659	131.8	2009	—	—

10.5 Trend Values by the Method of Least Squares

In this method a straight line trend is assumed for trend values, where the line is

$$y = a + bt.$$

Here t = time taking values $1, 2, 3, \ldots, n$;

y = time series values for different values of t.

By the method of least squares, we have

$$b = \frac{\text{S.P. } (yt)}{\text{S.S. } (t)} \text{ and } a = \bar{y} - b\bar{t}.$$

$$\text{Here S.P. } (yt) = \Sigma yt - \frac{\Sigma y \Sigma t}{n},\ \bar{t} = \frac{1}{n}\Sigma t,\ \bar{y} = \frac{1}{n}\Sigma y.$$

EXAMPLE 10.2 Using the data of Example 10.1 fit a straight line trend of the number of dead patients and predict the number of dead patients in 2015.

Solution.

Year	t	*No. of dead patients, y*	t^2	*yt*	*Year*	t	*No. of dead patients, y*	t^2	*yt*
1990	1	125	1	125	2000	11	142	121	1562
1991	2	128	4	256	2001	12	141	144	1692
1992	3	127	9	381	2002	13	135	169	1755
1993	4	122	16	488	2003	14	136	196	1904
1994	5	130	25	650	2004	15	130	225	1950
1995	6	135	36	810	2005	16	125	256	2000
1996	7	128	49	896	2006	17	120	289	2040
1997	8	125	64	1000	2007	18	118	324	2124
1998	9	124	81	1116	2008	19	110	361	2090
1999	10	127	100	1270	2009	20	105	400	2100
Total	55	1271	385	6992	Total	155	1162	2486	19217

$$n = 20,\ \Sigma t = 55 + 155 = 210,\ \bar{t} = \frac{1}{n}\Sigma t = \frac{210}{20} = 10.5.$$

$$\Sigma y = 1271 + 1162 = 2433,\ \bar{y} = \frac{1}{n}\Sigma y = \frac{2433}{20} = 121.65.$$

$$\Sigma t^2 = 385 + 2486 = 2870,\ \Sigma yt = 6992 + 19217 = 26209.$$

$$\text{S.S.}\,(t) = \Sigma t^2 - \frac{(\Sigma t)^2}{n} = 2870 - \frac{(210)^2}{20} = 665.$$

$$\text{S.P.}\,(yt) = \Sigma yt - \frac{\Sigma y \Sigma t}{n} = 26209 - \frac{2433 \times 210}{20} = 662.5.$$

$$b = \frac{\text{S.P}\,(yt)}{\text{S.S.}\,(t)} = \frac{662.5}{665} = 0.996.$$

$$a = \bar{y} - b\bar{t} = 121.65 - 0.996 \times 10.5 = 111.192.$$

Therefore, the fitted trend line is $y = a + bt = 111.192 + 0.996t$.

Estimate of number of dead patients in 2015 ($t = 26$) is

$$y = 111.192 + 0.996 \times 26 = 137.088 \approx 137.$$

Chapter 11

Demography

11.1 Introduction

Birth, death, marriage, divorce, migration, etc. influence family, society, area as well as country. These events are again influenced by social, cultural, economical, socio-economical, socio-political, etc. factors. These events are known as vital events. Demography deals with these vital events and analyses how these vital events change the population size, structure, regional variation and also analyses the causes of changes in population structure.

Definition of demography. The statistical and mathematical analysis of change in population size, structure, regional variation and change in population components, namely vital events is known as demography.

According to Hauser and Duncan demographic analysis deals with variation and causes of variation of population. In the opinion of other researchers, demography deals with collection, compilation, statistical analysis and mathematical presentation of components which influence the population size. Whatever be the opinion, the vital events play a significant role in the study of population size and structure.

11.2 Method of Collection of Vital Statistics

The statistics of vital events are collected by

(a) Census method,

(b) Sample survey method, and

(c) Registration of vital events.

Census method. Census method is a technique to collect information on demographic events, economical and social data from all individuals/households living in a specific sovereign country after a certain period of time. The census data, after collection, are edited, compiled and presented in a lucid form so that administrative authority can use it for national development.

Survey method. Census is usually done after ten years and there may be chances of errors to be creeped in census data. To avoid this problem and to guess the population composition within intercensal period, alternative technique is used in a smaller scale using some selective localities from the country as a whole. This latter technique of collection of statistical data is known as survey method.

Registration of vital events. Birth, death, marriage, divorce, migration occur at every moment in every locality. These events are, usually, recorded in the offices of unions, municipalities and city corporations. These offices can maintain the records of these events properly and correctly, if trained and responsible persons are employed for the jobs. However, in our country vital registration system is not strong enough till now, specially in rural areas.

The vital registration system is necessary to study the changes in population composition. More specifically, to study the measure of fertility and mortality and hence, to project the future population.

11.3 Measure of Fertility

According to Barclay, fertility is the level of completion of ever born children to mothers of a country/area. The measure of fertility can throw light in the direction of change in the population. The measures of fertility are:

(i) Crude Birth Rate (CBR),
(ii) General Fertility Rate (GFR),
(iii) General Matrimonial Fertility Rate (GMFR),
(iv) Age-specific Fertility Rate (ASFR),
(v) Age-specific Matrimonial Fertility Rate (ASMFR),
(vi) Total Fertility Rate (TFR),
(vii) Standardized Birth Rate (SBR).

(i) **Crude birth rate (CBR).** Let

P = total mid-year population of a country/area in a year,

B = total live births in the country/area in that year.

Then CBR is measured by $\text{CBR} = \dfrac{B}{P} \times 1000.$

This measure provides a rough idea about the fertility level. But, it has some limitations. These are: (a) In its calculations the mid-year population of the country is used as denominator. As a result, the calculated rate is an underestimate of fertility rate. The denominator used should be the female population of age group 15-49 years as these females are capable to give birth.

(b) Due to differential population distribution by age and sex in different localities, the CBR is not suitable to compare the fertility rates of two or more localities/countries.

As an example of CBR, let

B = total live births of an area in 2011 = 3155600.

P = total mid-year population of that area = 156834000.

Then $\text{CBR} = \dfrac{3155600}{156834000} \times 1000 = 20.12.$

The birth rate per thousand people is 20.12. In Bangladesh, the CBR is in decreasing trend since 1991 as is observed from the record of BBS [Statistical Pocket Book, 98; Bangladesh Bureau of Statistics].

Year	1991	1992	1993	1994	1995	1996	1997	1998
CBR	31.6	30.8	28.8	27.8	26.5	25.6	21.0	19.9

(ii) **General fertility rate (GFR).** In calculating CBR, we have used total mid-year population as denominator. But the birth of babies depends on female population of child-bearing ages 15-49. Hence, birth rate should be calculated using mid-year female population of ages 15-49 as denominator.

Let B = total births of a country/locality in a year,

$W_{15\text{-}49}$ = total mid-year female population of child-bearing ages 15-49 of the country in that year.

Then GFR is measured by $\text{GFR} = \dfrac{B}{W_{15\text{-}49}} \times 1000.$

This measure avoids limitation of CBR due to the use of total mid-year population. However, it is also affected by age distribution of female population as in some localities female population of ages 15-49 may be more than that of other localities. Moreover, all females of ages 15-49 may not give birth as some of them may not be married, some may be widowed, some may be divorced. As a result GFR may be affected.

(iii) **General matrimonial fertility rate (GMFR).** To avoid the limitation in the calculation of GFR the currently married women of ages 15-49 may be used as denominator in calculation of GMFR, where

$$\text{GMFR} = \frac{B}{W^m_{15\text{-}49}} \times 1000,$$

where $W^m_{15\text{-}49}$ = mid-year currently married women of ages 15-49 of the country/locality in the year.

For example, let

B = total live births of an area in a year = 267435.

$W_{15\text{-}49}$ = total mid-year female population of the area in that year = 6380780.

$$\text{The GFR} = \frac{B}{W_{15\text{-}49}} \times 1000 = \frac{267435}{6380780} \times 1000 = 42.11.$$

For the same population in the same year, let $W^m_{15\text{-}49} = 5244963$.

$$\text{Then GMFR} = \frac{B}{W^m_{15\text{-}49}} \times 1000 = \frac{267435}{5244963} \times 1000 = 50.99.$$

(iv) **Age-specific fertility rate (ASFR$_x$).** In calculating GFR we have used mid-year female population of child-bearing ages 15-49 as denominator. Currently married women of these ages usually give births. But the birth rates during initial child-bearing period and last stage of child-bearing period differ from those of middle stage child-bearing period. In Bangladesh maximum births take place during the period 20-34 years. This is observed from the ASFR of Bangladesh as shown below:

Age / Year	15-19	20-24	25-29	30-34	35-39	40-44	45-49	TFR
	ASFR							
1987	86.0	247.0	243.0	161.0	105.0	31.0	10.0	4.42
1988	78.1	248.0	243.6	153.4	100.3	37.9	17.2	4.39
1998	47.0	174.0	171.0	103.0	81.0	14.0	5.0	2.98

[*Source:* Statistical Year Book of Bangladesh, 1992 : BBS]

It is seen that the birth rates of females of some specific age groups are more than others. Thus, it needs to measure the age-specific fertility rate (ASFR), where

$$\text{ASFR}_x = \frac{B_x}{W_x} \times 1000.$$

$$\text{Again, ASMFR}_x = \frac{B_x}{W_x^m} \times 1000.$$

Here B_x = total births to mothers of age group x in the country in a year,

W_x = total mid-year female population of age group x in the country in that year.

W_x^m = total mid-year married female population of age group x in the country in that year.

The usual age groups are 15-19, 20-24, 25-29, 30-34, 35-39, 40-44 and 44-49.

Let us observe the age-specific fertility rate and age-specific matrimonial fertility rate as shown below:

Age group	15-19	20-24	25-29	30-34	35-39	40-44	45-49
B_x	4528	90626	154556	176970	116441	5098	462
W_x	129675	386277	2096882	2290624	183641	186024	100960
$\frac{B_x}{W_x} \times 1000$	34.92	234.61	73.71	77.26	63.41	27.41	4.58
W_x^m	98245	305564	1809762	1962446	179644	175862	98625
$\frac{B_x}{W_x^m} \times 1000$	46.09	296.59	85.40	90.18	64.82	28.99	4.68

(vi) **Total fertility rate (TFR).** Total fertility rate is the sum of the age-specific fertility rate.

It is given by $\text{TFR} = \sum_{x=15}^{49} \text{ASFR}_x$, if $x = 15, 16, \ldots, 49$.

Again, $\text{TFR} = 5\sum_{x=1}^{7} \text{ASFR}_x$, if x is shown as age group 15-19, 20-24, . . ., 44-49.

In the above example of ASFR_x, we have

$$\begin{aligned} \text{TFR} &= 5\sum \text{ASFR}_x \\ &= 5\,[0.03492 + 0.23461 + 0.07371 + 0.07726 + 0.06341 + 0.02741 + 0.00458] \\ &= 2.58 \end{aligned}$$

This fertility rate indicates that per 1000 women of ages 15-49 are expected to give birth 2.58 children provided none will die during this child-bearing periods.

(vii) **Standardized birth rate (SBR).** The birth rate changes due to the changes in sex composition, age composition and population distribution. These compositions are different in different countries/localities. In such a situation the CBR is not used to compare the birth rates of two or more populations. The comparison of birth rates are done if standardized birth rate is calculated.

The direct method of standardized birth rate is calculated by the formula:

$$\text{DSBR} = 5\sum_{x=1}^{7} \text{ASFR}_x^P \frac{W_x^s}{P^s},$$

where ASFR_x^P = age-specific fertility rate of age group x for a population, P,

W_x^s = female population of age group x in standard population, s,

P^s = total population in standard population.

Let us consider the calculation of DSBR using data given below:

Age group	15-19	20-24	25-29	30-34	35-39	40-44	45-49
B_x^P	5062	95362	167738	207645	118960	4596	562
W_x^P	149466	417682	1736854	2093460	1536708	187845	113645
W_x^s	150000	400000	2500000	2600000	2000000	2000000	1200000
$\frac{B_x^P}{W_x^P}$	0.03387	0.22831	0.09657	0.09919	0.07741	0.02447	0.00494
$\frac{B_x^P}{W_x^P} \times W_x^s$	5080	91325	241439	257887	154824	48934	5934

Therefore, $\text{DSBR} = 5\sum \text{ASFR}_x^P \frac{W_x^s}{P^s}$, where $P^s = 25000000$

$$= 5 \times \frac{805423}{25000000} = 0.16108 \text{ or } 161.08 \text{ per 1000 women.}$$

Gross reproductive rate (GRR)

It has already mentioned that TFR indicates the total expected births per women during her child-bearing ages 15-49 years. The born baby may be son or daughter. The present female birth will be the future mother if she reaches up to the child-bearing age. The future mother will enhance the population size. Therefore, we need to study the separate female birth rate. This birth rate can be used as a measure of growth of population.

The gross reproduction rate is the rate of female birth who will be the future mother on the assumption that none of female babies will die before reaching their child-bearing period.

Let B_x^f = total female birth in a year in the country,

B_x = total birth in a year in the country,

W_x = mid-year female population of child-bearing ages 15-49 in the country in that year.

$$\text{Then GRR} = \sum_{x=15}^{49} \frac{B_x^f}{W_x} = \sum \frac{B_x}{W_x} \times \frac{B_x^f}{B_x}.$$

Here $x = 15, 16, 17, \ldots, 49$. B_x^f/B_x is the ratio of female birth to total birth (sometimes it is replaced by sex ratio). If x is expressed in age group 15-49, 20-24, ..., 44-49, then

$$\text{GRR} = 5\sum_{h=1}^{7} \frac{B_x^f}{W_x}.$$

Sometimes GRR is also calculated by the formula:

$$\text{GRR} = \text{TFR} \times \frac{\sum B_x^f}{\sum B_x}.$$

Comment on GRR

(i) GRR = 1 indicates that, if the present fertility rate is not changed and if none of female babies dies, then there will be no change in the number of child-bearing mothers in future.

(ii) GRR > 1 indicates that, the child-bearing mothers will increase in future.

(iii) GRR < 1 indicates that, the child-bearing mothers will decrease in future.

Net reproduction rate (NRR)

In calculating GRR, if it is assumed that none of the female babies will die before reaching child-bearing age or during child-bearing age 15-49. This is one of the limitation in calculation of GRR. By nature or by many other reasons any human being may die in any age group. Any female can also die during child-bearing age or before reaching child-bearing age. To avoid this problem, net reproduction rate is calculated using the female death rate. The female death rate is utilized from life table calculated only for female population.

Then, $\text{NRR} = 5\sum_{h=1}^{7} \frac{B_{hx}^f}{W_{hx}} \times \frac{L_x}{l_0}$, where

L_x = person years to survive by female population of an age group $[x - (i + x); i = 4]$

l_0 = initial female cohort in calculating female life table.

Comment on NRR

(i) NRR = 1 indicates that, in any population a mother will be replaced by one new mother, on an average. The population of the country will show a trend of constant size.

(ii) NRR > 1 indicates that, one present mother will be replaced, on an average, by more future mothers and there may be an increasing trend in the population size.

(iii) NRR < 1 indicates that, one present mother will be replaced, on an average, by less number of mother. The population size will show decreasing trend.

It is known that, for every age or age group $\frac{L_x}{l_0} < 1$ and hence, NRR < GRR.

As an example, let us calculate GRR and NRR using the data given below:

Age group	15-19	20-24	25-29	30-34	35-39	40-44	45-49
B_x	6572	100458	203865	259648	154967	6098	680
W_x	174099	583699	2144971	2617308	1836888	223764	124499
B_x^f	3547	56943	102624	130083	74361	4058	475
$\frac{L_x}{l_0}$	0.93	0.92	0.90	0.89	0.88	0.87	0.85

Calculation of GRR and NRR

Age group	15-19	20-24	25-29	30-34	35-39	40-44	45-49
$\frac{B_x^f}{W_x}$	0.02037	0.09756	0.04784	0.04970	0.04048	0.01814	0.00382
$\frac{B_x}{W_x}$	0.03775	0.17211	0.09504	0.09920	0.08436	0.02725	0.00546
$\frac{B_x^f}{W_x} \times \frac{L_x}{l_0}$	0.01895	0.08975	0.04306	0.04423	0.03562	0.01578	0.00324

Therefore, $\text{GRR} = 5 \times \sum_{h=1}^{7} \frac{B_{hx}^f}{W_{hx}} = 5 \times 0.27791 \approx 1.39.$

The result indicates that the child-bearing mothers will be increased in future compared to the present number of child-bearing mothers.

We have $\text{TFR} = 5 \sum_{h=1}^{7} \frac{B_{hx}}{W_{hx}} = 5 \times 0.52117 = 2.60585.$

Assuming sex ratio as $M : F :: 100 : 100.3$, we have

$$\text{GRR} = \text{TFR} \times \frac{M}{M+F} = 2.60585 \times \frac{100}{100 + 100.3} = 1.30.$$

Again, $\text{NRR} = 5 \sum_{h=1}^{7} \frac{B_{hx}^f}{W_{hx}} \times \frac{L_x}{l_0} = 5 \times 0.25063 = 1.25.$

The result indicates that the present number of mothers will be replaced by more number of mothers. There will be increasing trend in population.

11.4 Measure of Mortality

The mortality measurement is very important to the population policy planners. Some deaths occur by nature as man is mortal. However, some deaths occurs due to diseases, enviroment, natural calamity, accidents, etc. Deaths due to natural calamity cannot be controlled but number of deaths can be reduced, if proper precaution is taken against death. Deaths due to diseases may not be avoided, but can be reduced by proper medication and by proper health management system.

Thus, death analysis is also an important component to the population policy planners. The important aspect of death analysis is the measure of mortality. Some important mortality measures are:

(i) Crude Death Rate (CDR),

(ii) Age-specific Death Rate (ASDR)

This death rate can also be classified by

(a) Infant Mortality Rate (IMR),

(b) Child Mortality Rate (CMR),

(c) Maternal Mortality Rate (MMR),

(d) Old-age Death Rate (ODR)

(iii) Death Rate by Causes (DRC).

(i) **Crude death rate (CDR).** Let

D = total deaths of a country/area in a year,

P = total mid-year population of that country/area in that year.

The CDR is measured by $\text{CDR} = \frac{D}{P} \times 1000.$

This death rate is used as an index of death of a country and it measures the probability of death. However, it has limitations, as the probabilities of death are different in different ages. The distributions of population by age are not similar for every community and hence, probabilities of death are also not similar. Therefore, this CDR is not suitable for comparative study of death in two or more communities. Even it is not suitable to compare the probability of death in two regions of a country. As an example, we can mention the rural (5.4 per 1000) and urban (4.0 per 1000) death rates of Bangladesh in 1998. [Statistical Pocket Book of Bangladesh, 1998; BBS]. As another example, let us calculate the CDR using the following data:

P = total mid-year population of an area in 2011 = 53067511

D = total deaths in that area in 2011 = 356788

$$\therefore \text{CDR} = \frac{D}{P} \times 1000 = \frac{356788}{53067511} \times 1000 = 6.72.$$

(ii) **Age-specific death rate (ASDR).** The crude death rate cannot depict the death rates of people of different ages. Usually, infant death rate, child death rate and death rate at old age are more compared to those of people of young and middle ages. The age-specific death rate provides information on death of people of different age groups. This rate is calculated by

$$\text{ASDR}_x = \frac{D_x}{P_x} \times 1000, \text{ where}$$

D_x = total deaths of people of age group [x to $(x+h)$, h = age interval] of a country/area in a year,

D_x = mid-year population of age group x to $(x+h)$ of that country/area in that year.

Let us calculate the age-specific death rate using the following data:

Age group	0-5	5-10	10-15	15-20	20-25	25-30	30-40	40-50	50-60	60+
D_x	682	332	765	2308	2143	1406	1279	1802	1655	1290
P_x	20625	25962	401621	422677	508962	401888	328725	211960	88972	42325

The calculated ASDR_x are as follows:

Age group	0-5	5-10	10-15	15-20	20-25	25-30	30-40	40-50	50-60	60+
$\frac{D_x}{P_x} \times 1000$	33.1	12.8	7.5	5.5	4.2	3.5	3.9	8.5	18.6	30.5

(a) **Infant mortality rate (IMR).** Though there is decreasing trend in infant mortality rate in Bangladesh, still it is higher compared to the child mortality rate and mortality rates in other ages. Higher number of babies cannot reach their next birth day. This number is, again, higher in rural area. Thus, special attention is needed to check this mortality rate as higher infant mortality is directly or indirectly related to higher fertility leading mothers of poor health.

The IMR is calculated by

$$IMR = \frac{D_1}{B} \times 1000, \text{ where}$$

D_1 = total deaths of babies below age one year in a country/area,

B = total live births in the country/area in that year.

The IMR of Bangladesh in 1998 was 57 per 1000 live births. The corresponding figures in rural and urban areas were 66 and 41, respectively.

Let $B = 548681$ of a country in a year.

$D_1 = 24866$.

Then $IMR = \frac{D_1}{B} \times 1000 = \frac{24866}{548681} \times 1000 = 45.3.$

The infant mortality rate has some limitations. These are:

(i) A baby may born in the last part of the academic calendar and his/her next birth day will be in the next calendar year. In such a case B is inflated compared to D_1 in a year and D_1 will be inflated in the next year. Thus, IMR will be affected. However, if these cases are not too much, IMR will be a good measure.

(ii) Some babies may die just after birth. Information of such births and deaths may not be recorded and hence, IMR is affected.

The infant mortality is again, divided into two separate rates, namely (1) Neonatal Death Rate (NDR) and (2) Post-Neonatal Death Rate (PNDR).

Deaths which occur within 28 days of birth are known as neonatal deaths. NDR is calculated using these deaths (within 28 days of birth) as numerator. Deaths of babies after 28 days of birth but before next birth day are known as post neonatal deaths. These deaths are used as numerator in calculating PNDR.

(b) **Child mortality rate (CMR).** The babies who die after one year but before reaching 5 years are known as child deaths. Using these deaths in a year child mortality rate is calculated, where

$$CMR = \frac{D_{1-4}}{B_{1-4}} \times 1000.$$

Here $D_{1\text{-}4}$ = total deaths of babies of ages 1-4 years in a year in any country/area.

$B_{1\text{-}4}$ = number of babies of ages 1-4 in mid-year of a year in a country/area.

Let us calculate CMR using the data given below of a year, where $D_{1\text{-}4} = 1005$ and $B_{1\text{-}4} = 35496$. Then

$$CMR = \frac{1005}{35496} \times 1000 = 28.31.$$

(c) **Maternal mortality rate (MMR).** The child-bearing period of female is 15-49 years. During this period some mothers die during delivery period or after delivery. The total deaths of these mothers is $D_{15\text{-}49}$ and the babies born to them is $B_{15\text{-}49}$. The ratio of this D and B is known as maternal mortality rate. Thus,

$$MMR = \frac{D_{15\text{-}49}}{B_{15\text{-}49}} \times 1000.$$

The MMR in Bangladesh in the year 1998 is 3.00 per 1000 mothers.

(d) **Old age death rate (OADR).** As man is mortal, some body will die, even if he is not sick, at some stage of his/her old age. Usually, a death is considered as old age death, if a man dies at the age of 60 and above. Let

D_{60}^{+} = total deaths of people at the age of 60 and above in a country in a year,
P_{60}^{+} = total population of people of that age in that country in that year.

Then $\text{OADR} = \dfrac{D_{60}^{+}}{P_{60}^{+}} \times 1000.$

Besides these death rates, one standardized death (SDR) is calculated to compare the death rates of two or more countries or localities, where

$$\text{SDR} = \sum \text{ASDR}_x \frac{P_x^s}{P^s}.$$

Here $\text{ASDR}_x = \dfrac{D_x}{P_x}$

D_x = total deaths of age group x to $(x + h)$ in a country/area in a year,
P_x = total mid-year population of age group x to $(x + h)$ in the country in that area,
P_x^s = total people of age group x to $(x + h)$ of standard population in a year,
P^s = total people of standard population in a year.

As an example, let us consider the calculation of SDR using the following data:

Age group	0-5	5-10	10-15	15-20	20-30	30-40	40-50	50-60	60⁺
D_x	850	458	968	2790	1855	1585	2278	1878	2050
P_x	25350	30182	95672	148965	185666	166794	225770	100560	88960
P_x^s	15000	17000	40000	44000	48000	60000	165000	150000	100000

Calculation of SDR

Age group	0-5	5-10	10-15	15-20	20-30	30-40	40-50	50-60	60⁺
$\text{ASDR}_x = \dfrac{D_x}{P_x}$	0.03353	0.01517	0.01012	0.00191	0.00999	0.00950	0.01009	0.01868	0.02304
$\dfrac{D_x}{P_x} \times P_x^s$	503	258	40	84	480	570	1665	2801	2304

$P^s = 639000.$

Therefore, $\text{SDR} = \sum \text{ASDR}_x \times \dfrac{P_x^s}{P^s} = \dfrac{8705}{639000} = 0.01362$ or 13.62 per 1000.

11.5 Life Table

The life table is the process of calculating probability of death and the process of calculation of expectation of life of a cohort who will be decreased gradually by death. The calculation of probability of death and expectation of life is calculated at each age or age interval. Here cohort is a group of people who were born in a day. Usually, the initial cohort is assumed 1000 or 100000 and the cohort is decreased only by death. The cohort will not be increased or decreased by birth or migration. The probability calculation is continued until every member of the cohort dies.

The life table is, basically, of two types, namely

(i) Complete life table : It is calculated for the cohort of each age.

(ii) Abridged life table : It is calculated for the cohort of different age interval. However, for the initial cohort (age 0) and cohort of age 1-4 years are calculated separately.

The life table is completed by calculating a column, titled expectation of life (e_x). It is based on many other columns. A short description of the tables are discussed below:

(a) $x =$ the age of a member of cohort at his last birth day.

(b) $l_x =$ number of members of cohort survived up to exact age x years. The initial cohort is $l_0 = 1000$ or 100000.

(c) $d_x =$ number of deaths at age x, where $d_x = l_x - l_{x+1}$.

(d) $q_x = \frac{d_x}{l_x} =$ rate of mortality of a cohort at age x.

(e) $p_x = 1 - q_x$

(f) $L_x =$ total years lived by l_x within age interval x to $x+1$. This is called total person years lived.

$L_x = \frac{1}{2}[l_x + l_{x+1}]$. However,

$L_0 = 0.3l_0 + 0.7l_1, L_1 = 0.4l_1 + 0.6l_2$.

(g) $T_x = L_x + L_{x+1} + L_{x+2} + \cdots =$ total person years lived by l_x who survive up to age x.

(h) $e_x = \frac{T_x}{l_x}$. Moreover, $m_x = \frac{d_x}{L_x}$ and $q_x = \frac{2m_x}{2+m_x}$.

In calculating abridged life table the l_x, L_x, T_x and e_x have similar meaning except some changes due to the age interval x to $(x+n)$ years. Here an abridged life table of Bangladesh of 1990 is shown below:

Age	q_x	l_x	e_x	Age	q_x	l_x	e_x	Age	q_x	l_x	e_x
0	0.113	1000	56.4	5-9	0.016	840	62.0	40-44	0.024	760	31.5
1	0.016	887	62.6	10-14	0.006	826	58.0	45-49	0.044	742	27.2
2	0.017	873	62.6	15-19	0.011	820	53.4	50-54	0.050	709	23.3
3	0.013	858	62.7	20-24	0.012	812	48.9	55-59	0.065	673	19.5
4	0.009	847	62.4	25-29	0.015	802	44.5	60-64	0.102	630	15.6
1-4	0.052	887	62.6	30-34	0.016	790	40.1	65-69	0.182	566	12.1
0-4	0.169	1000	56.7	35-39	0.022	777	35.8	70-74	0.279	463	9.3
								75^+	1.000	333	6.9

11.6 Population Projection

The population of a country is counted after every 10 years by census. However, the researchers or the population planners or the development planners need estimate of intercensal or post census population. The count of intercensal population is called *population estimation* and the estimation of population of one or more years after census is called *population projection.*

There are different methods of intercensal estimation of population and of population projection. Some of the methods are:

Intercensal Estimation Methods:

(a) Linear Method,

(b) Exponential Model Method.

Methods of Population Projection:

(c) Logistic Curve,

(d) Geometric Growth Function,

(e) Modified Exponential Function,

(f) Gompertz Curve.

(a) **Linear method.** In this method the estimation of population is done by the formula:

$$P_t = P_0(1 + rt), \text{ where}$$

P_0 = population of initial census,

P_t = population after t years of initial census

$$r = \text{growth rate of population} = \frac{P_t - P_0}{P_0 t}.$$

For example, the enumerated population of Bangladesh in 2001 is 124355 thousands. This enumerated population in 2011 is 142319 thousands [Population and Housing Census 2011, Preliminary Report; BBS].

Let $P_0 = 124355$, $P_t = 142319$. Then

$$r = \frac{P_t - P_0}{P_0 t} = \frac{142319 - 124355}{124355 \times 10}, \text{ where } t = 10$$
$$= 0.0144.$$

Now, we can estimate population of 2012, where

$$P_{2012} = P_{2011}(1 + rt) = 142319(1 + 0.0144 \times 1) = 144368.$$

(b) **Exponential model method.** In this method the estimated population is given by

$$P_t = P_0 e^{rt}, \text{ where}$$
$$r = \frac{1}{t}\left[\log_{10}\frac{P_t}{P_0}\right]\frac{1}{\log_{10} e}, \text{ or, } r = \frac{1}{t}\ln\frac{P_t}{P_0}.$$

If $P_{2011} = P_t = 142319$, $P_{2001} = P_0 = 124355$, then

$$r = \frac{1}{10}\left[\log_{10}\frac{142319}{124355}\right]\frac{1}{\log_{10} 2.7183} = 0.01349.$$

Therefore,

$$P_{2015} = P_{2011}e^{rt} = 142319 \times e^{0.01349 \times 4} = 150210.$$

(c) **Logistic curve.** The logistic curve is given by

$$P_t = \frac{1}{A + Be^{-kt}}, \text{ where}$$
$$A = \frac{1}{a}, \; P_0 = \frac{1}{A + B}.$$

Here A and B are constants.

(d) **Geometric growth function.** The geometric growth function for population projection is

$$P_t = P_0(1+r)^t, \text{ where}$$
P_0 = population of initial census,
P_t = population after t years of initial census,
r = growth rate of population
$$= \text{Antilog}\left[\frac{1}{t}\log\frac{P_t}{P_0}\right] - 1.$$

For example, the population of Bangladesh in 1991 is $P_0 = 106315$ thousands and the population of Bangladesh in 2001 is $P_t = 124355$ thousands. Let us estimate the population of 1996. Here

$$r = \text{Antilog}\left[\frac{1}{t}\log\frac{P_t}{P_0}\right] - 1$$
$$= \text{Antilog}\left[\frac{1}{10}\log\frac{124355}{106315}\right] - 1 = 0.015796.$$

Now, population of 1996 is $P_{1996} = P_0(1+r)^t = 106315(1+0.015796)^5 = 114981.$

(e) **Modified exponential function.** The modified exponential function is given by

$$P_t = a + bc^t$$

where P_t is the population of the time t; a, b and c are three parameters. The values of a, b and c can be estimated by the method of least squares. However, an alternative technique is used to estimate the values of a, b and c.

Tl alternative technique is used if population of multiple of 3 years are given. In that case, let

S_1 = total population of first one-third years,
S_2 = total population of second one-third years,
S_3 = total population of third one-third years.

Then, let us calculate $d_2 = S_3 - S_2$, $d_1 = S_2 - S_1$, $c^m = \dfrac{d_2}{d_1}$, where $m = n/3$ and n is the number of years for which populations are given

$$a = \frac{1}{m}\left[S_1 - \frac{d_1}{c^m - 1}\right], \quad b = \frac{d_1(c-1)}{(c^m-1)^2}.$$

Using these values of a, b and c any value of P_t for any number of years t can be estimated.

As an application of the method, let us consider the estimation of population of 2009 using the data of population of 1996-2004, where the population of these years are estimated using geometric growth function as discussed in (d) above.

Year	1996	1997	1998	1999	2000	2001	2002	2003	2004
Population	114981	116797	118642	120517	122420	124354	126318	128314	130340
Sum of population of 3 years ($m = 3$)	$S_1 = 350420$			$S_2 = 367291$			$S_3 = 384972$		

$$d_1 = S_2 - S_1 = 367291 - 350420 = 16871.$$

$$d_2 = S_3 - S_2 = 384972 - 367291 = 17681.$$

$$c^m = c^3 = \frac{d_2}{d_1} = \frac{17681}{16871} = 1.04801, \ c = 1.01575.$$

$$a = \frac{1}{m}\left[S_1 - \frac{d_1}{c^m - 1}\right] = \frac{1}{3}\left[350420 - \frac{16871}{(1.04801 - 1)}\right] = -986.$$

$$b = \frac{d_1(c-1)}{(c^m - 1)^2} = \frac{16871(1.01575 - 1)}{(1.04801 - 1)^2} = 115281.$$

$$\therefore \ P_t = a + bc^t = -986 + 115281(1.01575)^t.$$

For 2009, we have $t = 5$.

Therefore, $P_{2009} = -986 + 115281(1.0175)^5 = 123664$ (thousand).

(f) **Gompertz curve.** The curve to project the population is given by

$$P_t = ab^{c^t}, \text{ where}$$

$$P_t = \text{population of } t \text{ years},$$

a, b and c are three parameters.

Taking log on both sides of the curve, we have $\log P_t = \log a + c^t \log b$.

Let $Y_t = \log P_t$, $A = \log a$, $B = \log b$; then $Y_t = A + Bc^t$.

This curve is similar to the modified exponential function. Hence, the values of A, B and C are to be estimated similarly as these are estimated in fitting modified exponential function except that population of different years are transformed to logarithmic value.

11.7 Survival Analysis

Survival analysis is a collection of statistical procedures for data analysis for which the outcome variable of interest is time until an event occurs. For example, the time from birth until death, the time from start of cancer until death, the time to come back to the job after a major surgical procedure, the time from the detection of HIV until death, etc. Here the events are death, disease, relapse, recovery, etc. and survival time is the time that an individual has survived over some follow-up period. The time interval between starting point and a subsequent event is often called failure, known as survival time. The analysis involved in case of such data generally focuses on estimation of probability that an individual will survive for a given length of time.

Most survival analysis must consider a key analytical problem, called censoring. In essence, censoring occurs when we have some information about individual survival time, but we do not know the survival time exactly.

As an example of censoring, let us consider a case of cancer patient who is followed until he goes out of remission. **For this patient the study ends before he gets the event. The survival time of this patient is not known. The survival time of this patient is considered censored. There are three reasons why censoring may occur. These are:**

(i) a person does not experience the event before the study ends;

(ii) a person is lost to follow-up during the study period;

(iii) a person withdraws from the study because of death or some other reason (here death is not the event of interest).

A distribution of survival times can be characterized by a survival function represented by $S(t)$, where

$$S(t) = P(T > t).$$

Here T is a continuous random variable indicating survival time. The graph of $S(t)$ against t is known as survival curve as shown below:

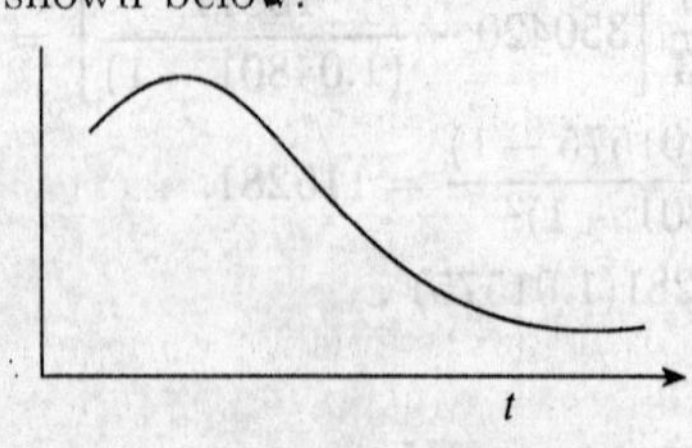

Theoretically, as t ranges from 0 up to infinity, the survivor function is a smooth curve. The function has the following characteristics:

(i) the function is nonincreasing and as t increases it moves downward,

(ii) at time $t = 0$, $S(t) = S(0) = 1$; that is, at the start of the study the value of $S(0) = 1$ indicates that no event has occurred and the probability of surviving past time zero is 1.

(iii) at time $t = \infty$, $S(t) = S(\infty) = 0$ indicates that, if the study period is increased without limit, eventually no body will survive and survivor curve must eventually fall to zero.

Though $S(t)$ is a curve, in practice the graph may be a step function rather than smooth curve. The step function may be as follows:

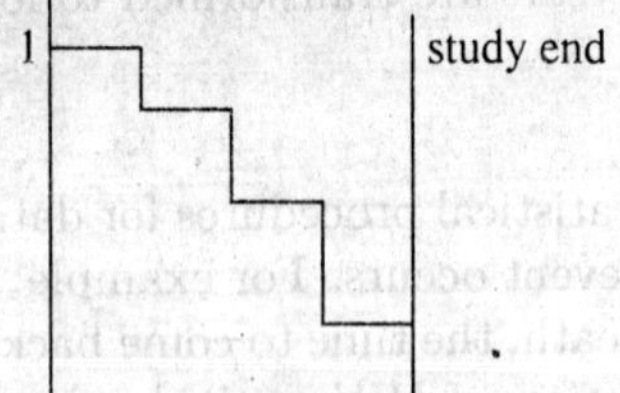

Moreover, the study period is never infinite and hence, $S(\infty)$ will not be zero. It means that, in some cases, every unit may not get the events.

Method of estimating survival curve

There are two important methods of estimating survival curve. These are:

(i) Life Table Method

(ii) The Product Limit Method or Kaplan-Meir Method

Life table method

The classical method of estimating a survival curve is life table method, wherein calculating life table we calculate mortality rate q_t, where

$$q_t = \frac{2m_t}{2 + m_t}, \quad m_t = \frac{d_t}{L_t},$$

if complete life table is constructed.

For abridged life $q_t = \dfrac{n \cdot n_{m_t}}{1 + n \cdot a_t \cdot n_{m_t}}$, where

n_{m_t} = death rate in the interval t to $(n+t)$ years

a_t = the average survival time in each year of a person who survives within t to $(n+t)$ years

n = age interval.

The survival probability is calculated by

$$S(t) = \frac{l_t}{l_0}, \text{ where}$$

l_0 = initial cohort who will be diminished only by death as time passes on. There will be no increase of cohort by birth or migration.

$$l_t = l_t - q_t \times l_{t-1}[l_1 = l_0 - l_0 q_0, l_2 = l_1 - q_1 l_1]$$

As an example, let us calculate survival probability $S(t)$ using the mortality rate (q_t) from abridged life table of Bangladesh calculated in 1990 as shown below:

Age t to $(t+n)$	n_{q_t}	l_t	$S(t) = \frac{l_t}{l_0}$
0-4	0.169	100000	1.000
5-9	0.016	83100	0.831
10-14	0.006	81770	0.818
15-19	0.011	81279	0.813
20-24	0.012	80385	0.804
25-29	0.015	79179	0.792
30-34	0.016	77991	0.780
35-39	0.022	76743	0.767
40-44	0.024	75055	0.751
45-49	0.044	73254	0.733
50-54	0.065	66529	0.665
55-59	0.102	62205	0.622
60-64	0.182	55860	0.559
65+	1.000	45693	0.457

Let us consider the calculation of survival probability of some cancer patients who survived some years after the detection of disease by life table method.

No. of patients	1	2	3	4	5	6	7	8	9	10	11
Survival years	0.25	0.50	1	1	2	2	3	3	3	4	4

Life table method of estimation of survival probability:

Survival time t *to* $(t+n)$ *years*	${}^{n}q_t$	l_t	d_t	$S(t) = \frac{l_t}{l_0}$
0-0.25	0.0000	11	0	1.0000
0.25-0.50	0.0909	11	1	1.0000
0.50-0.75	0.1000	10	1	0.9091
0.75-1.00	0.0000	9	0	0.8182
1.00-1.25	0.2222	9	2	0.8182
1.25-1.50	0.0000	7	0	0.6364
1.50-1.75	0.0000	7	0	0.6364
1.75-2.00	0.0000	7	0	0.6364
2.00-2.25	0.2857	7	2	0.6364
2.25-2.50	0.0000	5	0	0.4545
2.50-2.75	0.0000	5	0	0.4545
2.75-3.00	0.0000	5	0	0.4545
3.00-3.25	0.6000	5	3	0.1818
3.25-3.50	0.0000	2	0	0.1818
3.50-3.75	0.0000	2	0	0.1818
3.75-4.00	0.0000	2	0	0.1818
4.00-4.25	1.0000	2	2	0.0000

Here first patient survived up to $t = 0.25$ year and he died within 0.25-0.50 year. One out of 11 died within 0.25-0.50 year. The probability of death is

$$q_1 = \frac{1}{11} = 0.0909.$$

With the period 0-0.25 year none died and probability of death is $q_0 = 0.0000$. After 0.25 year 10 patients were left and second patient died within the period 0.50-0.75 year. The probability of death of this patient is

$$q_2 = \frac{1}{10} = 0.1000.$$

After 0.50 year up to 1 year 9 patients were left. None died within 0.75-1.00 year. The probability of death during this period is $q_3 = 0.0000$. During 1.00-1.25 years 2 out of 9 patients died.

The probability of death is

$$q_4 = \frac{2}{9} = 0.2222.$$

Among the remaining 7 patients none died during 1.25-1.50 years, 1.50-1.75 years and 1.75 years to 2.00 years. The probabilities of these 3 periods are $q_5 = q_6 = q_7 = 0.0000$. Out of the 7 patients 2 died during the period 2.00-2.25 years. The probability of death is

$$q_8 = \frac{2}{7} = 0.2857.$$

None died during 2.25-2.50 years, 2.50-2.75 years, 2.75-3.00 years. The probabilities of death during these periods are $q_9 = q_{10} = q_{11} = 0.0000$. Out of the remaining 5 patients 3 died during the period 3.00-3.25 years. The probability of death is

$$q_{12} = \frac{3}{5} = 0.6000.$$

No body died during the periods 3.25-3.50 years, 3.50-3.75 years and 3.75-4.00 years. The probabilities of death of these periods are $q_{13} = q_{14} = q_{15} = 0.0000$.. The last 2 patients died during the period 4.00-4.25 years. The probability of death is

$$q_{16} = \frac{2}{2} = 1.0000.$$

As both of them died, their survival probability is 0.0000.

The product limit method

The survival probability calculated by life table method changes only during the time interval in which at least one death occurs. In this method, if initial cohort is taken large and the members of cohort diminish by death at every stages of life, the changes in survival probabilities are observed. For smaller cohort like the patients of a particular disease who will die after the detection of the disease, there may be many time intervals in which there will be no death. In such cases survival probability will be similar for many time intervals. To avoid the problem *product-limit method* of estimating survival probability can be used. The method is also called *Kaplan-Meier method.*

In this method the failure time or the time after which the event occurs or the time after which the patient died is considered. For example, let us consider that 10 patients are under study after the detection of their disease. At initial time none dies. After 2 months, say, 1 patient out of 10 dies. So, the probability of death after 2 months is

$$q_1 = \frac{1}{10} = 0.1.$$

As none dies initially, $q_0 = 0.0$.

After 2 months 1 dies. Therefore $(10 - 1) = 9$ patients are left for follow-up study. Consider that, out of 9 patients, 2 die after 4 months. The probability of death after 4 months is

$$q_2 = \frac{2}{9} = 0.2222.$$

After 4 months the number of patients left for follow-up study are $(9 - 2) = 7$. Assume that, after 8 months one patient dies. The probability of death of this patient is

$$q_3 = \frac{1}{7} = 0.1429.$$

In a similar way the probabilities of death (event) of individuals under study are calculated and the survivors (l_t) are also found out. Then, the survival probability is

$$S(t) = \frac{l_t}{l_0}.$$

The method can be explained for some patients suffering from AIDS. The patients and their survival time after detection of the disease are given below:

Patients' number	1	2	3	4	5	6	7	8	9	10
Survival time in years	3	4	6	6	6	7	7	7	8	10

The calculation of survival probability $S(t)$ by *product limit method.*

Time *[Death after]*	q_t	l_t	$S(t) = \frac{l_t}{l_0}$
0	0.0000	10	1.0000
3	0.10000	9	0.9000
4	0.1111	8	0.8000
6	0.3750	5	0.5000
7	0.6000	2	0.2000
8	0.5000	1	0.1000
10	1.0000	0	0.0000

Initially none died. The probability of death is 0.0000. After 3 years 1 out of 10 died. The probability of death is

$$q_1 = \frac{1}{10} = 0.1000.$$

Next one died after 4 years. After 3 years 9 were left and hence, the probability of death of next one is

$$q_2 = \frac{1}{9} = 0.1111.$$

In a similar way, the death rate (q_t) and the survival number of patients (l_t) are calculated. Then $S(t)$ is the ratio of l_t to l_0.

11.8 Hazard Function

It is a function $h(t)$ which gives the instantaneous potential per unit time for the event to occur given that the individual has survived up to time t.

In contrast to the survival function, which focuses on not failing, the hazard function focuses on failing. In other words, as $S(t)$ goes up, $h(t)$ goes down, and vice versa. This function $h(t)$ can be considered as giving the negative side of the information given by the survival function $S(t)$.

Cox's proportional hazard model

It is used for presenting survival data, where survival of an individual depends on many variables, say, $x_1, x_2, \ldots, x_p$. Out of these variables some are time dependent and some are time independent. For example, sex is time independent variable but age, height, weight are time dependent variable.

Cox's proportional hazard model is given by

$$h(t) = h_0(t)e^{\sum_{i=1}^{p} B_i x_i}$$

$$h_0(t) = \text{base line hazard [It occurs, if } x_1 = x_2 = \cdots = x_p = 0]$$

The hazard is a rate, rather than a probability. Thus, the values of the hazard functions range between zero and infinity.

Chapter 12

Meta-Analysis

12.1 Introduction

Meta-analysis is a statistical technique for combining the results of similar experiments conducted in different environmental conditions by different researchers or by same researcher using different experimental units to study the significance of the experimental results. For example, a manufacturer produces three different proteins suitable for babies and needs a decision which protein should be marketed successfully. The manufacturer may request the doctors of urban area as well as rural area to prescribe the protein. If a doctor prescribes a protein, he/she can follow-up the impacts of use of protein. The same protein may not be effective for body-weight change of all babies in a similar way. In such a case the manufacturer can collect the information about the impacts of a protein and can analyse the data collected from doctors of both urban and rural localities and finally, a conclusion can be drawn on the basis of combined analysis of the data. Such combined analysis of data is known as meta-analysis.

In medical science, the impact of drug against hypertension may vary from person to person due to variaton in body-mass index or due to variation in ages. But the medical practitioner or researcher needs to identify one drug which is suitable for almost all patients. To study the suitability of a drug it can be applied on patients of different ages or patients of different body-mass indices through different experiments. Finally, conclusion can be drawn by combining the results of all experiments.

Meta-analysis are often, but not always, important components of a systematic study, specially, for clinical trials of a medical treatment, in an effort to conclude efficiently how well a treatment works. Thus, the analysis has some advantages. These are:

(i) To investigate the variation in the results due to sample diversity. Combined analytical results are more acceptable, if least diversity is observed due to sample diversity.

(ii) Common conclusion can be drawn on a parameter, if it is tested in different experimental situations.

(iii) Results of sample studies can be generalized for populaton in studies.

(iv) Helps to control the variation of results of inter-experiments.

(v) Statistical conclusion from combined analysis is more powerful.

(vi) Due to combination of results of several experiments local biases are less influential in conclusion.

The above advantages are achieved, if proper decision is taken to include the experiments for combined analysis. Similar type of experiments conducted in homogeneous environment or conducted on homogeneous experimental units are to be combined for better conclusive results. Here combinaton of results will not be effected by the variation in sample sizes. But variation in experimental conditions may distort the final conclusion if proper analytical methods are not followed. Due to change in experimental conditions, weighted analysis may be helpful for final conclusion. .

Therefore, inclusion of experiments for meta-analysis is very important. For example, if any researcher is interested to identify a best drug against a disase, drug can be prescribed for patients of similar type including a group of control units.

12.2 Application of Meta-Analysis

Repeated experiments are conducted in the field of analysis of variance, correlation analysis, regression analysis, test of hypothesis, estimation of parameters from sample data, etc. Thus, meta-analysis can be applied in the following analytical procedures to conclude about population parameter. The parameter can be estimated by weighted analysis and conlusion can be drawn by combination of P-values of test statistics. The fields in which it is used are:

(a) Combined analysis in design of experiments,

(b) Combined analysis in sample survey,

(c) Combined analysis in fitting regression model,

(d) Combined analysis in correlation study,

(e) Combined analysis in test of association of attributes,

(f) Combined analysis in test of hypothesis, etc.

Analytical procedures are shown by examples.

12.3 Meta-Analysis in Analysis of Variance

EXAMPLE 12.1 Three doctors prescribed 4 different types of medicines for patients of hypertension and asked the patients to record their systolic blood pressure after 7 days of using medicines. Medicines were prescribed for patients of different age-groups at random. The systolic blood pressure levels of the patients are as follows. The experiment is conducted through randomized block design assuming patients of one age-group as block.

Patients of Doctor-1

Systolic blood pressure levels (x_{1ij} mmHg) after using medicines

Age groups	*Medicines*				*Total*
(in years)	M_1	M_2	M_3	M_4	x_{1i}
<25	70	70	72	65	277
25-40	75	78	80	80	313
40-50	80	80	75	70	305
50-60	85	86	80	80	331
60^+	85	90	82	80	337
Total, x_{1j}	395	404	389	375	1563

Patients of Doctor-2

Systolic blood pressure levels (x_{2ij} mmHg) *after using medicines*

Age groups (in years)	*Medicines*				*Total*
	M_1	M_2	M_3	M_4	x_{2i}
<25	68	65	65	70	268
25-40	75	75	70	70	290
40-50	80	80	74	72	306
50-60	90	90	85	85	350
60^+	85	85	90	88	348
Total	398	395	384	385	1562

Patients of Doctor-3

Systolic blood pressure levels (x_{3ij} mmHg) *after using medicines*

Age groups (in years)	*Medicines*				*Total*
	M_1	M_2	M_3	M_4	x_{3i}
<25	65	65	65	65	260
25-40	78	80	70	72	300
40-50	85	85	80	75	325
50-60	80	80	85	90	335
60^+	90	90	95	85	360
Total	398	400	395	387	1580

(i) Are the medicines similar?

(ii) Identify the best medicine, if there is any.

(iii) Consider the patients of ages below 25 as conrol group and differentiate other groups from control group.

Solution. Let p = number of doctors = 3,

q = number of age-groups = 5,

r = number of medicines = 4

$[h = 1, 2, \ldots, p = 3; j = 1, 2, \ldots, r = 4; i = 1, 2, \ldots, q = 5]$.

The blood pressure levels of patients of different age-groups and of different doctors $[x_{hi}]$

Age groups (in years)	*Doctors*			*Total*
	1	2	3	x_i
<25	277	268	260	805
25-40	313	290	300	903
40-50	305	306	325	936
50-60	331	350	335	1016
60^+	337	348	360	1045
Total, x_h	1563	1562	1580	4705

x_{hi} = total of i-th block (Number of blocks = q = 5) of h-th experiment.

[h-th doctor and number of doctors, p = 3]

G = grand total = 4705, total number of observations = $pqr = 60$.

$$\text{C.T.} = \frac{G^2}{pqr} = \frac{(4705)^2}{3 \times 5 \times 4} = 368950.4167.$$

$$\text{Total S.S.} = \Sigma\Sigma\Sigma x_{hij}^2 - \text{C.T.} = 372859 - 368950.4167 = 3908.5833.$$

$$\text{S.S. (experiments)} = \text{S.S. (doctors)} = \Sigma \frac{x_h^2}{qr} - \text{C.T.}$$

$$= \frac{(1563)^2 + (1562)^2 + (1580)^2}{5 \times 4} - 368950.4167 = 10.2333$$

S.S. (blocks within doctors)

$$= \text{S.S. (age within doctors)} = \Sigma \left[\frac{\Sigma x_{hi}^2}{r} - \frac{x_h^2}{q \times r} \right]$$

$$= \left[\frac{x_{11}^2 + x_{12}^2 + x_{13}^2 + x_{14}^2 + x_{15}^2}{r} - \frac{x_1^2}{qr} \right] + \left[\frac{x_{21}^2 + x_{22}^2 + x_{23}^2 + x_{24}^2 + x_{25}^2}{r} - \frac{x_2^2}{qr} \right]$$

$$+ \left[\frac{x_{31}^2 + x_{32}^2 + x_{33}^2 + x_{34}^2 + x_{35}^2}{r} - \frac{x_3^2}{qr} \right]$$

$$= \left[\frac{(277)^2 + (313)^2 + (305)^2 + (331)^2 + (337)^2}{4} - \frac{(1563)^2}{5 \times 4} \right]$$

$$+ \left[\frac{(268)^2 + (290)^2 + (306)^2 + (350)^2 + (348)^2}{4} - \frac{(1562)^2}{5 \times 4} \right]$$

$$+ \left[\frac{(260)^2 + (300)^2 + (325)^2 + (335)^2 + (360)^2}{4} - \frac{(1580)^2}{5 \times 4} \right]$$

$$= 564.8 + 1298.8 + 1442.5 = 3306.1.$$

Blood pressure due to the use of medicine prescribed by the doctors $[x_{hj}]$:

Doctors	*Medicines*				*Total* x_h
	M_1	M_2	M_3	M_4	
D_1	395	404	389	375	1563
D_2	398	395	384	385	1562
D_3	398	400	395	387	1580
Total, x_j	1191	1199	1168	1147	4705

$$\text{S.S. (medicines)} = \frac{\Sigma x_j^2}{pq} - \text{C.T.}$$

$$= \frac{(1191)^2 + (1199)^2 + (1168)^2 + (1147)^2}{3 \times 5} - 368950.4167 = 110.5833.$$

$$\text{S.S. (doctors} \times \text{Medicines)} = \frac{\Sigma\Sigma x_{hj}^2}{q} - \text{C.T.} - \text{S.S. (doctors)} - \text{S.S. (medicines)}$$

$$= \frac{1}{5} [(395)^2 + (404)^2 + (389)^2 + (375)^2 + (398)^2$$

$$+ (395)^2 + (384)^2 + (385)^2 + (398)^2 + (400)^2 + (395)^2$$

$$+ (387)^2] - 368950.4167 - 10.2333 - 110.5833$$

$$= 27.7667.$$

S.S. (error) = total S.S. − S.S. (doctors) − S.S. (ages within doctors)
−S.S. (medicines) − S.S. (medicines×doctors)
= 3908.5833 − 10.2333 − 3306.1 − 110.5833 − 27.7667 = 453.9.

ANOVA Table

Sources of variation	*d.f.*	*S.S.*	$M.S. = \frac{S.S.}{d.f.}$	*Calculated F*	*5% value of F*
Doctors	2	$S_1 = 10.2333$	$s_1 = 5.1167$	$F_1 = \frac{s_1}{s_5} = 0.45$	3.23
Ages within doctors	8	$S_2 = 3306.1$	$s_2 = 413.2625$	$F_2 = \frac{s_2}{s_5} = 36.42$	2.18
Medicines	3	$S_3 = 110.5833$	$s_3 = 36.8611$	$F_3 = \frac{s_3}{s_5} = 3.25$	2.84
Doctors×Medicines	6	$S_4 = 27.7667$	$s_4 = 4.6275$	$F_4 = \frac{s_4}{s_5} = 0.41$	2.34
Error	40	$S_5 = 453.9$	$s_5 = 11.3475$		—
Total	59	—	—	—	—

(i) H_0 : All medicines are similar,
H_A : All medicines are not similar.
Comment. As $F_3 >$ 5% value of F (P-value < 0.05), H_0 is rejected. All medicines are not similar.

(ii) To identify the best medicine we can apply least significant difference [LSD], where

$$\text{LSD} = 5\% \text{ value of } t \text{ with d.f. of error} \sqrt{\frac{2 \text{ M.S. (error)}}{pq}}$$

$$= 2.021\sqrt{\frac{2 \times 11.3475}{3 \times 5}} = 1.23.$$

The mean blood pressure levels due to the use of medicines are:

$$\bar{M}_1 = 79.4,\ \bar{M}_2 = 79.93,\ \bar{M}_3 = 77.87,\ \bar{M}_4 = 76.47.$$

If $|\bar{M}_l \quad \bar{M}_k| \geq \text{LSD}$ $(l \neq k = 1, 2, 3, 4)$, $\bar{M}_l$ and $\bar{M}_k$ are different.

$|\bar{M}_1 - \bar{M}_2| = |79.4 - 79.93| = 0.53 < \text{LSD}$; M_1 and M_2 are similar.

$|\bar{M}_1 - \bar{M}_3| = |79.4 - 77.87| = 1.53 > \text{LSD}$; M_1 and M_3 are different. M_3 is better than M_1

$|\bar{M}_1 - \bar{M}_4| = |79.4 - 76.47| = 2.93 > \text{LSD}$; M_1 and M_4 are different. M_4 is better than M_1

$|\bar{M}_2 - \bar{M}_3| = |79.93 - 77.87| = 2.06 > \text{LSD}$; M_2 and M_3 are different. M_3 is better than M_2

$|\bar{M}_2 - \bar{M}_4| = |79.93 - 76.47| = 3.46 > \text{LSD}$; M_2 and M_4 are different. M_4 is better than M_2

$|\bar{M}_3 - \bar{M}_4| = |77.87 - 76.47| = 1.40 > 1.23$; M_3 and M_4 are different. M_4 is better than M_3.

M_3 and M_4 are better than M_1 and M_2. But M_4 is better than M_3. So, M_4 is the best medicine to reduce blood pressure level.

(iii) As the patients of ages below 25 are considered as control group, other groups of patients can be compared with control group by Dunnett's test, where the test statistic is

$$D = 5\% \text{ value of } d \text{ with d.f. of error for } q = 5 \text{ means } \sqrt{\frac{2 \text{ M.S. (error)}}{pr}}.$$

This 5% value of d is from Dunnett's table 2.54.

$$\therefore D = 2.54\sqrt{\frac{2 \times 11.3475}{12}} = 3.493.$$

Averages of i-th age-group are:

$$\bar{A}_i = \frac{1}{pr}x_i \; [x_i = \text{total of } i\text{-th age (block)}]$$

$$\bar{A}_1 = \frac{805}{12} = 67.083, \; \bar{A}_2 = \frac{903}{12} = 75.25, \; \bar{A}_3 = \frac{936}{12} = 78.00$$

$$\bar{A}_4 = \frac{1016}{12} = 84.667, \; \bar{A}_5 = \frac{1045}{12} = 87.083.$$

We need to compare $\bar{A}_1$ with other means. If $|\bar{A}_1 - \bar{A}_i| \geq D[i = 2, 3, 4, 5]$, there is significant difference between A_1 and other age-groups.

$|\bar{A}_1 - \bar{A}_2| = |67.083 - 75.25| = 8.167 > D$; age-group 1 and age-group 2 are different. Blood pressure level of group 1 is less than that of group 2.

$|\bar{A}_1 - \bar{A}_3| = |67.083 - 78.00| = 10.917 > D$; age-group 1 and age-group 3 are different. Blood pressure level of age-group 1 is different from that of age-group 3.

$|\bar{A}_1 - \bar{A}_4| = |67.083 - 84.667| = 17.584 > D$; age-group 1 and age-group 4 are different. Blood pressure level of age-group 1 is different from that of age-group 4.

$|\bar{A}_1 - \bar{A}_5| = |67.083 - 87.083| = 20.00 > D$; age-group 1 and age-group 5 are different. Blood pressure level of age-group 1 is different from that of age-group 5.

Here meta-analysis (combined analysis) of three experiments indicates that the medicines are not similar. But, individual analysis of three experiments does not signify the similar differentials in medicines. The analysis of variance tables of three experiments conducted by three doctors are shown below:

ANOVA Table [*Experiment*-1]

Sources of variation	*d.f.*	*S.S.*	*M.S.* $= \frac{S.S.}{d.f.}$	*Calculated* F	*5% value of* F
Ages	4	564.80	$s_1 = 141.2$	$F_1 = \frac{s_1}{s_3} = 15.29$	3.26
Medicines	3	88.95	$s_2 = 29.65$	$F_2 = \frac{s_2}{s_3} = 3.21$	3.49
Error	12	110.8	$s_3 = 9.233$		
Total	19				

As $F_2 <$ 5% value of F, medicines are similar, observed from the first experiment conducted by Doctor-1.

ANOVA Table [*Experiment-2*]

Sources of variation	*d.f.*	*S.S.*	$M.S. = \frac{S.S.}{d.f.}$	*Calculated F*	*5% value of F*
Ages	4	1298.4	324.7	$F_1 = 36.35$	3.26
Medicines	3	29.8	9.933	$F_2 = 1.11$	3.49
Error	12	107.2	8.933		
Total	19				

The experiment conducted by the second doctor also indicates that medicines are similar [$F_2 <$ 5% value of F].

ANOVA Table [*Experiment-3*]

Sources of variation	*d.f.*	*S.S.*	$M.S. = \frac{S.S.}{d.f.}$	*Calculated F*	*5% value of F*
Ages	4	1442.5	360.625	$F_1 = 18.34$	3.26
Medicines	3	19.6	6.533	$F_2 = 0.33$	3.49
Error	12	235.9	19.658		
Total	19				

Here also, it is seen that all medicines are similar.

But, if combined analysis is done, medicines are significantly different. In that sense the combined analysis (meta-analysis) is better than individual analysis of the experiment.

12.4 Meta-Analysis Applied in Testing the Significance of Population Mean

EXAMPLE 12.2 The following are the birth-weights (x_{ij} in kg) of different mothers giving births in different clinics in a day

Clinics	*Birth-weights (x_{ij} in kg) of mothers*					*Total* x_i	*Mean* $\bar{x}_i$
1	3.8	3.0	3.2	3.6	2.8	16.4	3.28
2	2.5	2.7	3.0	3.0	2.5	13.7	2.74
3	4.2	3.6	2.9	2.5	3.8	17.0	3.40
4	4.5	4.5	3.8	3.9	3.8	20.5	4.10

Do you think that the average birth-weight of babies born in different clinics is 3.0 kg?

Solution. We need to test hypothesis:

$$H_0 : M = 3.0, \text{ against } H_A : M \neq 3.0, \text{ where}$$
$$M = \text{mean birth-weight.}$$

For first clinic: [For each sample $n = 5$].

$$\bar{x}_1 = \frac{1}{n}\Sigma x_{1j} = \frac{16.4}{5} = 3.28.$$

$$s_1^2 = \frac{1}{n-1}\left[\Sigma x_{1j}^2 - \frac{(\Sigma x_{1j})^2}{n}\right] = \frac{1}{5-1}\left[54.48 - \frac{(16.4)^2}{5}\right] = 0.172.$$

$$\therefore t = \frac{\bar{x}_1 - M}{\sqrt{\frac{s_1^2}{n}}} = \frac{3.28 - 3.0}{\sqrt{\frac{0.172}{5}}} = 1.51.$$

The P-value of t is 0.206 > 0.05, H_0 is accepted. The average birth-weight of babies born in clinic-1 is 3.0 kg.

For second clinic:

$$\bar{x}_2 = \frac{1}{n}\Sigma x_{2j} = \frac{13.7}{5} = 2.74.$$

$$s_2^2 = \frac{1}{n-1}\left[\Sigma x_{2j}^2 - \frac{(\Sigma x_{2j})^2}{n}\right] = \frac{1}{5-1}\left[37.79 - \frac{(13.7)^2}{5}\right] = 0.063.$$

$$t = \frac{\bar{x}_2 - M}{\sqrt{\frac{s_2^2}{n}}} = \frac{2.74 - 3.0}{\sqrt{\frac{0.063}{5}}} = -2.316.$$

The P-value of t is 0.081 > 0.05, H_0 is accepted. The average birth-weight of babies born in clinic-2 is 3.0 kg.

For third clinic:

$$\bar{x}_3 = \frac{1}{n}\Sigma x_{3j} = \frac{17.0}{5} = 3.40.$$

$$s_3^2 = \frac{1}{n-1}\left[\Sigma x_{3j}^2 - \frac{(\Sigma x_{3j})^2}{n}\right] = \frac{1}{5-1}\left[59.7 - \frac{(17.0)^2}{5}\right] = 0.475.$$

$$t = \frac{\bar{x}_3 - M}{\sqrt{\frac{s_3^2}{n}}} = \frac{3.40 - 3.0}{\sqrt{\frac{0.475}{5}}} = 1.298.$$

The P-value of t is 0.264 > 0.05, H_0 is accepted. The average birth-weight of babies born in clinic-3 is 3.0 kg.

For fourth clinic:

$$\bar{x}_4 = \frac{1}{n}\Sigma x_{4j} = \frac{20.5}{5} = 4.10.$$

$$s_4^2 = \frac{1}{n-1}\left[\Sigma x_{4j}^2 - \frac{(\Sigma x_{4j})^2}{n}\right] = \frac{1}{5-1}\left[84.59 - \frac{(20.5)^2}{5}\right] = 0.135.$$

$$t = \frac{\bar{x}_4 - M}{\sqrt{\frac{s_4^2}{n}}} = \frac{4.10 - 3.0}{\sqrt{\frac{0.135}{5}}} = 6.694.$$

The P-value of t is 0.003 < 0.05, H_0 is rejected. The average birth-weight of babies born in clinic-4 is not 3.0 kg.

We observed 4 different results for babies born in 4 clinics. By meta-analysis, we can conclude about the hypothesis for all babies in a day. One of the technique of meta-analysis is the combination of P-values observed from four t-test statistics, where the combination of probabilities is done by

$$-2 \ln P = -2 \ln[p_1 \cdot p_2 \cdot p_3 \cdot p_4],$$

where p_i = p-values from i-th test. This $-2 \ln P$ is distributed as chi-square (χ^2) with $2k$ d.f. (Here $k = 4$). So, we have

$$-2 \ln P = -2 \ln[0.206 \times 0.081 \times 0.264 \times 0.003] = 22.47.$$

Five per cent value of χ^2 with $2 \times 4 = 8$ d.f. is 15.507. The calculated value is greater than tabulated value. Therefore, H_0 is rejected. The average birth-weight of all babies born in clinics is not 3.0 kg. There is variation in birth-weight due to the variation of mothers. This phenomenon can also be observed if we use the analysis of variance technique to observe the variations in birth-weights, where the analytical results are as follows:

We have $k = 4$ clinics, in each clinic there are $n = 5$ births.

Grand total of birth-weight is $G = 67.6$.

$$\text{C.T.} = \frac{G^2}{nk} = \frac{(67.6)^2}{5 \times 4} = 228.488.$$

$$\text{Total S.S.} = \Sigma\Sigma x_{ij}^2 - \text{C.T.} = 236.56 - 228.488 = 8.072.$$

$$\text{S.S. (clinics)} = \frac{\Sigma x_i^2}{n} - \text{C.T.} = \frac{1165.9}{5} - 228.488 = 4.692.$$

$$\text{S.S. (error)} = \text{total S.S.} - \text{S.S. (clinics)} = 8.072 - 4.692 = 3.38.$$

ANOVA Table

Sources of variation	*d.f.*	*S.S.*	*M.S.* = $\frac{S.S.}{d.f.}$	*Calculated F*	*5% value of F*
Clinics	3	4.692	1.564	7.404	0.002
Error	16	3.380	0.211		
Total	19				

As P-value < 0.05, birth-weights vary from clinic to clinic. All mothers giving births in clinics are not similar. If birth-weights of 4 different clinics are considered as 4 different samples, the above analysis of variance is also a meta-analysis.

12.5 Meta-Analysis for Common Estimate of Population Mean Using Different Simple Random Samples

EXAMPLE 12.3 Three different simple random samples are selected from 3 populations of patients of sizes $N_1 = 300$, $N_2 = 400$, $N_3 = 250$ visiting a diabetic centre in 3 different days to study their blood sugar levels. These patients are under treatment. The objective of the study is to verify whether treatment is effective. If treatment is effective, their blood sugar level should be 6.3 mmol/l, on an average. From the first population 36 patients, from the second population 48 patients and from the third population 30 patients are selected and their blood sugar levels (x_{ij}; $i = 1, 2, 3$; $j = 1, 2, \ldots, n_i$; where $n_1 = 36$, $n_2 = 48$ and $n_3 = 30$) are recorded.

	Blood sugar levels of samples (x_{ij} fasting)
First sample	7.5, 6.8, 6.4, 5.8, 8.2, 10.6, 5.5, 4.8, 6.5, 6.2, 6.0, 5.0, 5.9, 6.1, 6.0, 6.7, 4.7, 4.8, 5.0, 5.5, 6.5, 5.8, 6.7, 6.7, 4.5, 7.5, 7.6, 7.5, 4.8, 4.8, 5.5, 6.7, 7.0, 6.5, 6.0, 6.8
Second Sample	9.5, 9.0, 6.4, 6.2, 6.0, 5.8, 6.0, 7.5, 9.6, 9.0, 6.5, 6.2, 8.5, 4.8, 4.8, 5.8, 5.5, 5.6, 5.7, 6.8, 7.0, 7.5, 6.5, 7.0, 4.9, 5.5, 6.2, 6.3, 6.1, 6.0, 7.0, 7.8, 8.5, 10.6, 6.0, 7.5, 4.5, 5.3, 6.1, 6.0, 8.5, 9.6, 9.0, 7.0, 7.2, 6.4, 6.0, 6.0
Third sample	7.0, 7.0, 6.0, 6.4, 6.2, 5.5, 5.8, 5.0, 8.6, 10.4, 9.3, 9.0, 8.5, 4.5, 5.8, 4.6, 6.0, 5.9, 6.2, 6.1, 6.0, 6.2, 6.3, 5.5, 4.8, 4.0, 5.8, 6.0, 6.7, 7.5.

(i) Is the assumption of average blood sugar level 6.3 mmol/l correct for patients of every day and for all the populations?

(ii) Find combined average of blood sugar (fasting) levels of all patients of all samples.

Solution.

(i) We have $N_1 = 300$, $N_2 = 400$, $N_3 = 250$, $n_1 = 36$, $n_2 = 48$ and $n_3 = 30$.

For first sample:

$$\bar{x}_1 = \frac{1}{n_1}\Sigma x_{1j} = -\frac{224.9}{36} = 6.2472.$$

$$s_1^2 = \frac{1}{n_1 - 1}\left[\Sigma x_{1j}^2 - \frac{(\Sigma x_{1j})^2}{n_1}\right] = \frac{1}{36-1}\left[1455.85 - \frac{(224.9)^2}{36}\right] = 1.45285.$$

$$v(\bar{x}_1) = \frac{N_1 - n_1}{N_1 n_1}s_1^2 = \frac{300-36}{300 \times 36}1.45285 = 0.0355.$$

$$W_1 = \frac{1}{v(\bar{x}_1)} = \frac{1}{0.0355} = 28.1578.$$

We need to test $H_0 : M = M_0 = 6.3, H_A : M \neq M_0$.

The test statistic is

$$z_1 = \frac{\bar{x}_1 - M_0}{\sqrt{\frac{s_1^2}{n_1}}} = \frac{6.2472 - 6.3}{\sqrt{\frac{1.45285}{36}}} = -0.263, \ P\text{-value}, p_1 = 0.794.$$

As P-value > 0.05, H_0 is true. The average blood sugar level of patients selected from population -1 is 6.3.

For second sample:

$$\bar{x}_2 = \frac{1}{n_2}\Sigma x_{2j} = -\frac{327.2}{48} = 6.8167.$$

$$s_2^2 = \frac{1}{n_2 - 1}\left[\Sigma x_{2j}^2 - \frac{(\Sigma x_{2j})^2}{n_2}\right] = \frac{1}{48-1}\left[2328.26 - \frac{(327.2)^2}{48}\right] = 2.08184.$$

$$v(\bar{x}_2) = \frac{N_2 - n_2}{N_2 n_2}s_2^2 = \frac{400-48}{400 \times 48} \times 2.08184 = 0.03817.$$

$$W_2 = \frac{1}{v(\bar{x}_2)} = \frac{1}{0.03817} = 26.2005.$$

The test statistic is

$$z_2 = \frac{\bar{x}_2 - M_0}{\sqrt{\frac{s_2^2}{n_2}}} = \frac{6.8167 - 63}{\sqrt{\frac{2.08184}{48}}} = 2.481, \ P\text{-value}, p_2 = 0.017.$$

For third sample:

$$\bar{x}_3 = \frac{1}{n_3}\Sigma x_{3j} = \frac{192.6}{30} = 6.42.$$

$$s_3^2 = \frac{1}{n_3 - 1}\left[\Sigma x_{3j}^2 - \frac{(\Sigma x_{3j})^2}{n_3}\right] = \frac{1}{30-1}\left[1299.86 - \frac{(192.6)^2}{30}\right] = 2.1851.$$

$$v(\bar{x}_3) = \frac{N_3 - n_3}{N_3 n_3} s_3^2 = \frac{250 - 30}{250 \times 30} \times 2.1851 = 0.0641.$$

$$W_3 = \frac{1}{v(\bar{x}_3)} = \frac{1}{0.0641} = 15.6015.$$

The test statistic is

$$z_3 = \frac{\bar{x}_3 - M_0}{\sqrt{\frac{s_3^2}{n_3}}} = \frac{6.42 - 6.3}{\sqrt{\frac{2.1851}{30}}} = 0.445, \quad P\text{-value}, p_3 = 0.66.$$

We need to test the hypothesis $H_0 : M = M_0 = 6.3$ for all the sample observations. This can be done, as a meta-analysis, by combining the P-values of 3 samples. The combination of probabilities is done by

$$\begin{aligned} -2\ln P &= -2\ln(p_1 \times p_2 \times p_3) = -2\ln[0.794 \times 0.017 \times 0.66] \\ &= 9.44. \end{aligned}$$

Here $-2\ln P$ is distributed as χ^2 with $2k$ d.f., where k is the number of samples ($k = 3$). The 5% value of χ^2 with 6 d.f. is 12.592 and it is more than our calculated value (9.44). Therefore, null hypothesis of $H_0 : M = M_0 = 6.3$ is true for all samples together, though for second sample observations null hypothesis is not true.

(ii) The combined average blood sugar (fasting) level is as follows:

$$\begin{aligned} \bar{x}\,(\text{combined mean}) &= \frac{W_1\bar{x}_1 + W_2\bar{x}_2 + W_3\bar{x}_3}{W_1 + W_2 + W_3} \\ &= \frac{28.1578 \times 6.2472 + 26.2005 \times 6.8167 + 15.6015 \times 6.42}{28.1578 + 26.2005 + 15.6015} \\ &= 6.499. \end{aligned}$$

A common conclusion about the equality of mean blood sugar level for all three samples can also be done by analysis of variance. Here, we have

$$\begin{aligned} n &= n_1 + n_2 + n_3 = 36 + 48 + 30 \\ &= 114 \text{ (total sample observations of } k = 3 \text{ samples)}. \end{aligned}$$

Grand total is $G = \Sigma x_{1j} + \Sigma x_{2j} + \Sigma x_{3j} = 224.9 + 327.2 + 192.6 = 744.7.$

$$\text{C.T.} = \frac{G^2}{n} = \frac{(744.7)^2}{114} = 4864.7201.$$

$$\begin{aligned} \text{Total S.S.} &= \Sigma\Sigma x_{ij}^2 - \text{C.T.} \\ &= \Sigma x_{1j}^2 + \Sigma x_{2j}^2 + \Sigma x_{3j}^2 - \text{C.T.} \\ &= 1455.85 + 2328.26 + 1299.86 - 4864.7201 \\ &= 219.2499. \end{aligned}$$

$$\text{S.S. (samples)} = \frac{(\Sigma x_{1j}^2)^2}{n_1} + \frac{(\Sigma x_{2j}^2)^2}{n_2} + \frac{(\Sigma x_{3j})^2}{n_3} - \text{C.T.}$$

$$= \frac{(224.9)^2}{36} + \frac{(327.2)^2}{48} + \frac{(192.6)^2}{30} - 4864.7201 = 7.1855.$$

S.S. (error) = Total S.S. − S.S. (samples) = 219.2499 − 7.1855 = 212.0644.

ANOVA Table

Sources of variation	*d.f.*	*S.S.*	$M.S. = \frac{S.S.}{d.f.}$	*Calculated F*	*P-value*
Samples	2	7.1855	3.593	1.88	0.157
Error	111	212.0644	1.91		
Total	113				

As the P-value of F-statistic is 0.157 > 0.05, H_0 of similarity of average blood sugar (fasting) levels for all 3 samples is true.

12.6 Meta-Analysis in Testing Equality of Several Proportions

EXAMPLE 12.4 It is seen that [Example 12.3] for some of the persons blood sugar level is normal [≤ 6.3 mmol/l]. for some persons blood sugar level is not alarming and can be controlled by exercise and diet control [≤ 7.00 mmol/l] and for some persons treatment is needed [> 7.00 mmol/l]. This is true for all 3 samples.

(i) Are all 3 groups of persons in any sample equally distributed?

(ii) Is there any differential in 3 groups of persons in 3 samples?

Solution. We have $n_1 = 36$, $n_2 = 48$ and $n_3 = 30$. For each sample we need to test

H_0 : Three groups of persons of each sample are equally distributed.
[$H_0 : P_1 = P_2 = P_3 = P$, where P_i proportion of persons of i-th group]

H_A : Three groups of persons of each sample are not equally distributed.

The hypothesis can be tested by χ^2-test.

First sample:

Group	1	2	3	*Total*
	≤ 6.3	≤ 7.0	> 7.0	
Number of persons	19	10	7	36

The test statistic is

$$\chi_1^2 = \Sigma \frac{O_i^2}{E_i} - n; \qquad \begin{array}{l} O_i = \text{observed frequency of } i\text{-th group } (i = 1, 2, 3) \\ E_i = \text{frequency under hypothesis } (E_1 = E_2 = E_3 = 12) \end{array}$$

$$= \frac{1}{12}(19^2 + 10^2 + 7^2) - 36 = 6.5.$$

The P-value of χ_1^2 is 0.039 [χ^2 has 2 d.f.] < 0.05. The hypothesis of equal proportions of 3 groups is not true.

Second sample:

Group	1	2	3	*Total*
	≤ 6.3	≤ 7.0	> 7.0	
Number of persons	24	9	15	48

The χ^2-statistic is $\chi_2^2 = \frac{1}{16}[24^2 + 9^2 + 15^2] - 48 = 7.125$ $[E_1 = E_2 = E_3 = 16]$

The P-vaue of χ_2^2 is $0.028 < 0.05$. The hypothesis of equal proportions of 3 groups is not true.

Third sample:

Group	1	2	3	Total
	≤ 6.3	≤ 7.0	> 7.0	
Number of persons	20	4	6	30

The χ^2-statistic is $\chi_3^2 = \frac{1}{10}[20^2 + 4^2 + 6^2] - 30 = 15.2$ $[E_1 = E_2 = E_3 = 10]$.

The P-value of χ_3^2 is $0.001 < 0.05$. The null hypothesis is not true. A common conclusion can be done for all 3 samples by combining the P-values of all $3\chi^2$-test (as a meta-analysis) or by combining the χ^2-values.

The combined P-values follows χ^2-distribution with $2k$ d.f., where $k = 3$ samples. The combined value is

$$\begin{aligned} -2\ln P &= -2\ln p_1p_2p_3, \ (p_1 = 0.039, \ p_2 = 0.028, \ p_3 = 0.001) \\ &= -2\ln(0.039 \times 0.028 \times 0.001) = 27.45. \end{aligned}$$

Five per cent value of χ^2 with $2 \times 3 = 6$ d.f. is $12.592 < 27.45$. It indicates that the null hypothesis of equal proportions of 3 groups in all samples is not true.

The combined analysis can also be done by addition of the χ^2 values of 3 samples, where the added value is

$$\chi^2 = \chi_1^2 + \chi_2^2 + \chi_3^2 = 6.5 + 7.125 + 15.2 = 28.825.$$

It has $2 + 2 + 2 = 6$ d.f. and the calculated value is greater than 5% value of $\chi^2 = 12.592$. So, the null hypothesis of equal proportions of 3 groups of patients in 3 samples is not true.

12.7 Meta-Analysis Applied in Testing Independence of Two Characters

EXAMPLE 12.5 The following data represent the number of ever born children to mothers of different levels of education recorded by 3 persons from 3 localities.

Levels of education	Locality											
	L_1				L_2				L_3			
	Ever born children											
	O_{ij}				O_{ij}				O_{ij}			
	≤ 2	3-4	5^+	*Total*	≤ 2	3-4	5^+	*Total*	≤ 2	3-4	5^+	*Total*
Illiterate	15	28	12	55	15	28	32	75	13	50	25	88
Primary	20	17	8	45	25	30	13	68	12	48	12	72
Secondary	18	42	10	70	12	43	10	65	48	48	19	115
Higher	18	7	5	30	20	16	6	42	12	7	6	25
Total	71	94	35	200	72	117	61	250	85	153	62	300

(i) Are levels of education and levels of ever born children associated for each locality?

(ii) Draw a conclusion for all localities together.

Solution. We need to test the hypothesis:

H_0 : Levels of ever born children are independent of levels of education of mothers

H_A : Levels of ever born children are not independent of levels of education of mothers.

Under H_0 the test statistic is

$$\chi^2 = \Sigma\Sigma \frac{O_{ij}^2}{E_{ij}} - n, \text{ where}$$

O_{ij} = observation of i-th level of education corresponding to j-th level of ever born children for each locality,

$E_{ij} = \frac{R_i \times C_j}{n}$,

R_i = total of i-th level of education

C_j = total of j-th level of ever born children,

n = total mothers of each locality.

The E_{ij} values for different localities are shown below:

Levels of education	*Locality*											
	L_1				L_2				L_3			
	Ever born children											
	≤ 2	3-4	5^+	*Total*	≤ 2	3-4	5^+	*Total*	≤ 2	3-4	5^+	*Total*
Illiterate	19.525	25.85	9.625	55	21.6	35.1	18.3	75	24.933	44.88	18.187	88
Primary	15.975	21.15	7.875	45	19.584	31.824	16.592	68	20.4	36.72	14.88	72
Secondary	24.85	32.9	12.25	70	18.72	30.42	15.86	65	32.583	58.65	23.767	115
Higher	10.65	14.1	5.25	30	12.096	19.656	10.248	42	7.084	12.75	5.166	25
Total	71	94	35	200	72	117	61	250	85	153	62	300

Locality-1: $$\chi_1^2 = \frac{(15)^2}{19.525} + \frac{(28)^2}{25.85} + \frac{(12)^2}{9.625} + \frac{(20)^2}{15.975} + \frac{(17)^2}{21.15} + \frac{(8)^2}{7.875} + \frac{(18)^2}{24.85} + \frac{(42)^2}{32.9} + \frac{(10)^2}{12.25} + \frac{(18)^2}{10.65} + \frac{(7)^2}{14.1} + \frac{(5)^2}{5.25} - 200$$

$$= 17.122 \quad (P\text{-value}, p_1 = 0.009).$$

As the P-value of this χ^2 is $p_1 = 0.009 < 0.05$, H_0 is rejected. Levels of education and number of ever born children are associated.

Locality-2: $$\chi_2^2 = \frac{(15)^2}{21.6} + \frac{(28)^2}{35.1} + \frac{(32)^2}{18.3} + \frac{(25)^2}{19.584} + \frac{(30)^2}{31.824} + \frac{(13)^2}{16.592} + \frac{(12)^2}{18.72} + \frac{(43)^2}{30.42} + \frac{(10)^2}{15.86} + \frac{(28)^2}{12.096} + \frac{(16)^2}{19.656} + \frac{(6)^2}{10.248} - 250$$

$$= 65.22 \quad (P\text{-value}, p_2 = 0.0001).$$

As $p_2 < 0.05$, H_0 is rejected.

Locality-3: $\chi_3^2 = \frac{(13)^2}{24.933} + \frac{(50)^2}{44.88} + \frac{(25)^2}{18.189} + \frac{(12)^2}{20.4} + \frac{(48)^2}{36.72} + \frac{(12)^2}{14.88} + \frac{(48)^2}{32.583}$

$$+ \frac{(48)^2}{58.65} + \frac{(19)^2}{23.767} + \frac{(12)^2}{7.084} + \frac{(7)^2}{12.75} + \frac{(6)^2}{5.166} - 300$$

$$= 32.654 \quad (P\text{-value}, p_3 = 0.0001).$$

Here also H_0 is rejected.

For conclusion using data of all localities, meta-analysis can be performed by combining the P-values of 3 chi-squares, where the combined P-value is given by

$$\begin{aligned} -2\ln P &= -2\ln(p_1 \times p_2 \times p_3) \\ &= -2\ln[0.009 \times 0.0001 \times 0.0001] \\ &= 46.26. \end{aligned}$$

This $-2\ln P$ is distributed as χ^2 with $2k = 2 \times 3$ ($k = 3$, number of localities) d.f. The tabulated value of χ^2 at 5% level with 6 d.f. is 12.592; which is less than calculated $\chi^2(-2\ln P)$. So, H_0 of independence of levels of education and levels of ever born children is rejected.

The conclusion can also be done by combining the values of calculated χ^2 for 3 localities, where combined χ^2 is

$$\begin{aligned} \chi^2 &= \chi_1^2 + \chi_2^2 + \chi_3^2 \\ &= 17.122 + 65.22 + 32.654 \\ &= 114.996. \end{aligned}$$

This χ^2 has $6 + 6 + 6 = 18$ d.f. The tabulated χ^2 at 5% level with 18 d.f. is 28.869, which is less than calculated χ^2. This indicates that for all localities the levels of ever born children depend on levels of education of mothers.

12.8 Meta-Analysis Applied in Correlation and Regression

EXAMPLE 12.6 The following are the gestation period (x in months) of mothers and the body-weight (y in kg) to those mothers of rural and urban areas:

Urban	x	8.5	9.0	8.4	8.0	9.1	8.8	7.5	7.6	7.8	7.5	8.8
	y	3.8	4.2	3.0	3.5	3.5	4.2	2.8	2.5	3.0	3.2	4.0
Rural	x	8.4	8.5	9.0	8.4	8.7	8.5	7.5	8.0	9.0	7.0	8.5
	y	3.0	3.5	4.0	3.4	4.0	3.4	3.0	3.0	3.4	2.5	3.2

(i) Fit regression line of y on x for urban and rural areas separately and test the significance of regression in each case.

(ii) Is the regression significant for mothers of both urban area and rural area together?

(iii) Estimate a common regression coefficient for all mothers and test the significance of the regression coefficient.

(iv) Calculate correlation coefficient of x and y for urban and rural mothers separately and test the significance of correlation.

(v) Find a common correlation coefficient of all mothers and test the significance of correlation.

(vi) Fit a common regression line of y on x for both urban and rural areas together.

Solution. Let the fitted regression line be $\hat{y} = a + bx$, where $b = \dfrac{\text{S.P. }(xy)}{\text{S.S. }(x)}$, $a = \bar{y} - b\bar{x}$.

We have $n = 11$ for both groups of mothers.

(i) *Urban area:* $\text{S.S. }(x) = \Sigma x^2 - \dfrac{(\Sigma x)^2}{n} = 756.6 - \dfrac{(91)^2}{11} = 3.7818, \bar{x} = \dfrac{91}{11} = 8.27.$

$$\text{S.S. }(y) = \Sigma y^2 - \frac{(\Sigma y)^2}{n} = 132.55 - \frac{(37.7)^2}{11} = 3.3418, \bar{y} = \frac{37.7}{11} = 3.43.$$

$$\text{S.P. }(xy) = 314.71 - \frac{91 \times 37.7}{11} = 2.8282.$$

$$b = \frac{\text{S.P. }(xy)}{\text{S.S. }(x)} = \frac{2.8282}{3.7818} = 0.748.$$

$$a = \bar{y} - b\bar{x} = 3.43 - 0.748 \times 8.27 = -2.756.$$

$\therefore$ the fitted line is $\hat{y} = -2.756 + 0.748x$.

We need to test $H_0 : B = 0$, against $H_A : B \neq 0$.

We have $s^2 = \dfrac{1}{n-2}[\text{S.S. }(y) - b\text{S.P. }(xy)] = \dfrac{1}{11-2}[3.3418 - 0.748 \times 2.8282] = 0.1363.$

The test statistic is $t = \dfrac{b}{\sqrt{\frac{s^2}{\text{S.S. }(x)}}} = \dfrac{0.748}{\sqrt{\frac{0.1363}{3.7818}}} = 3.94$, P-value, $p_1 = 0.003$.

As P-value < 0.05, H_0 is rejected. The regression coefficient is significant.

Here $\text{S.S. (regression)} = b\,\text{S.P. }(xy) = 0.748 \times 2.8282 = 2.1155.$

$$\text{S.S. (error)} = \text{S.S. }(y) - \text{S.S. (regression)} = 3.3418 - 2.1155 = 1.2263.$$

$$\text{M.S. (error)} = \frac{\text{S.S. (error)}}{\text{d.f.}} = \frac{1.2263}{9} = 0.1363.$$

$$\text{S.S. (regression)} = \frac{\text{S.S. (regression)}}{\text{d.f.}} = \frac{2.1155}{1} = 2.1155.$$

$$F = \frac{\text{M.S. (regression)}}{\text{M.S. (error)}} = \frac{2.1155}{0.1363} = 15.52, \; P\text{-value}, p_1 = 0.003.$$

As P-value is < 0.05, the regression is significant.

Rural area: $\text{S.S. }(x) = \Sigma x^2 - \dfrac{(\Sigma x)^2}{n} = 764.81 - \dfrac{(91.5)^2}{11} = 3.6964, \; \bar{x} = \dfrac{91.5}{11} = 8.32.$

$$\text{S.S. }(y) = \Sigma y^2 - \frac{(\Sigma y)^2}{n} = 122.42 - \frac{(36.4)^2}{11} = 1.9691, \; \bar{y} = \frac{36.4}{11} = 3.31.$$

$$\text{S.P. }(xy) = \Sigma xy - \frac{\Sigma x \Sigma y}{n} = 305.01 - \frac{91.5 \times 364}{11} = 2.2282.$$

$$b = \frac{\text{S.P. }(xy)}{\text{S.S. }(x)} = \frac{2.2282}{3.6964} = 0.603.$$

$$a = \bar{y} - b\bar{x} = 3.31 - 0.603 \times 8.32 = -1.707.$$

$\therefore$ the fitted line is $\hat{y} = -1.707 + 0.603x$.

$$\text{S.S. (regression)} = b\,\text{S.P. }(xy) = 0.603 \times 2.2282 = 1.344.$$

$$\text{S.S. (error)} = \text{S.S. }(y) - \text{S.S. (regression)} = 1.9691 - 1.344 = 0.6251.$$

$$\text{M.S. (error)} = \frac{\text{S.S. (error)}}{n-2} = \frac{0.6251}{9} = 0.069 = s^2.$$

The test statistic for $H_0 : B = 0$ is

$$t = \frac{b}{\sqrt{\frac{s^2}{\text{S.S. }(x)}}} = \frac{0.603}{\sqrt{\frac{0.069}{3.6964}}} = \frac{0.603}{0.1366} = 4.41; \ P\text{-value}, \ p_2 = 0.002.$$

As P-value < 0.05, H_0 is rejected. The regression coefficient is significant.

$$\text{M.S. (regression)} = \frac{\text{S.S. (regression)}}{\text{d.f.}} = \frac{1.344}{1} = 1.344$$

$$F = \frac{\text{M.S. (regression)}}{\text{M.S. (error)}} = \frac{1.344}{0.069} = 19.48, \ P\text{-value}, \ p_2 = 0.002.$$

As P-value < 0.05, the regression is significant.

(ii) To test the significance of regression for both urban and rural area, we can do the meta-analysis by combining the P-values of both F-statistic, where the combined test statistic is

$$-2 \ln P = -2 \ln p_1 \times p_2 = -\ln(0.003 \times 0.002) = 24.05.$$

This test statistic is distributed as χ^2 with $2k = 4\,(k = 2)$ d.f. The tabulated value of χ^2 with 4 d.f. is 9.488, which is less than calculated $\chi^2 = -2 \ln P$. The regression is significant for urban area and rural area taken together.

(iii) We have $v(b) = \frac{s^2}{\text{S.S. }(x)} = \frac{0.1363}{3.7818} = 0.036$ for urban area.

Let $W_1 = \frac{1}{v(b)} = \frac{1}{0.036} = 27.78.$

Again, $v(b) = \frac{s^2}{\text{S.S. }(x)} = \frac{0.069}{3.6964} = 0.0187$ for rural area.

Let $W_2 = \frac{1}{v(b)} = \frac{1}{0.0187} = 53.48.$

Estimate of common regression coefficient is

$$\frac{W_1 b_1 + W_2 b_2}{W_1 + W_2}, \text{ where } b_1 = \text{estimate of } B \text{ for urban area} = 0.748$$

$$= \frac{27.78 \times 0.748 + 53.48 \times 0.603}{27.48 + 53.48}, \ b_2 = \text{estimate of } B \text{ for rural area} = 0.603$$

$$= 0.655.$$

The significance of the common regression coefficient can be concluded by combining the P-values of two t-test statistics, where the combined P value is

$$-2 \ln P = -2 \ln p_1 \times p_2 = -2 \ln 0.003 \times 0.022 = 24.05.$$

This value of $\chi^2 >$ 5% value of χ^2 with $2k = 4(k = 2)$ d.f.
The regression coefficient is significant.

(iv) The correlation coefficient for urban area is

$$r_1 = \frac{\text{S.P. }(xy)}{\sqrt{\text{S.S. }(x)\,\text{S.S. }(y)}} = \frac{2.8282}{\sqrt{3.7818 \times 3.3418}} = 0.796.$$

We need to test $H_0 : P = 0$, against $H_A : P \neq 0$, where P is the population correlation coefficient. The test statistic is

$$t_1 = \frac{r_1\sqrt{n-2}}{\sqrt{1-r_1^2}} = \frac{0.796\sqrt{11-2}}{\sqrt{1-(0.796)^2}} = \frac{2.388}{0.6053} = 3.94.$$

P-value of t_1 is $p_1 = 0.003$. As P-value < 0.05, the correlation coefficient is significant. The correlation coefficient for rural area is

$$r_2 = \frac{\text{S.P. }(xy)}{\sqrt{\text{S.S. }(x)\,\text{S.S. }(y)}} = \frac{2.2282}{\sqrt{3.6964 \times 1.9641}} = 0.826.$$

The test statistic is $t_2 = \dfrac{r_2\sqrt{n-2}}{\sqrt{1-r_2^2}} = \dfrac{0.826\sqrt{11-2}}{\sqrt{1-(0.826)^2}} = 4.41.$

P-value of t_2 is $0.002 < 0.05$. The correlation coefficient is significant.

(v) Let $v(r_1) = \dfrac{1-r_1^2}{n-2} = \dfrac{1-(0.796)^2}{11-2} = 0.0407.$

Let $W_1 = \dfrac{1}{v(r_1)} = 24.56.$

$v(r_2) = \dfrac{1-r_2^2}{n-2} = \dfrac{1-(0.826)^2}{11-2} = 0.0353.$

Let $W_2 = \dfrac{1}{v(r_2)} = 28.33.$

The common estimate of correlation coefficient is

$$r = \frac{W_1 r_1 + W_2 r_2}{W_1 + W_2} = \frac{24.56 \times 0.796 + 28.33 \times 0.826}{24.56 + 28.33} = 0.812.$$

A common conclusion can be drawn by combining the P-values of t_1 and t_2, where the combined P-value is

$$-2\ln P = -2\ln p_1 \times p_2 = -2\ln 0.003 \times 0.002 = 24.05.$$

As this calculated value is greater than tabulated value of χ^2 at 5% level with $2k = 4$ ($k = 2$) d.f. (9.488), the correlation coefficient is significant both for urban and rural areas together.

(vi) The value of a for urban area is, say, $a_1 = -2.756$ and for rural area is, say, $a_2 = -1.707$.

The variance of a is given by $v(a) = \dfrac{s^2}{n - \frac{(\Sigma x)^2}{\Sigma x^2}}$.

This value for urban area is

$$V(a) = \frac{0.1363}{11 - \frac{(91)^2}{756.6}} = 1.36. \text{ Let } V_1 = \frac{1}{v(a)} = \frac{1}{1.36} = 0.735.$$

The value for rural area is

$$V(a) = \frac{0.069}{11 - \frac{(91.5)^2}{764.81}} = 1.15. \text{ Let } V_2 = \frac{1}{v(a)} = \frac{1}{1.15} = 0.869.$$

The combined estimate of a is

$$a = \frac{V_1 a_1 + V_2 a_2}{V_1 + V_2} = \frac{0.735(-2.756) + 0.869(-1.707)}{0.735 + 0.869} = -2.18$$

Therefore, the common regression line for both urban and rural areas is

$$\hat{y} = -2.18 + 0.655x.$$

EXERCISE – 12

1. From $N = 500$ patients of a hospital $n = 22$ patients are selected at random and their heights (x cm) and weights (y kg) are measured and given below:

x :	155	154	156	156	155	157	157	158	152	152	153	154	155	156
y :	60	60	65	66	65	58	62	68	60	62	64	65	66	67

x :	153	155	154	156	155	157	157	158
y :	54	55	58	62	62	67	68	70

From $N = 300$ patients of another hospital $n = 16$ patients are selected at random and their heights (x cm) and weights (y kg) are measured and given below:

x	158	155	157	152	150	151	156	154	158	158	157	158	154	155	156	157
y	65	67	68	58	55	54	56	70	68	67	62	64	65	66	67	58

(i) Is the average height of both groups of patients 154 cm?
(ii) Is the average weight of both groups of patients 60 kg?
(iii) Find a common average of heights of both groups of patients.
(iv) Are the regression coefficients of weight on height for two groups similar?
(v) Find a common regression coefficient for both groups of patients.
(vi) Are the correlation coefficients of height and weight of both groups of patients similar?
(vii) Find a common correlation coefficient of height and weight of both groups of patients.

2. (a) The following information on duration of breastfeeding (x in months) of 3 different groups of mothers observed by 3 independent research workers

Research worker	*Sample size*	*Total duration of breastfeeding* Σx	Σx^2
1	$n_1 = 40$	588	8900
2	$n_2 = 35$	555	8967
3	$n_3 = 50$	672	9345

(i) Is the average breastfeeding period 15 months? Make a common conclusion from all samples.
(ii) Find a common average of breastfeeding periods.

(b) The above investigated mothers are classified by their level of education and duration of breastfeeding

Level of education	*Duration of breastfeeding (in months)*											
	Sample-1				*Sample-2*				*Sample-3*			
	< 6	6-12	12^+	*Total*	< 6	6-12	12^+	*Total*	< 6	6-12	12^+	*Total*
Illiterate	8	7	5	20	4	5	3	12	10	12	6	28
Primary	5	4	3	12	4	5	4	13	4	5	3	12
Secondary and above	4	2	2	8	4	3	3	10	3	4	3	10
Total	17	13	10	40	12	13	10	35	17	21	12	50

Is duration of breastfeeding associated with level of education?

3. Four investigators independently have undertaken 4 investigations on correlation of age and blood pressure levels of some persons. They have calculated correlation coefficients from different sample sizes. The investigated results are given below:

Sample sizes (n_i)	*Correlation coefficients* (r_i)
50	0.70
60	0.62
50	0.48
40	0.55

Are the correlation coefficients similar? Calculate one common correlation coefficient of all observations.

4. Three investigators conducted independently 3 surveyes to study the association of suffering from cancer and smoking habit. The investigated results are shown below:

Smoking habit	*Investigators*								
	1			2			3		
	Stages of cancer								
	Initial	*Last*	*Total*	*Initial*	*Last*	*Total*	*Initial*	*Last*	*Total*
Yes	18	17	35	20	18	38	25	20	45
No	10	7	17	14	8	22	18	12	30
Total	28	24	52	34	26	60	43	32	75

Is smoking habit associated with suffering from cancer?
Use (i) Mantel-Haenszel test, (ii) Meta-analysis for common conclusion.

5. Four investigators independently recorded the blood pressure levels (mmHg) of different persons. The recorded information are as follows:

Investigators	*Recorded blood pressure levels*
1	75, 80, 82, 85, 80, 75, 70, 85, 88, 90, 85, 80, 72, 70, 80
2	82, 80, 75, 70, 65, 70, 72, 88, 90, 95, 88, 85, 80, 82, 80
3	70, 75, 78, 82, 80, 85, 87, 86, 90, 95, 100, 90, 80, 85
4	80, 80, 80, 70, 75, 65, 68, 72, 70, 70, 65, 75, 80, 80

Is median blood pressure level 75?

6. The following are the information on amount of s. triglyceride (x mg/dl) and amount of s. cholesterol (y mg/dl) of some persons recorded by two investigators:

Investigator-1	*Investigator*-2
$n_1 = 50, \Sigma x = 3780, \Sigma x^2 = 285862$	$n_2 = 40, \Sigma x = 3655, \Sigma x^2 = 334508$
$\Sigma y = 4585, \Sigma y^2 = 420588$	$\Sigma y = 4077, \Sigma y^2 = 415725$
$\Sigma xy = 346725$	$\Sigma xy = 372665$

Are the regression coefficients of y on x observed by two investigators similar?

7. Three investigators have conducted 3 experiments to study the impacts of same medicine produced by 4 different companies to reduce the blood sugar levels of diabetic patients. The fasting blood sugar levels of different patients after the use of medicines are observed and given below:

Investigator	*Company*	*Fasting blood sugar* (mmol/l) *levels of patients* (x_{ijl})					*Total*
1	C_1	6.8	7.5	7.0	7.8	8.0	37.1
	C_2	6.5	6.3	6.3	6.3	6.0	31.4
	C_3	5.8	5.5	5.2	6.0	6.2	28.7
	C_4	7.5	8.0	8.0	7.8	7.5	38.8
	Total, x_1						136.0
2	C_1	7.0	7.0	7.5	7.5	7.2	36.2
	C_2	5.8	5.5	6.0	6.0	5.4	28.7
	C_3	5.5	5.0	5.2	5.6	5.5	26.8
	C_4	7.8	8.2	8.2	7.0	7.0	38.2
	Total, x_2						129.9
3	C_1	6.3	6.4	6.5	6.5	6.5	32.2
	C_2	5.5	5.4	5.0	5.0	5.0	25.9
	C_3	4.8	5.2	5.0	5.2	5.5	25.7
	C_4	7.8	8.5	8.0	8.2	7.5	40.0
	Total, x_3						123.8

(i) Are the medicines of different companies similarly effective?

(ii) Is there any difference between companies C_1 and C_4?

(iii) The mean blood sugar level by the medicine of company C_3 is observed minimum. Do you think that the medicine produced by C_3 is better than that of other companies?

Chapter 13

Multivariate Analysis

13.1 Introduction

Applied research work is mainly based on data analysis, where data are collected from different types of sampling units covering social, physical, biological, medical and other aspects which need scientific inquiry. Data analysis, while interesting with one variable, becomes truely fascinating and challenging when several variables are involved. Researchers in the field of biological, medical, physical and social sciences frequently collect information on several variables. The numerical analysis of the data collected on some variables from same sample points is known as multivariate analysis.

The multivariate analysis has two main aspects, viz., (1) dependence analysis, and (2) interdependence analysis. As the variables are observed from same sample point, these are expected to be related. The study of the relationship of variable is a topic of dependence analysis. The multivariate regression analysis, discriminant analysis and multivariate analysis of variance [MANOVA] are the topics of dependence analysis. In these analyses, investigation of dependence among variables is performed to study the nature of relationship among variables. Under certain restrictions the analytical results are accepted to validate the prior assumptions regarding the relationship of the variables. As the variables are observed, in most cases, from homogeneous sample points, the values of a variable are expected not to be very heterogenous. Moreover, as the variables are interrelated, some functions of the variables may be sufficient to explain the variability in the data set. We need to suggest fewer components compared to a large number of variables under study so that most of the variabilities of the data set can be explained by fewer components. The selection of fewer components is a data reduction technique, where reduced data set is sufficient to explain most of the variablilities in the data set. Thus, interdependence analysis is a data reduction technique. Among different techniques of data reduction principal component analysis, factor analysis, cluster analysis, etc. are very important.

13.2 Principal Component Analysis

It has already been mentioned that social and medical data of sample points are interrelated as many variables are under study from same sample points. As a result, practical problems in respect of number of correlation due to a large number of variables are arisen. A comprehensive idea from a large number of correlation is beyond imagination. Thus, the researcher needs to identify some important components which can explain almost all variabilities in the data set and which can provide a comprehensive idea about population characteristic. Such component or components are the linear function of the original variables under study. The principal component analysis is a technique to select components so that 90% or more of the variability in the data set can be explained by the new selected components.

At the early stage of scientific investigation there was very little knowledge about theoretical aspect of selection of components which can explain most of the variability in the data set. Moreover, computing facilities were also not available to every researcher. But,. nowadays theory is developed further and computing facilities are easily available to everybody. So, data reduction technique can be easily applied so that one can summarise systematically the most of the variabilities in the data set. However, reduction in number of variables by new components is done in such a way that the process maintains as much of original information as possible.

Let there be p variables $x_1, x_2, \ldots, x_p$. Values of each variable is recorded from n sample points, where the n values of p variables can be presented as follows:

$$X = \begin{bmatrix} X_1 & X_2 & \cdots & X_p \\ x_{11} & x_{21} & \cdots & x_{p1} \\ x_{12} & x_{22} & \cdots & x_{p2} \\ \vdots & \vdots & \vdots & \vdots \\ x_{1n} & x_{2n} & \cdots & x_{pn} \end{bmatrix}.$$

Here arrangement of np values of p variables in rows and columns is called a matrix denoted by X. The matrix X has p-vectors $X_1, X_2, \ldots, X_p$ each one has n observations recorded from n sample points. Assume that the mean vector of X is zero and variance-covariance matrix of X is Σ, where

$$\Sigma = \begin{bmatrix} v(x_1) & \text{cov}(x_1x_2) & \text{cov}(x_1x_3) & \cdots & \text{cov}(x_1x_p) \\ \text{cov}(x_1x_2) & v(x_2) & \text{cov}(x_2x_3) & \cdots & \text{cov}(x_2x_p) \\ \vdots & \vdots & \vdots & \vdots & \vdots \\ \text{cov}(x_1x_p) & \text{cov}(x_2x_p) & \cdots & \cdots & v(x_p) \end{bmatrix}$$

Here Σ is a real positive definite matrix. Let us consider that the non-zero eigenvalues of Σ are

$$\lambda_1 > \lambda_2 > \cdots > \lambda_p$$

and the corresponding eigenvectors are $\gamma_1, \gamma_2, \ldots, \gamma_p$, where

$$\Gamma = [\gamma_1 \gamma_2 \cdots \gamma_p].$$

This Γ is a matrix of p eigenvectors $\gamma_1, \gamma_2, \ldots, \gamma_p$. Let us consider an orthogonal transformation of X vector to Y vector, where

$$Y = \Gamma' X$$

and $Y = [Y_1, Y_2, \ldots, Y_p]$ is a vector of new components and is known as *principal components.*

An overall measure of variability of X can be taken as $\text{tr}\Sigma$ or the generalized variance of determinant of Σ. Here

$$\text{tr}\Sigma = \sum_{i=1}^{p} \lambda_i.$$

The overall measure of variability of Y is same as that of X. Here the j-th component explains the variability of X by an amount

$$\frac{\lambda_i}{\Sigma \lambda_i} \times 100.$$

Due to positive definite matrix of Σ the strict positivity of λ_i is guaranteed. Those components are extracted for further analysis if these explain 90% or more [Mardia et al., 1979] of the variation of the data set.

In practice, the analysis is performed on the basis of variance-covariance or correlation matrix of the variables under study. Let us assume that l_i is the i-th eigenvalue of the correlation matrix of X and g_{ij} is the j-th value of eigenvector corresponding to i-th eigenvalue. Then

$$y_j = \sum_{i}^{p} g_{ij} x_i.$$

The importance of x_i in extracting y_j is measured by the amount

$$\frac{g_{ij}\sqrt{l_i}}{\sqrt{S_{ii}}},$$

where S_{ii} is the i-th diagonal element of variance-covariance matrix of the data set. The value of S_{ii} is 1, if analysis is done from correlation matrix.

13.3 Factor Analysis

It has already been stated that many variables can be observed from each sample point to infer about population parameter vectors. But all the variables may not be needed in the analysis as some of these are interrelated. For example, let us consider the case of selection of doctors for some hospitals. The appointment of doctors or the selection of doctors may be done by the authority after reviewing the result of last MBBS examination (x_1), the result of written test (x_2) and the marks of oral examination (x_3). All these results depend on general ability (f) of the candidates seeking appointment. This general ability (f) of the candidates can be expressed as follows:

$$x_1 = a_1 f + u_1,\ x_2 = a_2 f + u_2 \text{ and } x_3 = a_3 f + u_3.$$

Here f is the general ability of the candidates and is known as *common factor*. This f can be observed from the sample points. Factor analysis is a multivariate technique to identify this common factor which is a linear function of other 3 variables x_1, x_2 and x_3. Here the functions of x_1, x_2 and x_3 are known as factor models.

In the factor model, $a_i (i = 1, 2, 3)$ is the factor loading of the common factor f and u_i is random component. Here the variance of u_i will be minimum if x_i and f is highly correlated.

The sources of variance of u_i are two. One is due to the difference in general ability in a particular topic, and another is due to the inefficient evaluation of the ability of the candidate. The objective of the factor analysis is to identify the common factor (f) on the basis of the variables under study. One of the technique of identification of the common factor or factors is the principal factor method which is similar to principal component analysis.

Before performing the factor analysis, it is important to investigate whether a particular data set is suitable for the analysis or not. This can be done by Kaiser-Meyer-Olkin (KMO) statistic, where the statistic is

$$\text{KMO} = \frac{\Sigma\Sigma r_{ij}^2}{\Sigma\Sigma r_{ij}^2 + \Sigma\Sigma a_{ij}^2}.$$

Here $r_{ij} (i \neq j)$ is the correlation coefficient of i-th and j-th variables and a_{ij} is the partial correlation coefficient of those two variables. The value of KMO becomes 1, if $a_{ij} = 0$. In such a case the factor analysis is fruitful. It implies that factor analysis will not be accepted, if KMO is small. Kaiser (1974) has mentioned that, the factor analysis is excellent, if KMO is greater than or equal to 0.90; it is meritorious, if $0.80 \leq$ KMO ≤ 0.90; middle, if $0.70 \leq$ KMO ≤ 0.80, mediocre, if $0.60 \leq$ KMO ≤ 0.70; miserable, if $0.50 \leq$ KMO ≤ 0.60 and unacceptable, if KMO < 0.50.

Similar KMO statistic can also be computed to decide about the inclusion of i-th variable in a factor analysis. This statistic is known as 'Measure of Sampling Adequacy' (MSA$_i$) of i-th variable, where

$$\text{MSA}_i = \frac{\sum_j r_{ij}^2}{\sum r_{ij}^2 + \sum a_{ij}^2}, \; i \neq j.$$

Here also the inclusion of i-th variable in factor analysis is not acceptable, if MSA$_i$ < 0.50. The dropping of variable can also be done, if the multiple correlation coefficient R_i of i-th variable with other variables is small ($i \neq j$).

As an example of factor analysis, let us extract the factors for the data discussed below:

The following are the total expenditure (y in 000 taka) per day in clinics, where the patients are served by a number of doctors (x_1), a number of nurses (x_2), service persons (x_3), medicine supplied to the patients (x_4) in each clinic, the area of the clinic (x_5; $x_5 = 0$, if the clinic is in rural area, $x_5 = 1$, if it is situated in semi-urban area and $x_5 = 2$, if it is in urban area) and a number of patients visited in the clinics in a day (x_6).

Sl. No.	x_1	x_2	x_3	x_4	x_5	x_6	y	*Sl. No.*	x_1	x_2	x_3	x_4	x_5	x_6	y
01	5	3	8	44	0	77	20.6	16	6	3	12	40	1	95	36.5
02	4	2	7	32	0	62	18.5	17	5	4	14	35	1	78	33.0
03	3	2	5	23	0	50	15.8	18	6	3	13	28	1	70	28.5
04	6	5	10	38	0	70	22.7	19	5	5	15	38	1	85	35.5
05	5	4	8	25	0	55	18.0	20	4	3	10	30	1	65	24.0
06	4	3	8	35	0	60	24.5	21	5	3	10	30	1	70	28.8
07	5	5	12	48	0	88	32.0	22	5	4	12	40	1	78	33.4
08	4	3	10	30	0	66	25.6	23	6	6	15	48	2	100	45.0
09	4	2	8	18	0	52	16.5	24	8	6	12	50	2	125	48.0
10	5	2	10	15	0	45	17.0	25	10	8	16	52	2	130	50.5
11	6	4	12	30	0	55	27.5	26	12	8	15	70	2	150	52.5
12	8	6	15	40	1	75	38.5	27	10	5	20	82	2	150	55.5
13	6	4	10	45	1	70	40.0	28	8	6	17	66	2	128	48.0
14	7	4	10	55	1	75	42.0	29	10	8	15	58	2	125	50.5
15	8	4	12	42	1	68	38.0	30	7	6	12	45	2	78	45.0

The factor analysis is done using the values of the variable corresponding to $x_5 = 2$. The KMO statistic of these variables is $0.457 < 0.50$. Though the factor analysis is not acceptable (as KMO < 0.50), it is done as an application of the technique and to explain the importance of the variables in extracting factors. The 'Principal Component Method' is used to extract the factors using the correlation matrix of the six variables x_1, x_2, x_3, x_4, x_6 and y. The correlation matrix provides eigenvalues

$$\lambda_1 = 4.005, \lambda_2 = 1.478, \lambda_3 = 0.296, \lambda_4 = 0.130, \lambda_5 = 0.085, \lambda_6 = 0.006.$$

Among these eigenvalues the first two eigenvalues (λ_1 and λ_2) can explain 91.39% of variation of data set, where the percentage of variation by λ_1 is

$$\frac{\lambda_1}{\Sigma\lambda_i} = \frac{4.005}{6.00} = 0.66755 \text{ or } 66.755\%$$

The variation by λ_2 is

$$\frac{\lambda_2}{\Sigma\lambda_i} = \frac{1.478}{6.00} = 0.24634 \text{ or } 24.634.$$

Total variation can be explained by λ_1, and λ_2 is $66.755 + 24.634 = 91.39\%$.

As first two eigenvalues can explain 91.39% variation of the data set, two common factors are extracted by 'Principal Component Method'. The factors are linear function of the variables x_1, x_2, x_3, x_4, x_6 and y, where the weights (coefficients) of the variables are the elements of eigenvector corresponding to eigenvalues λ_1 and λ_2. The elements of the eigenvector corresponding to λ_1 are

$$[0.852,\ 0.189,\ 0.766,\ 0.911,\ 0.941,\ 0.970].$$

Thus, the factor-1 is

$$[f_1 = 0.852x_1 + 0.189x_2 + 0.766x_3 + 0.911x_4 + 0.941x_6 + 0.970y].$$

It is seen that the variable y is most important to explain the variation in the data set followed by x_6.

The elements of the eigenvector corresponding to λ_2 are

$$[0.490,\ 0.945,\ -0.477,\ -0.336,\ 0.071,\ 0.009].$$

The factor-2 is

$$[f_2 = 0.490x_1 + 0.945x_2 - 0.477x_3 - 0.336x_4 + 0.071x_6 + 0.009y].$$

13.4 Cluster Analysis

The cluster analysis has wide application in biology, medicine, agriculture, marketing, etc. The numerical taxonomy in the field of biology is used to classify the animals into class, order and family. Different species of plants have different characteristics. Therefore, plant specimens can be classified into homogeneous groups. In agriculture, the land fertility of a particular region may not be homogeneous for any type of crop. Then the pieces of land sharing similar fertility for a particular crop may be grouped together. In medicine, the patients of similar disease may be clustered together because of creating facility of similar treatment to a cluster of patients. For example, let us think that a group of people are identified as diabetic patients, but all of them are at initial stage. So, they can be clustered in one group and similar treatment may be prescribed for all of the patients.

Since similar objects/units form a cluster, all the sample points in any cluster will provide similar information about the population characteristic. Therefore, for further analysis one or fewer units of a cluster may be included in the study.

Basic steps in cluster analysis

In cluster analysis the sample objects are clustered on the basis of some characteristics. Therefore, to start with the analysis, a number of decisions must be made regarding the characteristics to be considered, the number of variables to be included in the analysis, the measurement of distance between objects and the criterion to group the objects. The selection of variables for cluster analysis is important, since the exclusion of important variables will provide poor or misleading findings. For example, in medical treatment the patients of hypertension are needed to be clustered. In clustering the patients the blood pressure levels of patients is very important characteristic. This variable and age of patients must be included in the analysis. After selection of variables, the next important point to be considered is the measure of similarity and the measure of distance between objects. Two objects will be included in two groups if their distance is maximum and they will be included in the same cluster if they are close to each other. One of the methods of measuring the distance of two objects is the calculation of Minkowski Metric, where the elements of the metric is given by

$$d_{ij} = \left[\sum_k (X_{ik} - X_{jk})^2\right]^{\frac{1}{2}}, \quad i \neq j.$$

If $i = j$, the d_{ij} is $d_{ii} = 0$. The d_{ij} value is calculated for all k variables. This d_{ij} indicates the Ecludian distance between i-th and j-th objects. The elements of d_{ij} can be shown by an example for 5 objects using $k = 2$ for the variables x and y, where

x	3.5	3.5	3.8	3.5
y	1.3	4.0	3.2	1.0

$$\begin{aligned} \text{Here } d_{11} &= \left[(x_{11} - x_{11})^2 + (y_{11} - y_{11})\right]^{\frac{1}{2}} = 0 \\ d_{12} &= \left[(x_{11} - x_{12})^2 + (y_{11} - y_{12}\right]^{\frac{1}{2}} \\ &= \left[(3.5 - 3.5)^2 + (1.3 - 4.0)^2\right]^{\frac{1}{2}} \\ &= \left[(-2.7)^2\right]^{\frac{1}{2}} = 2.7. \end{aligned}$$

Similarly, other d_{ij} values are calculated. Then, the d matrix with elements d_{ij} is as follows:

$$d = \begin{bmatrix} 0.00 & 2.70 & 1.92 & 0.30 \\ 2.70 & 0.00 & 0.85 & 3.00 \\ 1.92 & 0.85 & 0.00 & 2.22 \\ 0.30 & 3.00 & 2.22 & 0.00 \end{bmatrix}$$

From this matrix, it is observed that the fourth object is closer to the first one as the distance between first and fourth objects is minimum. The second and third objects are closer to each other. Therefore, the first and fourth object is to be included in one cluster. The similar process is to be followed in clustering the other objects following the other steps of analysis.

13.5 Discriminant Analysis

In medical science a doctor needs to diagnise a new patient by observing his different characteristics, viz., age, height, weight, blood pressure level, blood sugar level, amount of haemoglobin, blood urea, etc. The study of these variables helps the doctor to diagnise a patient. The patients may be grouped into several classes according to their disease. Then a new patient visits the doctor. Accordingly, the doctor will ask the patient for investigation of the above variables. The investigated results will help the doctor to enlist this patient into any one of the groups of patients earlier identified. This process of enlistment of the new patient into any of the earlier groups is done by a multivariate technique known as *Discriminant Analysis.*

The objective of the distriminant analysis is to classify the sample units (objects) into two or more groups. This is done with the help of linear combination of predictor or explanatory variables in such a way that misclassification error is minimum. The distriminant rules are

(i) ML Method of Discrimination,

(ii) Bayes Discriminant Rule and

(iii) Fisher's Discriminant Rule.

Let us explain the method of discrimination by an example. Consider that we have three sets of observations collected from 3 populations, where each set contains 20 observations of two variables, say x_1 and x_2. The mean vectors of two variables for 3 sets of observations are:

$$\bar{Y}_1 = \begin{bmatrix} 47.500 \\ 2.915 \end{bmatrix}, \quad \bar{Y}_2 = \begin{bmatrix} 48.45 \\ 4.25 \end{bmatrix}, \quad \bar{Y}_3 = \begin{bmatrix} 42.25 \\ 2.20 \end{bmatrix}, \quad \left[\bar{Y}_i = \begin{bmatrix} \bar{x}_1 \\ \bar{x}_2 \end{bmatrix}\right]$$

The variance-covariance matrix of two variables are:

$$S_1 = \begin{bmatrix} 52.65 & 5.8825 \\ 5.8825 & 0.9013 \end{bmatrix}, \quad S_2 = \begin{bmatrix} 42.1475 & 9.4325 \\ 9.4325 & 4.3085 \end{bmatrix}, \quad S_3 = \begin{bmatrix} 47.6475 & 5.595 \\ 5.595 & 0.835 \end{bmatrix}$$

$$\left[S_i = \begin{bmatrix} v(x_1) & \text{cov}(x_1 x_2) \\ \text{cov}(x_1 x_2) & v(x_2) \end{bmatrix}, \; i = 1, 2, 3 \right]$$

We have $k = 3, n_1 = n_2 = n_3 = 20 \; (i = 1, 2, 3)$.

We have $S_u = \sum n_i S_i / (n - k), \; n = \sum n_i = n_1 + n_2 + n_3 = 60$.

$$S_u = \begin{bmatrix} 49.9807 & 7.3368 \\ 7.3368 & 2.1210 \end{bmatrix}.$$

$$S_u^{-1} = \frac{1}{2.1210 \times 49.9807 - (7.3368)^2)} \begin{bmatrix} 2.1210 & -7.3368 \\ -7.3368 & 49.9807 \end{bmatrix}$$

$$= \begin{bmatrix} 0.04065 & -0.14060 \\ -0.14060 & 0.95784 \end{bmatrix}.$$

By ML method of discrimination, we need to calculate a statistic

$$D_{ij} = (\bar{Y}_i - \bar{Y}_j)' S_u^{-1} x - \frac{1}{2} \bar{Y}_i' S_u^{-1} \bar{Y}_i + \frac{1}{2} \bar{Y}_j S_u^{-1} \bar{Y}_i,$$

where $i \neq j = 1, 2, 3$.

The above discriminant functions are linearly related by $D_{12} + D_{23} = D_{13}$.

The rule to allocate x to a population is that

x is allocated to Population-1, if $D_{12} > 0$ and $D_{13} > 0$

x is allocated to Population-2, if $D_{12} < 0$ and $D_{13} > 0$

x is allocated to Population-3, if $D_{12} < 0$ and $D_{13} < 0$.

In the above example, let us allocate a new observation $x = [6.2 \;\; 3.2]$ to a population. Here $D_{13} = 1.77 > 0$ and $D_{12} = 6.46 > 0$. Therefore, x belongs to Population-1.

APPENDICES

Appendix—1

Table of Random Digits

1	2	3	4	5	6	7	8	9	10	11
1	51449	39284	85527	67168	91284	19954	91166	70918	85957	19492
2	16144	56830	67507	97275	25982	69294	32841	20861	83114	12531
3	48145	48230	99481	13050	81818	25282	66466	24461	97021	21072
4	83780	48351	85422	42978	26088	17869	94245	26622	48318	73850
5	95329	38482	93510	39170	63683	40587	80451	43058	81923	97072
6	11179	69004	34273	36062	26234	58601	47159	82248	95968	99722
7	94631	52413	31524	02316	27611	15888	13525	43809	40014	30667
8	64275	10294	35027	25604	65695	36014	17988	02734	31732	29911
9	72125	19232	10782	30615	42005	90419	32447	53688	36125	28456
10	16463	42028	27927	48403	88963	79615	41218	43290	53618	68082
11	10036	66273	69506	19610	01479	92338	55140	81097	73071	61544
12	85356	51400	88502	98267	73943	25828	38219	13268	09016	77465
13	84076	82087	55053	75370	71030	92275	55497	97123	40919	57479
14	76731	39755	78537	51937	11680	78820	50082	56068	36908	55399
15	19032	73472	79399	05549	14772	32746	38841	45524	13535	03113
16	72791	59040	61529	74437	74482	76619	05232	28616	98690	24011
17	11553	00135	28306	65571	34465	47423	39198	54456	95283	54637
18	71405	70352	76763	64002	62461	41982	15933	46942	36941	93412
19	17594	10116	55483	96219	85493	96955	89180	59690	82170	77643
20	09584	23476	09243	65568	89128	36747	63692	09986	47687	46448
21	81677	62634	52794	01466	85938	14565	79993	44956	82254	65223
22	45849	01177	13773	43523	69825	03222	58458	77463	58521	07273
23	97252	92257	90419	01241	52516	66293	14536	23870	78402	41759
24	26232	77422	76289	57587	42831	87047	20092	92676	12017	43554
25	87799	33602	01931	66913	63008	03745	93939	07178	70003	18158
26	46120	62298	69126	07862	76731	58527	39342	42749	57050	91725
27	53292	55652	11834	47581	25682	64085	26587	92289	41853	38354
28	81606	56009	06021	98392	40450	87721	50917	16978	39472	23505
29	67819	47314	96988	89931	49395	37071	72658	53947	11996	64631
30	50458	20350	87362	83996	86422	58694	71813	97695	28804	58523
31	59772	27000	97805	25042	09916	77569	71347	62667	09330	02152
32	94752	91056	08939	93410	59204	04644	44336	55570	21106	76588
33	01885	82054	45944	55308	55487	56455	56940	68787	36591	29914
34	85190	91941	86714	76593	77199	39724	99548	13827	84961	76740
35	97747	67607	14549	08215	95408	46381	12449	03672	40325	77312
36	43318	84469	26047	86003	34786	38931	34846	28711	42833	93019
37	47874	71365	76603	57440	49514	17335	71969	58055	99136	73589
38	24259	48079	71198	95859	94212	55402	93392	31965	94622	11673
39	31947	64805	34133	03245	24546	48934	41730	47831	26531	02230
40	37911	93224	87153	54541	57529	38299	65659	00202	07054	40168
41	82714	15799	93126	74180	94171	97117	31431	00323	62793	11995
42	82927	37884	74411	45887	36713	52339	68421	35968	67714	05883
43	65934	21782	35804	36676	35404	69987	52268	19894	81977	87764
44	56953	04356	68903	21369	35901	86797	83901	68581	02397	55359
45	16278	17165	67843	49349	90163	97337	35003	34915	91485	33814
46	96339	95028	48468	12279	81039	56531	10759	19579	00015	22829
47	84110	49661	13988	75909	35580	18426	29038	79111	56049	96451
48	49017	60748	03412	09880	94091	90052	43596	21424	16584	67970
49	43560	05552	54344	69418	01327	07771	25364	77373	34841	75927
50	25206	15177	63049	12464	16149	18759	96184	15968	89446	07168

Appendix—2

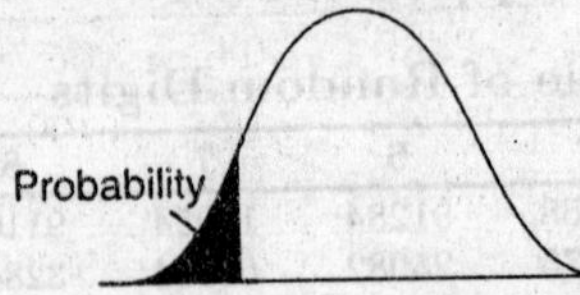

The table gives the probability that the standard normal random variable will fall at or below z

The Standard Normal Probability Distribution

z	.00	.01	.02	.03	.04	.05	.06	.07	.08	.09
−3.4	.0003	.0003	.0003	.0003	.0003	.0003	.0003	.0003	.0003	.0002
−3.3	.0005	.0005	.0005	.0004	.0004	.0004	.0004	.0004	.0004	.0003
−3.2	.0007	.0006	.0006	.0006	.0006	.0006	.0006	.0005	.0005	.0005
−3.1	.0010	.0009	.0009	.0009	.0008	.0008	.0008	.0008	.0007	.0007
−3.0	.0013	.0013	.0013	.0012	.0012	.0011	.0011	.0011	.0010	.0010
−2.9	.0019	.0018	.0018	.0017	.0016	.0016	.0015	.0015	.0014	.0014
−2.8	.0026	.0025	.0024	.0023	.0023	.0022	.0021	.0021	.0020	.0019
−2.7	.0035	.0034	.0033	.0032	.0031	.0030	.0029	.0028	.0027	.0026
−2.6	.0047	.0045	.0044	.0043	.0041	.0040	.0039	.0038	.0037	.0036
−2.5	.0062	.0060	.0059	.0057	.0055	.0054	.0052	.0051	.0049	.0048
−2.4	.0082	.0080	.0078	.0075	.0073	.0071	.0069	.0068	.0066	.0064
−2.3	.0107	.0104	.0102	.0099	.0096	.0094	.0091	.0089	.0087	.0084
−2.2	.0139	.0136	.0132	.0129	.0125	.0122	.0119	.0116	.0113	.0110
−2.1	.0179	.0174	.0170	.0166	.0162	.0158	.0154	.0150	.0146	.0143
−2.0	.0228	.0222	.0217	.0212	.0207	.0202	.0197	.0192	.0188	.0183
−1.9	.0287	.0281	.0274	.0268	.0262	.0256	.0250	.0244	.0239	.0233
−1.8	.0359	.0351	.0344	.0336	.0329	.0322	.0314	.0307	.0301	.0294
−1.7	.0446	.0436	.0427	.0418	.0409	.0401	.0392	.0384	.0375	.0367
−1.6	.0548	.0537	.0526	.0516	.0505	.0495	.0485	.0475	.0465	.0455
−1.5	.0668	.0655	.0643	.0630	.0618	.0606	.0594	.0582	.0571	.0559
−1.4	.0808	.0793	.0778	.0764	.0749	.0735	.0721	.0708	.0694	.0681
−1.3	.0968	.0951	.0934	.0918	.0901	.0885	.0869	.0853	.0838	.0823
−1.2	.1151	.1131	.1112	.1093	.1075	.1056	.1038	.1020	.1003	.0985
−1.1	.1357	.1335	.1314	.1292	.1271	.1251	.1230	.1210	.1190	.1170
−1.0	.1587	.1562	.1539	.1515	.1492	.1469	.1446	.1423	.1401	.1379
−0.9	.1841	.1814	.1788	.1762	.1736	.1711	.1685	.1660	.1635	.1611
−0.8	.2119	.2090	.2061	.2033	.2005	.1977	.1949	.1922	.1894	.1867
−0.7	.2420	.2389	.2358	.2327	.2296	.2266	.2236	.2206	.2177	.2148
−0.6	.2743	.2709	.2676	.2643	.2611	.2578	.2546	.2514	.2483	.2451
−0.5	.3085	.3050	.3015	.2981	.2946	.2912	.2877	.2843	.2810	.2776
−0.4	.3446	.3409	.3372	.3336	.3300	.3264	.3228	.3192	.3156	.3121
−0.3	.3821	.3783	.3745	.3707	.3669	.3632	.3594	.3557	.3520	.3483
−0.2	.4207	.4168	.4129	.4090	.4052	.4013	.3974	.3936	.3897	.3859
−0.1	.4602	.4562	.4522	.4483	.4443	.4404	.4364	.4325	.4286	.4247
−0.0	.5000	.4960	.4920	.4880	.4840	.4801	.4761	.4721	.4681	.4641
−3.5	.0002									
−4.0	.00003									
−4.5	.000003									
−5.0	.0000003									
−6.0	.000000001									

Appendix—2 (Continued)

z	.00	.01	.02	.03	.04	.05	.06	.07	.08	.09
0.0	.5000	.5040	.5080	.5120	.5160	.5199	.5239	.5279	.5319	.5359
0.1	.5398	.5438	.5478	.5517	.5557	.5596	.5636	.5675	.5714	.5753
0.2	.5793	.5832	.5871	.5910	.5948	.5987	.6026	.6064	.6103	.6141
0.3	.6179	.6217	.6255	.6293	.6331	.6368	.6406	.6443	.6480	.6517
0.4	.6554	.6591	.6628	.6664	.6700	.6736	.6772	.6808	.6844	.6879
0.5	.6915	.6950	.6985	.7019	.7054	.7088	.7123	.7157	.7190	.7224
0.6	.7257	.7291	.7324	.7357	.7389	.7422	.7454	.7486	.7517	.7549
0.7	.7580	.7611	.7642	.7673	.7704	.7734	.7764	.7794	.7823	.7852
0.8	.7881	.7910	.7939	.7967	.7995	.8023	.8051	.8078	.8106	.8133
0.9	.8159	.8186	.8212	.8238	.8264	.8289	.8315	.8340	.8365	.8389
1.0	.8413	.8438	.8461	.8485	.8508	.8531	.8554	.8577	.8599	.8621
1.1	.8643	.8665	.8686	.8708	.8729	.8749	.8770	.8790	.8810	.8830
1.2	.8849	.8869	.8888	.8907	.8925	.8944	.8962	.8980	.8997	.9015
1.3	.9032	.9049	.9066	.9082	.9099	.9115	.9131	.9147	.9162	.9177
1.4	.9192	.9207	.9222	.9236	.9251	.9265	.9279	.9292	.9306	.9319
1.5	.9332	.9345	.9357	.9370	.9382	.9394	.9406	.9418	.9429	.9441
1.6	.9452	.9463	.9474	.9484	.9495	.9505	.9515	.9525	.9535	.9545
1.7	.9554	.9564	.9573	.9582	.9591	.9599	.9608	.9616	.9625	.9633
1.8	.9641	.9649	.9656	.9664	.9671	.9678	.9686	.9693	.9699	.9706
1.9	.9713	.9719	.9726	.9732	.9738	.9744	.9750	.9756	.9761	.9767
2.0	.9772	.9778	.9783	.9788	.9793	.9798	.9803	.9808	.9812	.9817
2.1	.9821	.9826	.9830	.9834	.9838	.9842	.9846	.9850	.9854	.9857
2.2	.9861	.9864	.9868	.9871	.9875	.9878	.9881	.9884	.9887	.9890
2.3	.9893	.9896	.9898	.9901	.9904	.9906	.9909	.9911	.9913	.9916
2.4	.9918	.9920	.9922	.9925	.9927	.9929	.9931	.9932	.9934	.9936
2.5	.9938	.9940	.9941	.9943	.9945	.9946	.9948	.9949	.9951	.9952
2.6	.9953	.9955	.9956	.9957	.9959	.9960	.9961	.9962	.9963	.9964
2.7	.9965	.9966	.9967	.9968	.9969	.9970	.9971	.9972	.9973	.9974
2.8	.9974	.9975	.9976	.9977	.9977	.9978	.9979	.9979	.9980	.9981
2.9	.9981	.9982	.9982	.9983	.9984	.9984	.9985	.9985	.9986	.9986
3.0	.9987	.9987	.9987	.9988	.9988	.9989	.9989	.9989	.9990	.9990
3.1	.9990	.9991	.9991	.9991	.9992	.9992	.9992	.9992	.9993	.9993
3.2	.9993	.9993	.9994	.9994	.9994	.9994	.9994	.9995	.9995	.9995
3.3	.9995	.9995	.9996	.9996	.9996	.9996	.9996	.9996	.9996	.9997
3.4	.9997	.9997	.9997	.9997	.9997	.9997	.9997	.9997	.9997	.9998
3.5	.9998									
3.6	.99997									
3.7	.999997									
3.8	.9999997									
3.9	.999999999									

Appendix—3

Critical Values for Chi-Squared Tests

Degrees of Freedom	10% Level	5% Level	1% Level	0.1% Level
1	2.706	3.841	6.635	10.828
2	4.605	5.991	9.210	13.816
3	6.251	7.815	11.345	16.266
4	7.779	9.488	13.277	18.467
5	9.236	11.071	15.086	20.515
6	10.645	12.592	16.812	22.458
7	12.017	14.067	18.475	24.322
8	13.362	15.507	20.090	26.124
9	14.684	16.919	21.666	27.877
10	15.987	18.307	23.209	29.588
11	17.275	19.675	24.725	31.264
12	18.549	21.026	26.217	32.909
13	19.812	22.362	27.688	34.528
14	21.064	23.685	29.141	36.123
15	22.307	24.996	30.578	37.697
16	23.542	26.296	32.000	39.252
17	24.769	27.587	33.409	40.790
18	25.989	28.869	34.805	42.312
19	27.204	30.144	36.191	43.820
20	28.412	31.410	37.566	45.315
21	29.615	32.671	38.932	46.797
22	30.813	33.924	40.289	48.268
23	32.007	35.172	41.638	49.728
24	33.196	36.415	42.980	51.179
25	34.382	37.652	44.314	52.620
26	35.563	38.885	45.642	54.052
27	36.741	40.113	46.963	55.476
28	37.916	41.337	48.278	56.892
29	39.087	42.557	49.588	58.301
30	40.256	43.773	50.892	59.703
31	41.422	44.985	52.191	61.098
32	42.585	46.194	53.486	62.487
33	43.745	47.400	54.776	63.870
34	44.903	48.602	56.061	65.247
35	46.059	48.802	57.342	66.619
36	47.212	50.998	58.619	67.985
37	48.363	52.192	59.893	69.346
38	49.513	53.384	61.162	70.703
39	50.660	54.572	62.428	72.055
40	51.805	55.758	63.691	73.402
41	52.949	56.942	64.950	74.745
42	54.090	58.124	66.206	76.084
43	55.230	59.304	67.459	77.419
44	56.369	60.481	68.710	78.749
45	57.505	61.656	69.957	80.077
46	58.641	62.830	71.201	81.400
47	59.774	64.001	72.443	82.720
48	60.907	65.171	73.683	84.037
49	62.038	66.339	74.919	85.351
50	63.167	67.505	76.154	86.661

Appendix—3 (Continued)

Degrees of Freedom	10% Level	5% Level	1% Level	0.1% Level
51	64.295	68.669	77.386	87.968
52	65.422	69.832	78.616	89.272
53	66.548	70.993	79.843	90.573
54	67.673	72.153	81.069	91.872
55	68.796	73.311	82.292	93.167
56	69.919	74.468	83.513	94.461
57	71.040	75.624	84.733	95.751
58	72.160	76.778	85.950	97.039
59	73.279	77.931	87.166	98.324
60	74.397	79.082	88.379	99.607
61	75.514	80.232	89.591	100.888
62	76.630	81.381	90.802	102.166
63	77.745	82.529	92.010	103.442
64	78.860	83.675	93.217	104.716
65	79.973	84.821	94.422	105.988
66	81.085	85.965	95.626	107.258
67	82.197	87.108	96.828	108.526
68	83.308	88.250	98.028	109.791
69	84.418	89.391	99.228	111.055
70	85.527	90.531	100.425	112.317
71	86.635	91.670	101.621	113.577
72	87.743	92.808	102.816	114.835
73	88.850	93.945	104.010	116.091
74	89.956	95.081	105.202	117.346
75	91.061	96.217	106.393	118.599
76	92.166	97.351	107.583	119.850
77	93.270	98.484	108.771	121.100
78	94.374	99.617	109.958	122.348
79	95.476	100.749	111.144	123.594
80	96.578	101.879	112.329	124.839
81	97.680	103.010	113.512	126.083
82	98.780	104.139	114.695	127.324
83	99.880	105.267	115.876	127.565
84	100.980	106.395	117.057	129.804
85	102.079	107.522	118.236	131.041
86	103.177	108.648	119.414	132.277
87	104.275	109.773	120.591	133.512
88	105.372	110.898	121.767	134.745
89	106.469	112.022	122.942	135.978
90	107.565	113.145	124.116	137.208
91	108.661	114.268	125.289	138.438
92	109.756	115.390	126.462	139.666
93	110.850	116.511	127.633	140.893
94	111.944	117.632	128.803	142.119
95	113.038	118.752	129.973	143.344
96	114.131	119.871	131.141	144.567
97	115.223	120.990	132.309	145.789
98	116.315	122.108	133.476	147.010
99	117.407	123.225	134.642	148.230
100	118.498	124.342	135.807	149.449

Appendix—4

The t Table

Confidence level :								
Two-sided		80%	90%	95%	98%	99%	99.8%	99.9%
One-sided		90%	95%	97.5%	99%	99.5%	99.9%	99.95%
Hypothesis Test level :								
Two-sided		0.20	0.10	0.05	0.02	0.01	0.002	0.001
One sided		0.10	0.05	0.025	0.01	0.005	0.001	0.0005
For one Sample : n	**In general : Degree of freedom**	**Critical values**						
2	1	3.078	6.314	12.706	31.821	63.657	318.309	636.619
3	2	1.886	2.920	4.303	6.965	9.925	22.327	31.599
4	3	1.638	2.353	3.182	4.541	5.841	10.215	12.924
5	4	1.533	2.132	2.776	3.747	4.604	7.173	8.610
6	5	1.476	2.015	2.571	3.365	4.032	5.893	6.869
7	6	1.440	1.947	2.447	3.143	3.707	5.208	5.959
8	7	1.415	1.895	2.365	2.998	3.499	4.785	5.408
9	8	1.397	1.860	2.306	2.896	3.355	4.501	5.041
10	9	1.383	1.833	2.262	2.821	3.250	4.297	4.781
11	10	1.372	1.812	2.228	2.764	3.169	4.144	4.587
12	11	1.363	1.796	2.201	2.718	3.106	4.025	4.437
13	12	1.356	1.782	2.179	2.681	3.055	3.930	4.318
14	13	1.350	1.771	2.160	2.650	3.012	3.852	4.221
15	14	1.345	1.761	2.145	2.624	2.977	3.787	4.140
16	15	1.341	1.753	2.131	2.602	2.947	3.733	4.073
17	16	1.337	1.746	2.120	2.583	2.921	3.686	4.015
18	17	1.333	1.740	2.110	2.567	2.898	3.646	3.965
19	18	1.330	1.734	2.101	2.552	2.878	3.610	3.922
20	19	1.328	1.729	2.093	2.539	2.861	3.579	3.883
21	20	1.325	1.725	2.086	2.528	2.845	3.552	3.850
22	21	1.323	1.721	2.080	2.518	2.831	3.527	3.819
23	22	1.321	1.717	2.074	2.508	2.819	3.505	3.792
24	23	1.319	1.714	2.069	2.500	2.807	3.485	3.768
25	24	1.318	1.711	2.064	2.492	2.797	3.467	3.745
26	25	1.316	1.708	2.060	2.485	2.787	3.450	3.725
27	26	1.315	1.706	2.056	2.479	2.779	3.435	3.707
28	27	1.314	1.703	2.052	2.473	2.771	3.421	3.690
29	28	1.313	1.701	2.048	2.467	2.763	3.408	3.674
30	29	1.311	1.699	2.045	2.462	2.756	3.396	3.659
31	30	1.310	1.697	2.042	2.457	2.750	3.385	3.646
32	31	1.309	1.696	2.040	2.453	2.744	3.375	3.633
33	32	1.309	1.694	2.037	2.449	2.738	3.365	3.622
34	33	1.308	1.692	2.035	2.445	2.733	3.356	3.611
35	34	1.307	1.691	2.032	2.441	2.728	3.348	3.601
36	35	1.306	1.690	2.030	2.438	2.724	3.340	3.591
37	36	1.306	1.688	2.028	2.434	2.719	3.333	3.582
38	37	1.305	1.687	2.026	2.431	2.715	3.326	3.574
39	38	1.304	1.686	2.024	2.429	2.712	3.319	3.566
40	39	1.304	1.685	2.023	2.426	2.708	3.313	3.558
	Infinity	1.282	1.645	1.960	2.326	2.576	3.090	3.291

Appendix—5

F Distribution

Values of F Exceeded with Probabilities of 5 and 1 per cent

df (denominator)	1	2	3	4	5	6	7	8	9	10	11	12	14	16	20	24	30	40	50	75	100	200	500	∞
	df (numerator)																							
1	161	200	216	225	230	234	237	239	241	242	243	244	245	246	248	249	250	251	252	253	253	254	254	254
	4.052	4.999	5.403	5.625	5.764	5.859	5.928	5.981	6.022	6.056	6.082	6.106	6.142	6.169	6.208	6.234	6.261	6.286	6.302	6.323	6.334	6.352	6.361	6.366
2	18.51	19.00	19.16	19.25	19.30	19.33	19.36	19.37	19.38	19.39	19.40	19.41	19.42	19.43	19.44	19.45	19.46	19.47	19.47	19.48	19.49	19.49	19.50	19.50
	98.49	99.00	99.17	99.25	99.30	99.33	99.36	99.37	99.39	99.40	99.41	99.42	99.43	99.44	99.45	99.46	99.47	99.48	99.48	99.49	99.49	99.49	99.50	99.50
3	10.13	9.55	9.28	9.12	9.01	8.94	8.88	8.84	8.81	8.78	8.76	8.74	8.71	8.69	8.66	8.64	8.62	8.60	8.58	8.57	8.56	8.45	8.54	8.53
	34.12	30.82	29.46	28.71	28.24	27.91	27.67	27.49	27.34	27.23	27.13	27.05	26.92	26.83	26.69	26.60	26.50	26.41	26.35	26.27	26.23	26.18	26.14	26.12
4	7.71	6.94	6.59	6.39	6.26	6.16	6.09	6.04	6.00	5.96	5.93	5.91	5.87	5.84	5.80	5.77	5.74	5.71	5.70	5.68	5.66	5.65	5.64	5.63
	21.00	18.00	16.69	15.98	15.52	15.21	14.98	14.80	14.66	14.54	14.45	14.37	14.24	14.15	14.02	13.93	13.83	13.74	13.69	13.61	13.57	13.52	13.48	13.46
5	6.61	5.79	5.41	5.19	5.05	4.95	4.88	4.82	4.78	4.74	4.70	4.68	4.64	4.60	4.56	4.53	4.50	4.46	4.44	4.42	4.40	4.38	4.37	4.36
	16.26	13.27	12.06	11.39	10.97	10.67	10.45	10.29	10.15	10.05	9.96	9.89	9.77	9.68	9.55	9.47	9.38	9.29	9.24	9.17	9.13	9.07	9.04	9.02
6	5.99	5.14	4.76	4.53	4.39	4.28	4.21	4.15	4.10	4.06	4.03	4.00	3.96	3.92	3.87	3.84	3.81	3.77	3.75	3.72	3.71	3.69	3.68	3.67
	13.74	10.92	9.78	9.15	8.75	8.47	8.26	8.10	7.98	7.87	7.79	7.72	7.60	7.52	7.39	7.31	7.23	7.14	7.09	7.02	6.99	6.94	6.90	6.88
7	5.59	4.74	4.34	4.12	3.97	3.87	3.79	3.73	3.68	3.63	3.60	3.57	3.52	3.49	3.44	3.41	3.38	3.34	3.32	3.29	3.28	3.25	3.24	3.23
	12.25	9.55	8.45	7.85	7.46	7.19	7.00	6.84	6.71	6.62	6.54	6.47	6.35	6.27	6.15	6.07	5.98	5.90	5.85	5.78	5.75	5.70	5.67	5.65
8	5.32	4.46	4.07	3.84	3.69	3.58	3.50	3.44	3.39	3.34	3.31	3.28	3.23	3.20	3.15	3.12	3.08	3.05	3.03	3.00	2.98	2.96	2.94	2.93
	11.26	8.65	7.59	7.01	6.63	6.37	6.19	6.03	5.91	5.82	5.74	5.67	5.56	5.48	5.36	5.28	5.20	5.11	5.06	5.00	4.96	4.91	4.88	4.86
9	5.12	4.26	3.86	3.63	3.48	3.37	3.29	3.23	3.18	3.13	3.10	3.07	3.02	2.98	2.93	2.90	2.86	2.82	2.80	2.77	2.76	2.73	2.72	2.71
	10.56	8.02	6.99	6.42	6.06	5.80	5.62	5.47	5.35	5.26	5.18	5.11	5.00	4.92	4.80	4.73	4.64	4.56	4.51	4.45	4.41	4.36	4.33	4.31
10	4.96	4.10	3.71	3.48	3.33	3.22	3.14	3.07	3.02	2.97	2.94	2.91	2.86	2.82	2.77	2.74	2.70	2.67	2.64	2.61	2.59	2.56	2.55	2.54
	10.04	7.56	6.55	5.99	5.64	5.39	5.21	5.06	5.95	4.85	4.78	4.71	4.60	4.52	4.41	4.33	4.25	4.17	4.12	4.05	4.01	3.96	3.93	3.91
11	4.84	3.98	3.59	3.36	3.20	3.09	3.01	2.95	2.90	2.86	2.82	2.79	2.74	2.70	2.65	2.61	2.57	2.53	2.50	2.47	2.45	2.22	2.41	2.40
	9.65	7.20	6.22	5.67	5.32	5.07	4.88	4.74	4.63	4.54	4.46	4.40	4.29	4.21	4.10	4.02	3.94	3.86	3.80	3.74	3.70	3.86	3.62	3.60
12	4.75	3.88	3.49	3.26	3.11	3.00	2.92	2.85	2.80	2.76	2.72	2.69	2.64	2.60	2.54	2.50	2.46	2.42	2.40	2.36	2.35	2.32	2.31	2.30
	9.33	6.93	5.95	5.41	5.06	4.82	4.65	4.50	4.39	4.30	4.22	4.16	4.05	3.98	3.86	3.78	3.70	3.61	3.56	3.49	3.46	3.41	3.38	3.36
13	4.67	3.80	3.41	3.18	3.02	2.92	2.84	2.77	2.72	2.67	2.63	2.60	2.55	2.51	2.46	2.42	2.38	2.34	2.32	2.28	2.26	2.24	2.22	2.21
	9.07	6.70	5.74	5.20	4.86	4.62	4.44	4.30	4.19	4.10	4.02	3.96	3.85	3.78	3.67	3.59	3.51	3.42	3.37	3.30	3.27	3.21	3.18	3.16
14	4.60	3.74	3.34	3.11	2.96	2.85	2.77	2.70	2.65	2.60	2.56	2.53	2.48	2.44	2.39	2.35	2.31	2.27	2.24	2.21	2.19	2.16	2.14	2.13
	8.86	6.51	5.56	5.03	4.69	4.46	4.28	4.14	4.03	3.94	3.86	3.80	3.70	3.62	3.51	3.43	3.84	3.26	3.21	3.14	3.11	3.06	3.02	3.00
15	4.54	3.68	3.29	3.06	2.90	2.79	2.70	2.64	2.59	2.55	2.51	2.48	2.43	2.39	2.33	2.29	2.25	2.21	2.18	2.15	2.12	2.10	2.08	2.07
	8.68	6.36	5.42	4.89	4.56	4.32	4.14	4.00	3.89	3.80	3.73	3.67	3.56	3.48	3.36	3.29	3.20	3.12	3.07	3.00	2.97	2.92	2.89	2.87
16	4.49	3.63	3.24	3.01	2.85	2.74	2.66	2.59	2.54	2.4[illegible]	2.45	2.42	2.37	2.33	2.28	2.24	2.20	2.16	2.13	2.09	2.07	2.04	2.02	2.01
	8.53	6.23	5.29	4.77	4.44	4.20	4.03	3.89	3.78	3.69	3.61	3.55	3.45	3.37	3.25	3.18	3.10	3.01	2.96	2.98	2.86	2.80	2.77	2.75
17	4.45	3.59	3.20	2.96	2.81	2.70	2.62	2.55	2.50	2.45	2.41	2.38	2.33	2.29	2.23	2.19	2.15	2.11	2.08	2.04	2.02	1.99	1.97	1.96
	8.40	6.11	5.18	4.67	4.34	4.10	3.93	3.79	3.68	3.59	3.52	3.45	3.35	3.27	3.16	3.08	3.00	2.92	2.86	2.79	2.76	2.70	2.67	2.65
18	4.41	3.55	3.16	2.93	2.77	3.66	2.58	2.51	2.46	2.41	2.37	2.34	2.29	2.25	2.19	2.15	2.11	2.07	2.04	2.00	1.98	1.95	1.93	1.92
	8.28	6.01	5.09	4.58	4.25	4.01	3.85	3.71	3.60	3.51	3.44	3.37	3.27	3.19	3.07	3.00	2.91	2.83	2.78	2.71	2.68	2.62	2.59	2.57

Contd.

Appendix—5 (Continued)

	1	2	3	4	5	6	7	8	9	10	11	12	14	16	20	24	30	40	50	75	100	200	500	∞
19	4.38	3.52	3.13	2.90	2.74	2.63	2.55	2.48	2.43	2.38	2.34	2.31	2.26	2.21	2.15	2.11	2.07	2.02	2.00	1.96	1.94	1.91	1.90	1.88
	8.18	5.93	5.01	4.50	4.17	3.94	3.77	3.63	3.52	3.43	3.36	3.30	3.19	3.12	3.00	2.92	2.84	2.76	2.70	2.63	2.60	2.54	2.51	2.49
20	4.35	3.49	3.10	2.87	2.71	2.60	2.52	2.45	2.40	2.35	2.31	2.28	2.23	2.18	2.12	2.08	2.04	1.99	1.96	1.92	1.90	1.87	1.85	1.84
	8.10	5.85	4.94	4.43	4.10	3.87	3.71	3.56	3.45	3.37	3.30	3.23	3.13	3.05	2.94	2.86	2.77	2.69	2.63	2.56	2.53	2.47	2.44	2.42
21	4.32	3.47	3.07	2.84	2.68	2.57	2.49	2.42	2.37	2.32	2.28	2.25	2.20	2.15	2.09	2.05	2.00	1.96	1.93	1.89	1.87	1.84	1.82	1.81
	8.02	5.78	4.87	4.37	4.04	3.81	3.65	3.51	3.40	3.31	3.24	3.17	3.07	2.99	2.88	2.80	2.72	2.63	2.58	2.51	2.47	2.42	2.38	2.36
22	4.30	3.44	3.05	2.82	2.66	2.55	2.47	2.40	2.35	2.30	2.26	2.23	2.18	2.13	2.07	2.03	1.98	1.93	1.91	1.87	1.84	1.81	1.80	1.78
	7.94	5.72	4.82	4.31	3.99	3.76	3.59	3.45	3.35	3.26	3.18	3.12	3.02	2.94	2.83	2.75	2.67	2.58	2.53	2.46	2.42	2.37	2.33	2.31
23	4.28	3.42	3.03	2.80	2.64	2.53	2.45	2.38	2.32	2.28	2.24	2.20	2.14	2.10	2.04	2.00	1.96	1.91	1.88	1.84	1.82	1.79	1.77	1.76
	7.88	5.66	4.76	4.26	3.94	3.71	3.54	3.41	3.30	3.21	3.14	3.07	2.97	2.89	2.78	2.70	2.62	2.53	2.48	2.41	2.37	2.32	2.28	2.26
24	4.26	3.40	3.01	2.78	2.62	2.51	2.43	2.36	2.30	2.26	2.22	2.18	2.13	2.09	2.02	1.98	1.94	1.83	1.86	1.82	1.80	1.76	1.74	1.73
	7.82	5.61	4.72	4.22	3.90	3.67	3.50	3.36	3.25	3.17	3.09	3.03	2.93	2.85	2.74	2.66	2.58	2.49	2.44	2.36	2.33	2.27	2.23	2.21
25	4.24	3.38	2.99	2.76	2.60	2.49	2.41	2.34	2.28	2.24	2.20	2.16	2.11	2.06	2.00	1.96	1.92	1.87	1.84	1.80	1.77	1.74	1.72	1.71
	7.77	5.57	4.68	4.18	3.86	3.63	3.46	3.32	3.21	3.13	3.05	2.99	1.89	2.81	2.70	2.62	2.54	2.45	2.40	2.32	2.29	2.23	2.19	2.17
26	4.22	3.37	2.98	2.74	2.59	2.47	2.39	2.32	2.27	2.22	2.18	2.15	2.10	2.05	1.99	1.95	1.90	1.85	1.82	1.78	1.76	1.72	1.70	1.69
	7.72	5.53	4.64	4.14	3.82	3.59	3.42	3.29	3.17	3.09	3.02	2.96	1.86	2.77	2.66	2.58	2.50	2.41	2.36	2.28	2.25	2.19	2.15	2.13
27	4.21	3.35	2.96	2.73	2.57	2.46	2.37	2.30	2.25	2.20	2.16	2.13	2.08	2.03	1.97	1.93	1.88	1.84	1.80	1.76	1.74	1.71	1.68	1.67
	7.68	5.49	4.60	4.11	3.79	3.56	3.39	3.26	3.14	3.06	2.98	2.93	2.83	2.74	2.63	2.55	2.47	2.38	2.33	2.25	2.21	2.16	2.12	2.10
28	4.20	3.34	2.95	2.71	2.56	2.44	2.36	2.29	2.24	2.19	2.15	2.12	2.06	2.02	2.96	1.91	1.87	1.81	1.78	1.75	1.72	1.69	1.67	1.65
	7.64	5.45	4.57	4.07	3.76	3.53	3.36	3.23	3.11	3.03	2.95	2.90	2.80	2.71	2.60	2.52	2.44	2.35	2.30	2.22	2.18	2.13	2.09	2.06
29	4.18	3.33	2.93	2.70	2.54	2.43	2.35	2.28	2.22	2.18	2.14	2.10	2.05	2.00	1.94	1.90	1.85	1.80	1.77	1.73	1.71	1.68	1.65	1.64
	7.60	5.42	4.54	4.04	3.73	3.50	3.33	3.20	3.08	3.00	2.92	2.87	2.77	2.68	2.57	2.49	2.41	2.32	2.27	2.19	2.15	2.10	2.06	2.03
30	4.17	3.32	2.92	2.69	2.53	2.42	2.34	2.27	2.21	2.16	2.12	2.09	2.04	1.99	1.93	1.89	1.84	1.79	1.76	1.72	1.69	1.66	1.64	1.62
	7.56	5.39	4.51	4.02	3.70	3.47	3.30	3.17	3.06	2.98	2.90	2.84	2.74	2.66	2.55	2.47	2.38	2.29	2.24	2.16	2.13	2.07	2.03	2.01
32	4.15	3.30	2.90	2.67	2.51	2.40	2.32	2.25	2.19	2.14	2.10	2.07	2.02	1.97	1.91	1.86	1.82	1.76	1.74	1.69	1.67	1.64	1.61	1.59
	7.50	5.34	4.46	3.97	3.66	3.42	3.25	3.12	3.01	2.94	2.86	2.80	2.70	2.62	2.51	2.42	2.34	2.25	2.20	2.12	2.08	2.02	1.98	1.96
34	4.13	3.28	2.88	2.65	2.49	2.38	2.30	2.23	2.17	2.12	2.08	2.05	2.00	1.95	1.89	1.84	1.80	1.74	1.71	1.67	1.64	1.61	1.59	1.57
	7.44	5.29	4.42	3.93	3.61	3.83	3.21	3.08	2.97	2.89	2.82	2.76	2.66	2.58	2.47	2.38	2.30	2.21	2.15	2.08	2.04	1.98	1.94	1.91
36	4.11	3.26	2.86	2.63	2.48	2.36	2.28	2.21	2.15	2.10	2.06	2.03	1.98	1.93	1.87	1.82	1.78	1.72	1.69	1.65	1.62	1.59	1.56	1.55
	7.39	5.25	4.38	3.89	3.58	3.35	3.18	3.04	2.94	2.86	2.78	2.72	2.62	2.54	2.43	2.35	2.26	2.17	2.12	2.04	2.00	1.94	1.90	1.87
38	4.10	3.25	2.85	2.62	2.46	2.35	2.26	2.19	2.14	2.09	2.05	2.02	1.96	1.92	1.85	1.80	1.76	1.71	1.67	1.63	1.60	1.57	1.54	1.53
	7.35	5.21	4.34	3.86	3.54	3.32	3.15	3.02	2.91	2.92	2.75	2.69	2.59	2.51	2.40	2.32	2.22	2.14	2.08	2.00	1.97	1.90	1.86	1.84
40	4.07	3.23	2.84	2.61	2.45	2.34	2.25	2.18	2.12	2.07	2.04	2.00	1.95	1.90	1.84	1.79	1.74	1.69	1.66	1.61	1.59	1.55	1.53	1.51
	7.31	5.18	4.31	3.83	3.51	3.29	3.12	2.99	2.88	2.80	2.73	2.66	2.56	2.49	2.37	2.29	2.20	2.11	2.05	1.97	1.94	1.88	1.84	1.81

Contd.

Appendix—5 (Continued)

Values of F Exceeded with Probabilities of 5 and 1 Precent

	df (numerator)																							
	1	2	3	4	5	6	7	8	9	10	11	12	14	16	20	24	30	40	50	75	100	200	500	∞
42	4.07	3.22	2.83	2.59	2.44	2.32	2.24	2.17	2.11	2.06	2.02	1.99	1.94	1.89	1.82	1.78	1.73	1.68	1.64	1.60	1.57	1.54	1.51	1.49
	7.27	5.15	4.29	3.80	3.49	3.26	3.10	2.96	2.86	2.77	2.70	2.64	2.54	2.46	2.35	2.26	2.17	2.08	2.02	1.94	1.91	1.85	1.80	1.78
44	4.06	3.21	2.82	2.58	2.43	2.31	2.23	2.16	2.10	2.05	2.01	1.98	1.92	1.88	1.81	1.76	1.72	1.66	1.63	1.58	1.56	1.52	1.50	1.48
	7.24	5.12	4.26	3.78	3.46	3.24	3.07	2.94	2.84	2.75	2.68	2.62	2.52	2.44	2.32	2.24	2.15	2.06	2.00	1.92	1.88	1.82	1.78	1.75
46	4.05	3.20	2.81	2.57	2.42	2.30	2.22	2.14	2.09	2.04	2.00	1.97	1.91	1.87	1.80	1.75	1.71	1.65	1.62	1.57	1.54	1.51	1.48	1.46
	7.21	5.10	4.24	3.76	3.44	3.22	3.05	2.92	2.82	2.73	2.66	2.60	2.50	2.42	2.30	2.22	2.13	2.04	1.98	1.90	1.86	1.80	1.76	1.72
48	4.04	3.19	2.80	2.56	2.41	2.30	2.21	2.14	2.08	2.03	1.99	1.96	1.90	1.86	1.79	1.74	1.70	1.64	1.61	1.56	1.53	1.50	1.47	1.45
	7.19	5.08	4.22	3.74	3.42	3.20	3.04	2.90	2.80	2.71	2.64	2.58	2.48	2.40	2.28	2.20	2.11	2.02	1.96	1.88	1.84	1.78	1.73	1.70
50	4.03	3.18	2.79	2.56	2.40	2.29	2.20	2.13	2.07	2.02	1.98	1.95	1.90	1.85	1.78	1.74	1.69	1.63	1.60	1.55	1.52	1.48	1.46	1.44
	7.17	5.06	4.20	3.72	3.41	3.18	3.02	2.88	2.78	2.70	2.62	2.56	2.46	2.39	2.26	2.18	2.10	2.00	1.94	1.86	1.82	1.76	1.71	1.68
55	4.02	3.17	2.78	2.54	2.38	2.27	2.18	2.11	2.05	2.00	1.97	1.93	1.88	1.83	1.76	1.72	1.67	1.61	1.58	1.52	1.50	1.46	1.43	1.41
	7.12	5.01	4.16	3.68	3.37	3.15	2.98	2.85	2.75	2.66	2.59	2.53	2.43	2.35	2.23	2.15	2.06	1.96	1.90	1.82	1.78	1.71	1.66	1.64
60	4.00	3.15	2.76	2.52	2.37	2.25	2.17	2.10	2.04	1.99	1.95	1.92	1.86	1.81	1.75	1.70	1.65	1.59	1.56	1.50	1.48	1.44	1.41	1.39
	7.08	4.98	4.13	3.65	3.34	3.12	2.95	2.82	2.72	2.63	2.56	2.50	2.40	2.32	2.20	2.12	2.03	1.93	1.87	1.79	1.74	1.68	1.63	1.60
65	3.99	3.14	2.75	2.51	2.36	2.24	2.15	2.08	2.02	1.98	1.94	1.90	1.85	1.80	1.73	1.68	1.63	1.57	1.54	1.49	1.46	1.42	1.39	1.37
	7.04	4.95	4.10	3.62	3.31	3.09	2.93	2.79	2.70	2.61	2.54	2,47	2,37	2.30	2.18	2.09	2.00	1.90	1.84	1.76	1.71	1.64	1.60	1.56
70	3.98	3.13	2.74	2.50	2.35	2.23	2.14	2.07	2.01	1.97	1.93	1.89	1.84	1.79	1.72	1.67	1.62	1.56	1.53	1.47	1.45	1.40	1.37	1.35
	7.01	4.92	2.08	3.60	3.29	3.07	2.91	2.77	2.67	2.59	2.51	2.45	2.35	2.23	2.15	2.07	1.98	1.88	1.82	1.74	1.69	1.62	1.56	1.53
80	3.96	3.11	2.72	2.48	2.33	2.21	2.12	2.05	1.99	1.95	1.91	1.88	1.82	1.77	1.70	1.65	1.60	1.54	1.51	1.54	1.42	1.38	1.35	1.32
	6.96	4.88	4.04	3.56	3.25	3.04	2.87	2.74	2.64	2.55	2.48	2.41	2.32	2.24	2.11	2.03	1.94	1.84	1.78	1.70	1.65	1.57	1.52	1.49
100	3.94	3.09	2.70	2.46	2.30	2.19	2.10	2.03	1.97	1.92	1.88	1.85	1.79	1.75	1.68	1.63	1.57	1.51	1.48	1.42	1.39	1.34	1.30	1.28
	6.90	4.82	3.98	3.51	3.20	2.99	2.82	2.69	2.59	2.51	2.43	2.36	2.26	2.19	2.06	1.98	1.89	1.79	1.73	1.64	1.59	1.51	1.46	1.43
125	3.92	3.07	2.68	2.44	2.29	2.17	2.08	2.01	1.95	1.90	1.85	1.83	1.77	1.72	1.65	1.60	1.55	1.49	1.45	1.39	1.36	1.31	1.27	1.25
	6.84	4.78	3.94	3.47	3.17	2.95	2.79	2.65	2.56	2.47	2.40	2.33	2.23	2.15	2.03	1.94	1.85	1.75	1.68	1.59	1.54	1.46	1.40	1.37
150	3.91	3.06	2.67	2.43	2.27	2.16	2.07	2.00	1.94	1.89	1.85	1.82	1.76	1.71	1.64	1.59	1.54	1.47	1.44	1.37	1.34	1.29	1.25	1.22
	6.81	4.75	3.91	3.44	3.14	2.92	2.76	2.62	2.53	2.44	2.37	2.30	2.20	2.12	2.00	1.91	1.83	1.72	1.66	1.56	1.51	1.43	1.37	1.33
200	3.89	3.04	2.65	2.41	2.26	2.14	2.05	1.98	1.92	1.87	1.83	1.80	1.74	1.69	1.62	1.57	1.52	1.45	1.42	1.35	1.32	1.26	1.22	1.19
	6.76	4.71	3.88	3.41	3.11	2.90	2.73	2.60	2.50	2.41	2.34	2.28	2.17	2.09	1.97	1.88	1.79	1.69	1.62	1.53	1.48	1.39	1.33	1.28
400	3.86	3.02	2.62	2.39	2.23	2.12	2.03	1.96	1.90	1.85	1.91	1.78	1.72	1.67	1.60	1.54	1.49	1.42	1.38	1.32	1.28	1.22	1.16	1.13
	6.70	4.66	3.83	3.36	3.06	2.85	2.69	2.55	2.46	2.37	2.29	2.23	2.12	2.04	1.92	1.84	1.74	1.64	1.57	1.47	1.42	1.32	1.24	1.19
1000	3.85	3.00	2.61	2.38	2.22	2.10	2.02	1.95	1.89	1.84	1.80	1.76	1.70	1.65	1.85	1.53	1.47	1.41	1.36	1.30	1.26	1.19	1.13	1.08
	6.66	4.62	3.80	3.34	3.04	2.82	2.66	2.53	2.43	2.34	2.26	2.20	2.09	2.01	1.98	1.81	1.71	1.61	1.54	1.44	1.38	1.28	1.19	1.11
∞	3.84	2.99	2.60	2.37	2.21	2.09	2.01	1.94	1.88	1.83	1.79	1.75	1.69	1.64	1.57	1.52	1.46	1.40	1.35	1.28	1.24	1.17	1.11	1.00
	6.64	4.60	3.78	3.32	3.02	2.80	2.64	2.51	2.41	2.32	2.24	2.18	2.07	1.99	1.87	1.79	1.69	1.59	1.52	1.41	1.36	1.25	1.15	1.00

Source : E.S. Pearson and H.O. Hartley Biometrika tables for statisticians, Vol. 1 (Cambridge, England : Cambridge University Press 1966), pp. 171–173.

Appendix—6

Dunnett Table

Number of Groups Including Control Group

dfe	alpha	2	3	4	5	6	7	8	9	10
5	0.05	2.57	3.03	3.29	3.48	3.62	3.73	3.82	3.9	3.97
	0.01	4.03	4.63	4.98	5.22	5.41	5.56	5.69	5.8	5.89
6	0.05	2.45	2.86	3.1	3.26	3.39	3.49	3.57	3.64	3.71
	0.01	3.71	4.21	4.51	4.71	4.87	5	5.1	5.2	5.28
7	0.05	2.36	2.75	2.97	3.12	3.24	3.33	3.41	3.47	3.53
	0.01	3.5	3.95	4.21	4.39	4.53	4.64	4.74	4.82	4.89
8	0.05	2.31	2.67	2.88	3.02	3.13	3.22	3.29	3.35	3.41
	0.01	3.36	3.77	4	4.17	4.29	4.4	4.48	4.56	4.62
9	0.05	2.26	2.61	2.81	2.95	3.05	3.14	3.2	3.26	3.32
	0.01	3.25	3.63	3.85	4.01	4.12	4.22	4.3	4.37	4.43
10	0.05	2.23	2.57	2.76	2.89	2.99	3.07	3.14	3.19	3.24
	0.01	3.17	3.53	3.74	3.88	3.99	4.08	4.16	4.22	4.28
11	0.05	2.2	2.53	2.72	2.84	2.94	3.02	3.08	3.14	3.19
	0.01	3.11	3.45	3.65	3.79	3.89	3.98	4.05	4.11	4.16
12	0.05	2.18	2.5	2.68	2.81	2.9	2.98	3.04	3.09	3.14
	0.01	3.05	3.39	3.58	3.71	3.81	3.89	3.96	4.02	4.07
13	0.05	2.16	2.48	2.65	2.78	2.87	2.94	3	3.06	3.1
	0.01	3.01	3.33	3.52	3.65	3.74	3.82	3.89	3.94	3.99
14	0.05	2.14	2.46	2.63	2.75	2.84	2.91	2.97	3.02	3.07
	0.01	2.98	3.29	3.47	3.59	3.69	3.76	3.83	3.88	3.93
15	0.05	2.13	2.44	2.61	2.73	2.82	2.89	2.95	3	3.04
	0.01	2.95	3.25	3.43	3.55	3.64	3.71	3.78	3.83	3.88
16	0.05	2.12	2.42	2.59	2.71	2.8	2.87	2.92	2.97	3.02
	0.01	2.92	3.22	3.39	3.51	3.6	3.67	3.73	3.78	3.83
17	0.05	2.11	2.41	2.58	2.69	2.78	2.85	2.9	2.95	3
	0.01	2.9	3.19	3.36	3.47	3.56	3.63	3.69	3.74	3.79
18	0.05	2.1	2.4	2.56	2.68	2.76	2.83	2.89	2.94	2.98
	0.01	2.88	3.17	3.33	3.44	3.53	3.6	3.66	3.71	3.75
19	0.05	2.09	2.39	2.55	2.66	2.75	2.81	2.87	2.92	2.96
	0.01	2.86	3.15	3.31	3.42	3.5	3.57	3.63	3.68	3.72
20	0.05	2.09	2.38	2.54	2.65	2.73	2.8	2.86	2.9	2.95
	0.01	2.85	3.13	3.29	3.4	3.48	3.55	3.6	3.65	3.69
24	0.05	2.06	2.35	2.51	2.61	2.7	2.76	2.81	2.86	2.9
	0.01	2.8	3.07	3.22	3.32	3.4	3.47	3.52	3.57	3.61
30	0.05	2.04	2.32	2.47	2.58	2.66	2.72	2.77	2.82	2.86
	0.01	2.75	3.01	3.15	3.25	3.33	3.39	3.44	3.49	3.52
40	0.05	2.02	2.29	2.44	2.54	2.62	2.68	2.73	2.77	2.81
	0.01	2.7	2.95	3.09	3.19	3.26	3.32	3.37	3.41	3.44
60	0.05	2	2.27	2.41	2.51	2.58	2.64	2.69	2.73	2.77
	0.01	2.66	2.9	3.03	3.12	3.19	3.25	3.29	3.33	3.37

Appendix—7

Critical Values $q'(p, \text{df}; 0.05)$ for Duncan's Multiple Range Test

df p →	2	3	4	5	6	7	8	9	10	11	12	13	14	15	16	17	18	19
1	17.969	17.969	17.969	17.969	17.969	17.969	17.969	17.969	17.969	17.969	17.969	17.969	17.969	17.969	17.969	17.969	17.969	17.969
2	6.085	6.085	6.085	6.085	6.085	6.085	6.085	6.085	6.085	6.085	6.085	6.085	6.085	6.085	6.085	6.085	6.085	6.085
3	4.501	4.516	4.516	4.516	4.516	4.516	4.516	4.516	4.516	4.516	4.516	4.516	4.516	4.516	4.516	4.516	4.516	4.516
4	3.926	4.013	4.033	4.033	4.033	4.033	4.033	4.033	4.033	4.033	4.033	4.033	4.033	4.033	4.033	4.033	4.033	4.033
5	3.635	3.749	3.796	3.814	3.814	3.814	3.814	3.814	3.814	3.814	3.814	3.814	3.814	3.814	3.814	3.814	3.814	3.814
6	3.460	3.586	3.649	3.680	3.694	3.697	3.697	3.697	3.697	3.697	3.697	3.697	3.697	3.697	3.697	3.697	3.697	3.697
7	3.344	3.477	3.548	3.588	3.611	3.622	3.625	3.625	3.625	3.625	3.625	3.625	3.625	3.625	3.625	3.625	3.625	3.625
8	3.261	3.398	3.475	3.521	3.549	3.566	3.575	3.579	3.579	3.579	3.579	3.579	3.579	3.579	3.579	3.579	3.579	3.579
9	3.199	3.339	3.420	3.470	3.502	3.523	3.536	3.544	3.547	3.547	3.547	3.547	3.547	3.547	3.547	3.547	3.547	3.547
10	3.151	3.293	3.376	3.430	3.465	3.489	3.505	3.516	3.522	3.525	3.525	3.525	3.525	3.525	3.525	3.525	3.525	3.525
11	3.113	3.256	3.341	3.397	3.435	3.462	3.480	3.493	3.501	3.506	3.509	3.510	3.510	3.510	3.510	3.510	3.510	3.510
12	3.081	3.225	3.312	3.370	3.410	3.439	3.459	3.474	3.484	3.491	3.495	3.498	3.498	3.498	3.498	3.498	3.498	3.498
13	3.055	3.200	3.288	3.348	3.389	3.419	3.441	3.458	3.470	3.478	3.484	3.488	3.490	3.490	3.490	3.490	3.490	3.490
14	3.033	3.178	3.268	3.328	3.371	3.403	3.426	3.444	3.457	3.467	3.474	3.479	3.482	3.484	3.484	3.484	3.484	3.484
15	3.014	3.160	3.250	3.312	3.356	3.389	3.413	3.432	3.446	3.457	3.465	3.471	3.476	3.478	3.480	3.480	3.480	3.480
16	2.998	3.144	3.235	3.297	3.343	3.376	3.402	3.422	3.437	3.449	3.458	3.465	3.470	3.473	3.476	3.477	3.477	3.477
17	2.984	3.130	3.222	3.285	3.331	3.365	3.392	3.412	3.429	3.441	3.451	3.459	3.465	3.469	3.472	3.474	3.475	3.475
18	2.971	3.117	3.210	3.274	3.320	3.356	3.383	3.404	3.421	3.435	3.445	3.454	3.460	3.465	3.469	3.472	3.473	3.474
19	2.960	3.106	3.199	3.264	3.311	3.347	3.375	3.397	3.415	3.429	3.440	3.449	3.456	3.462	3.466	3.469	3.472	3.473
20	2.950	3.097	3.190	3.255	3.303	3.339	3.368	3.390	3.409	3.423	3.435	3.445	3.452	3.459	3.463	3.467	3.470	3.472
21	2.941	3.088	3.181	3.247	3.295	3.332	3.361	3.385	3.403	3.418	3.431	3.441	3.449	3.456	3.461	3.465	3.469	3.471
22	2.933	3.080	3.173	3.239	3.288	3.326	3.355	3.379	3.398	3.414	3.427	3.437	3.446	3.453	3.459	3.464	3.467	3.470
23	2.926	3.072	3.166	3.233	3.282	3.320	3.350	3.374	3.394	3.410	3.423	3.434	3.443	3.451	3.457	3.462	3.466	3.469
24	2.919	3.066	3.160	3.226	3.276	3.315	3.345	3.370	3.390	3.406	3.420	3.431	3.441	3.449	3.455	3.461	3.465	3.469
25	2.913	3.059	3.154	3.221	3.271	3.310	3.341	3.366	3.386	3.403	3.417	3.429	3.439	3.447	3.454	3.459	3.464	3.468
26	2.907	3.054	3.149	3.216	3.266	3.305	3.336	3.362	3.382	3.400	3.414	3.426	3.436	3.445	3.452	3.458	3.463	3.468
27	2.902	3.049	3.144	3.211	3.262	3.301	3.332	3.358	3.379	3.397	3.412	3.424	3.434	3.443	3.451	3.457	3.463	3.467
28	2.897	3.044	3.139	3.206	3.257	3.297	3.329	3.355	3.376	3.394	3.409	3.422	3.433	3.442	3.450	3.456	3.462	3.467
29	2.892	3.039	3.135	3.202	3.253	3.293	3.326	3.352	3.373	3.392	3.407	3.420	3.431	3.440	3.448	3.455	3.461	3.466
30	2.888	3.035	3.131	3.199	3.250	3.290	3.322	3.349	3.371	3.389	3.405	3.418	3.429	3.439	3.447	3.454	3.460	3.466
31	2.884	3.031	3.127	3.195	3.246	3.287	3.319	3.346	3.368	3.387	3.403	3.416	3.428	3.438	3.446	3.454	3.460	3.465
32	2.881	3.028	3.123	3.192	3.243	3.284	3.317	3.344	3.366	3.385	3.401	3.415	3.426	3.436	3.445	3.453	3.459	3.465
33	2.877	3.024	3.120	3.188	3.240	3.281	3.314	3.341	3.364	3.383	3.399	3.413	3.425	3.435	3.444	3.452	3.459	3.465
34	2.874	3.021	3.117	3.185	3.238	3.279	3.312	3.339	3.362	3.381	3.398	3.412	3.424	3.434	3.443	3.451	3.458	3.464
35	2.871	3.018	3.114	3.183	3.235	3.276	3.309	3.337	3.360	3.379	3.396	3.410	3.423	3.433	3.443	3.451	3.458	3.464
36	2.868	3.015	3.111	3.180	3.232	3.274	3.307	3.335	3.358	3.378	3.395	3.409	3.421	3.432	3.442	3.450	3.457	3.464
37	2.865	3.013	3.109	3.178	3.230	3.272	3.305	3.333	3.356	3.376	3.393	3.408	3.420	3.431	3.441	3.449	3.457	3.463
38	2.863	3.010	3.106	3.175	3.228	3.270	3.303	3.331	3.355	3.375	3.392	3.407	3.419	3.431	3.440	3.449	3.456	3.463
39	2.861	3.008	3.104	3.173	3.226	3.268	3.301	3.330	3.353	3.373	3.391	3.406	3.418	3.430	3.440	3.448	3.456	3.463
40	2.858	3.005	3.102	3.171	3.224	3.266	3.300	3.328	3.352	3.372	3.389	3.404	3.418	3.429	3.439	3.448	3.456	3.463
48	2.843	2.991	3.087	3.157	3.211	3.253	3.288	3.318	3.342	3.363	3.382	3.398	3.412	3.424	3.435	3.445	3.453	3.461
60	2.829	2.976	3.073	3.143	3.198	3.241	3.277	3.307	3.333	3.355	3.374	3.391	3.406	3.419	3.431	3.441	3.451	3.460
80	2.814	2.961	3.059	3.130	3.185	3.229	3.266	3.297	3.323	3.346	3.366	3.384	3.400	3.414	3.427	3.438	3.449	3.458
120	2.800	2.947	3.045	3.116	3.172	3.217	3.254	3.286	3.313	3.337	3.358	3.377	3.394	3.409	3.423	3.435	3.446	3.457
240	2.786	2.933	3.031	3.103	3.159	3.205	3.243	3.276	3.304	3.329	3.350	3.370	3.388	3.404	3.418	3.432	3.444	3.455
Inf	2.772	2.918	3.017	3.089	3.146	3.193	3.232	3.265	3.294	3.320	3.343	3.363	3.382	3.399	3.41	3.428	3.442	3.454

Appendix—7 (Continued)

Critical Values $q'(p, \text{df}; 0.01)$ for Duncan's Multiple Range Test

df $p \rightarrow$	2	3	4	5	6	7	8	9	10	11	12	13	14	15	16	17	18	19
1	90.024	90.024	90.024	90.024	90.024	90.024	90.024	90.024	90.024	90.024	90.024	90.024	90.024	90.024	90.024	90.024	90.024	90.024
2	14.036	14.036	14.036	14.036	14.036	14.036	14.036	14.036	14.036	14.036	14.036	14.036	14.036	14.036	14.036	14.036	14.036	14.036
3	8.260	8.321	8.321	8.321	8.321	8.321	8.321	8.321	8.321	8.321	8.321	8.321	8.321	8.321	8.321	8.321	8.321	8.321
4	6.511	6.677	6.740	6.755	6.755	6.755	6.755	6.755	6.755	6.755	6.755	6.755	6.755	6.755	6.755	6.755	6.755	6.755
5	5.702	5.893	5.989	6.040	6.065	6.074	6.074	6.074	6.074	6.074	6.074	6.074	6.074	6.074	6.074	6.074	6.074	6.074
6	5.243	5.439	5.549	5.614	5.655	5.680	5.694	5.701	5.703	5.703	5.703	5.703	5.703	5.703	5.703	5.703	5.703	5.703
7	4.949	5.145	5.260	5.333	5.383	5.416	5.439	5.454	5.464	5.470	5.472	5.472	5.472	5.472	5.472	5.472	5.472	5.472
8	4.745	4.939	5.056	5.134	5.189	5.227	5.256	5.276	5.291	5.302	5.309	5.313	5.316	5.317	5.317	5.317	5.317	5.317
9	4.596	4.787	4.906	4.986	5.043	5.086	5.117	5.142	5.160	5.174	5.185	5.193	5.199	5.202	5.205	5.206	5.206	5.206
10	4.482	4.671	4.789	4.871	4.931	4.975	5.010	5.036	5.058	5.074	5.087	5.098	5.106	5.112	5.117	5.120	5.122	5.123
11	4.392	4.579	4.697	4.780	4.841	4.887	4.923	4.952	4.975	4.994	5.009	5.021	5.031	5.039	5.045	5.050	5.054	5.057
12	4.320	4.504	4.622	4.705	4.767	4.815	4.852	4.882	4.907	4.927	4.944	4.957	4.969	4.978	4.986	4.993	4.998	5.002
13	4.260	4.442	4.560	4.643	4.706	4.754	4.793	4.824	4.850	4.871	4.889	4.904	4.917	4.927	4.936	4.944	4.950	4.955
14	4.210	4.391	4.508	4.591	4.654	4.703	4.743	4.775	4.802	4.824	4.843	4.859	4.872	4.884	4.894	4.902	4.909	4.916
15	4.167	4.346	4.463	4.547	4.610	4.660	4.700	4.733	4.760	4.783	4.803	4.820	4.834	4.846	4.857	4.866	4.874	4.881
16	4.131	4.308	4.425	4.508	4.572	4.622	4.662	4.696	4.724	4.748	4.768	4.785	4.800	4.813	4.825	4.835	4.843	4.851
17	4.099	4.275	4.391	4.474	4.538	4.589	4.630	4.664	4.692	4.717	4.737	4.755	4.771	4.785	4.797	4.807	4.816	4.824
18	4.071	4.246	4.361	4.445	4.509	4.559	4.601	4.635	4.664	4.689	4.710	4.729	4.745	4.759	4.771	4.782	4.792	4.801
19	4.046	4.220	4.335	4.418	4.483	4.533	4.575	4.610	4.639	4.664	4.686	4.705	4.722	4.736	4.749	4.760	4.771	4.780
20	4.024	4.197	4.312	4.395	4.459	4.510	4.552	4.587	4.617	4.642	4.664	4.684	4.701	4.716	4.729	4.741	4.751	4.761
21	4.004	4.177	4.291	4.374	4.438	4.489	4.531	4.567	4.597	4.622	4.645	4.664	4.682	4.697	4.711	4.723	4.734	4.743
22	3.986	4.158	4.272	4.355	4.419	4.470	4.513	4.548	4.578	4.604	4.627	4.647	4.664	4.680	4.694	4.706	4.718	4.728
23	3.970	4.141	4.254	4.337	4.402	4.453	4.496	4.531	4.562	4.588	4.611	4.631	4.649	4.665	4.679	4.692	4.703	4.713
24	3.955	4.126	4.239	4.322	4.386	4.437	4.480	4.516	4.546	4.573	4.596	4.616	4.634	4.651	4.665	4.678	4.690	4.700
25	3.942	4.112	4.224	4.307	4.371	4.423	4.466	4.502	4.532	4.559	4.582	4.603	4.621	4.638	4.652	4.665	4.677	4.688
26	3.930	4.099	4.211	4.294	4.358	4.410	4.452	4.489	4.520	4.546	4.570	4.591	4.609	4.626	4.640	4.654	4.666	4.677
27	3.918	4.087	4.199	4.282	4.346	4.397	4.440	4.477	4.508	4.535	4.558	4.579	4.598	4.615	4.630	4.643	4.655	4.667
28	3.908	4.076	4.188	4.270	4.334	4.386	4.429	4.465	4.497	4.524	4.548	4.569	4.587	4.604	4.619	4.633	4.646	4.657
29	3.898	4.065	4.177	4.260	4.324	4.376	4.419	4.455	4.486	4.514	4.538	4.559	4.578	4.595	4.610	4.624	4.637	4.648
30	3.889	4.056	4.168	4.250	4.314	4.366	4.409	4.445	4.477	4.504	4.528	4.550	4.569	4.586	4.601	4.615	4.628	4.640
31	3.881	4.047	4.159	4.241	4.305	4.357	4.400	4.436	4.468	4.495	4.519	4.541	4.560	4.577	4.593	4.607	4.620	4.632
32	3.873	4.039	4.150	4.232	4.296	4.348	4.391	4.428	4.459	4.487	4.511	4.533	4.552	4.570	4.585	4.600	4.613	4.625
33	3.865	4.031	4.142	4.224	4.288	4.340	4.383	4.420	4.452	4.479	4.504	4.525	4.545	4.562	4.578	4.592	4.606	4.618
34	3.859	4.024	4.135	4.217	4.281	4.333	4.376	4.413	4.444	4.472	4.496	4.518	4.538	4.555	4.571	4.586	4.599	4.611
35	3.852	4.017	4.128	4.210	4.273	4.325	4.369	4.406	4.437	4.465	4.490	4.511	4.531	4.549	4.565	4.579	4.593	4.605
36	3.846	4.011	4.121	4.203	4.267	4.319	4.362	4.399	4.431	4.459	4.483	4.505	4.525	4.543	4.559	4.573	4.587	4.599
37	3.840	4.005	4.115	4.197	4.260	4.312	4.356	4.393	4.425	4.452	4.477	4.499	4.519	4.537	4.553	4.568	4.581	4.594
38	3.835	3.999	4.109	4.191	4.254	4.306	4.350	4.387	4.419	4.447	4.471	4.493	4.513	4.531	4.548	4.562	4.576	4.589

Bibliography

[1] Aczel, A.D. (1995) : Statistics : Concepts amd Applications, IRWIN.

[2] Bernard Rosner (2005) : Fundamentals of Biostatistics, Amazan Book.

[3] Bhuyan, K.C. (2009) : Methods of Statistics, 2nd Revised Edition, Sahitya Prokashani

[4] Bluman, A.G. (2000) : Elementary Statistics.

[5] Chap, T.L. (2003) : Introductory Biostatistics, Wiley Interscience.

[6] Chase, W. and Bown Fred (1992) : General Statistics, John Wiley and Sons, Inc.

[7] Elifsern,.K.W. (1998) : Fundamentals of Social Statistics.

[8] Everitt, B. and Dunn, G. (2001) : Applied Multivariate Data Analysis, 2nd Ed., Arnold.

[9] Frank, H. and Althoen, S.C. (1994) : Statistics : Concepts and Analysis, Cambridge University Press.

[10] Geoffrey, R. Norman and David L. Streiner (2007) : Biostatistics : The Bare Essentials, 3rd Ed., Amazon Book.

[11] Gereald van Belle, Llyod D. Fisher, Patrick J. Heagerly, Thomson Lumley (2004) : Biostatistics : A methodology for the Health Sciences, Wiley Interscience.

[12] Hogg, R.V. and Tannis, E.A. (1997) : Probability and Statistical Inference, 2nd Ed., Arnold.

[13] Jacob S. Siegel (2004) : The Methods and Materials of Demography, Amazan Book.

[14] Marcello Pagano and Kimberlee Gauvereau (2000) : Principles of Biostatistics, 2nd Ed., Amazan Book.

[15] Michael R. Chernick, Robert H. Friis (2003) : Introductory Biostatistics for the Health Sciences including Bootstrap, Wiley Interscience.

[16] Michael Borenstein, Larry V. Hedges, Julian, P.T., Higgins, Hannah Rothstein Chichester (2009) : Introduction to Meta Analysis, Wiley.

[17] Mario F,. Triola (2012) : Elementary Statistics, 7th Ed., John Wiley and Sons, Inc.

[18] Robert E. Potter (1983) : Fertility, Biology and Behaviour : An analysis of the Proximate Determinants, Amazan Book.

[19] Sr. Ralph D' Agostino, Lisa Sullivan and Alexa Beiser (2005) : Introductory Applied Biostatistics, Amazan Book.

[20] Tukey, J. (1947) : Exploratory Data Analysis, Addison Wesley.

[21] Wayne W. Daniel : Biostatistics : A foundation for analysis in Health sciences, 7th Ed., John Wiley and Sons, Inc.

Index

Some exciting gifts are up for grabs with this book!!

FREE Gifts Rs. 1000/-

ADVANCED BIOSTATISTICS

K. C. Bhuyan

DEAR READERS

By now you must have gone through this book in the entirety. So we invite your assessment on this title with absolute impartiality and candid observations. After all, we publishers faithfully depend on your benchmarking. Friends, please fill up the open spaces below with CAPITAL LETTERS. You may add some extra sheets of paper if the space provided is not sufficient for you.

i. What are the important topics of your syllabus which are not adequately covered in this book?

ii. In which chapters and on what topics the treatment of the subject-matter by the author is not satisfactory and why?

iii. Have you come across the typos, factual inaccuracies, inconsistencies, and other forms of mistakes in the book? If yes, then please specify the mistakes and page numbers.

iv. Name top three books on the same subject in your consideration. Could you provide some reasons?

Your 1st Preference: ....

Your 2nd Preference: ....

Your 3rd Preference: ....

v. Any suggestion to improve the quality of illustrations, diagrams, etc.?

vi. How did you like the cover design? Any suggestion on colours?

vii. Suggestions for further improvement of production-standard:

viii. Who recommended you this book:

☐ Teacher ☐ Friend ☐ Parents ☐ Bookseller ☐ Librarian

ix. Name of the person who has recommended:

Designation: ....

Address: ....

Phone No.: ....

e-mail: ....

x. Name and address of the bookseller from whom you have purchased this book:

....

....

....

xi. Name and address of the library where you would like to find the book:

....

....

xii. Name and address of your institution (Please mention the University or Board):

....

....

xiii. Your name, e-mail, and complete postal address:

....

....

....

xiv. A few other titles published by us that you would like to read:

....

....

The best assessment will be rewarded half-yearly. The award will be in the form of our publications, as decided by the Editorial and Marketing Operations Division, amounting to Rs. 1000/- (Rupees one thousand) only.

....

Please mail the filled up coupon at your earliest to :

Editorial & Marketing Operations Division

NEW CENTRAL BOOK AGENCY (P) Ltd

8/1 Chintamoni Das Lane, Kolkata 700 009

West Bengal, INDIA

Phone + Fax: 033 + 2241 8411, 2241 8573, 2241 9094

e-mail: ncbapvtltd@eth.net, ncba@cal2.vsnl.net.in

Website: www.centralgroup.com